Remote Sensing Applications in Marine Science and Technology

NATO ASI Series

Advanced Science Institutes Series

A series presenting the results of activities sponsored by the NATO Science Committee, which aims at the dissemation of advanced scientific and technological knowledge, with a view to strengthening links between scientific communities.

The series is published by an international board of publishers in conjunction with the NATO Scientific Affairs Division

A	Life Sciences	Plenum Publishing Corporation
B	Physics	London and New York
C	Mathematical and Physical Sciences	D. Reidel Publishing Company Dordrecht, Boston and Lancaster
D	Behavioural and Social Sciences	Martinus Nijhoff Publishers
E	Engineering and Materials Sciences	The Hague, Boston and Lancaster
F	Computer and Systems Sciences	Springer Verlag
G	Ecological Sciences	Heidelberg

Series C: Mathematical and Physical Sciences No. 106

Remote Sensing Applications in Marine Science and Technology

edited by

Arthur P. Cracknell

Carnegie Laboratory of Physics,
University of Dundee, Scotland, U.K.

D. Reidel Publishing Company

Dordrecht / Boston / Lancaster

Published in cooperation with NATO Scientific Affairs Division

Proceedings of the NATO Advanced Study Institute on
Remote Sensing Applications in Marine Science and Technology
Dundee, Scotland
1-21 August, 1982

Library of Congress Cataloging in Publication Data

NATO Advanced Study Institute (1982 : Dundee, Tayside)
Remote sensing applications in marine science and technology.

(NATO ASI series. Series C, Mathematical and physical sciences ; no. 106)
Includes index.
1. Oceanography—Remote sensing—Congresses.
I. Cracknell, Arthur P. II. Title. III. Series.
GC10.4.R4N37 1982 551.46'0028 83-9475
ISBN 90-277-1608-0

This summer school was co-sponsored by:
British Tourist Authority
Council of Europe
European Association of Remote Sensing Laboratories
European Space Agency

Published by D. Reidel Publishing Company
P.O. Box 17, 3300 AA Dordrecht, Holland

Sold and distributed in the U.S.A. and Canada
by Kluwer Academic Publishers,
190 Old Derby Street, Hingham, MA 02043, U.S.A.

In all other countries, sold and distributed
by Kluwer Academic Publishers Group,
P.O. Box 322, 3300 AH Dordrecht, Holland

D. Reidel Publishing Company is a member of the Kluwer Academic Publishers Group

Printed in The Netherlands

ORGANISATION

Organising Committee

Professor A.P. Cracknell, Director
Dr. W.M. Young, Secretary
Dr. R.K. Bullard
Mr. D.D. Hardy
Professor G. Østrem
Dr. B.E. Storey (Accommodation)
Dr. A.O. Tooke (Social Programme)
Mr. G.R. Whyte (Technical Arrangements)

Demonstrating, Secretarial and Technical Assistance

Mr. K.J. Arbuthnott
Mr. I.M. Ballingall
Mrs. M.M. Bulloch
Dr. J.A. Charlton
Mr. I. Durajczyk
Mr. G. Dutch
Miss C.L. Eggie
Mr. M.G. Flynn
Mrs. B. Johnston
Mr. J.N. Low
Mr. K.G. Miller
Miss P.M. Mitchell
Mr. K. Muirhead
Mrs. K.M. Payne
Mr. W.A. Reekie
Miss A.M. Shivers
Mr. S.B. Wilson

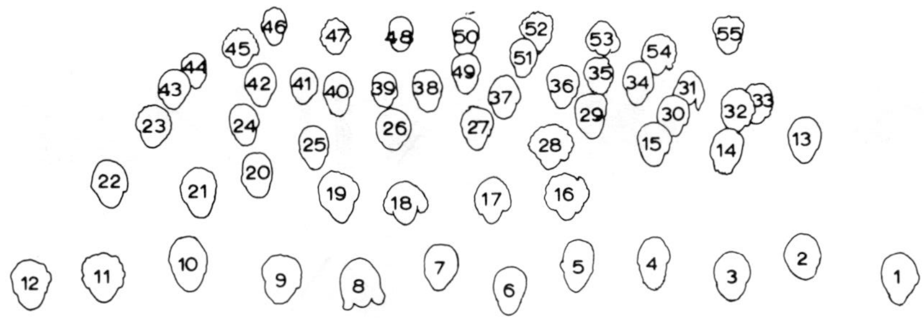

1. Jolly
2. Storey
3. Tooke
4. Offiler
5. Bullard
6. Sathyendranath
7. Cracknell
8. Wyatt
9. Yentsch
10. Young
11. Muralikrishna
12. Singh
13. Scharfen
14. Stott
15. Kurter
16. Economides
17. Macedo
18. Charmasson
19. Taccini
20. Sengele
21. Muirhead
22. McFarlane
23. Miskolczi
24. Wilson
25. Magnin
26. McHardy
27. Clement
28. Lacey
29. Maracci
30. Safak
31. Jayko
32. Nilsen
33. Offiler (Mrs.)
34. Gaspar
35. Font
36. Uysal
37. Nacu
38. Quader
39. Lin
40. Da Silva
41. Lambert
42. Bowyer
43. Reid
44. Langot
45. Hansen
46. Valeur
47. Alves
48. Sun
49. Nyfors
50. Calmet
51. Demarcq
52. Lawrence
53. Shivers
54. Payne
55. Flikke

TABLE OF CONTENTS

PREFACE

This summer school was a sequel to the summer school on Remote Sensing in Meteorology, Oceanography and Hydrology which was held in Dundee in 1980 and the proceedings of which were published by Ellis Horwood Ltd., Chichester, England. At the present summer school we concentrated on only part of the subject area that was covered in 1980. Although there was some repetition of material that was presented in 1980, because by and large we had a new set of participants, most subjects were treated in considerably greater detail than had been possible previously. The major topics covered in the present summer school were (i) the general principles of remote sensing with particular reference to marine applications, (ii) applications to physical oceanography, (iii) marine resources applications and (iv) coastal monitoring and protection.

The material contained in this volume represents the written texts of most of the lectures presented at the summer school. One important set of lecture notes was not available; this was for the lectures on active microwave techniques, principally synthetic aperture radar, by W. Alpers from Hamburg. For this material we would refer the reader to "Imaging Ocean Surface Waves by Synthetic Aperture Radar - A Review" by W. Alpers, which is to appear as chapter 6 in "Satellite Microwave Remote Sensing" edited by T.D. Allan (Ellis Horwood, Chichester) which is to be published in 1983.

This summer school was part of a continuing programme of activities in postgraduate education and training in remote sensing undertaken by the European Association of Remote Sensing Laboratories, with stimulation from an initiative taken by the Council of Europe. This activity was described in some detail in the Foreword to the proceedings of our 1980 summer school. In addition to the main financial support received from NATO, additional financial support was also received from the British Tourist Authority, the Council of Europe, the European Association of Remote Sensing Laboratories and the European Space Agency. The University of Dundee also contributed in kind, in numerous ways, to the success of the summer school. I am enormously grateful to all those colleagues for all the tremendous hard work and enthusiasm that they put into handling all aspects of the organisation of the summer school. I am also grateful to those who have assisted technically in the preparation of copy for these proceedings, especially Mrs. M. Bulloch who has typed the prelims and index and retyped one or two chapters, to Miss M.M. Benstead who has redrawn many of the very poor diagrams that were submitted and to Mrs. K.M. Payne who has compiled the index.

A. P. Cracknell (ed.), Remote Sensing Applications in Marine Science and Technology, xi–xii.

The success or failure of a summer school of this nature lies very much in the hands of the lecturers and the participants. That this particular school was extremely successful was due very largely to the hard work and enthusiasm of everyone who came and took part. We were very grateful to our lecturers, all of them very busy people with many other commitments, who took the time and trouble to prepare their lecture material, to come to Dundee to give their lectures and participate in formal and informal discussions, and (in most cases) to prepare the camera-ready copy for inclusion in these proceedings. And, of course, we were very grateful to the 56 "students" who came to participate in the summer school and who were very enthusiastic in that participation. Lecturers and students shared the common hospitality of residence in Chalmers Hall and many lively and constructive discussions took place. Barriers associated with differences in age or in professional and academic status were largely swept away in the discussions of matters of common scientific and technical interest. In addition to representatives from all the obvious European countries we also enjoyed the company of participants from as far afield as China, the Philippines, the Indian subcontinent, some African countries and, of course, North America. It is one of the minor consolations of the Director, after all the work and worry involved in putting together such a summer school, to see young people from countries on opposite sides of the world talking together like old friends within a few hours of having arrived in Dundee as complete strangers. In our own small way, dealing with people with common scientific interests and having the financial aspects smoothed by the assistance of our various sponsors, we can do our own little bit towards the advancement of understanding and friendship among the different peoples of the world.

Events such as this, which deal with subject matter at the leading edge of human knowledge and achievement, provide a very efficient and also very enjoyable way of communicating the knowledge and awareness of the current state of new developments widely and quickly in the scientific community. It is very much to be hoped that events such as this will be able to continue, in spite of the deep financial problems that seem to be affecting many academic institutions throughout the world at the moment.

Arthur P. Cracknell

CHAPTER 1

REMOTE MEASUREMENT OF THE OCEAN - AN OVERVIEW

John R. Apel

Applied Physics Laboratory
The Johns Hopkins University
Laurel, MD 20707 USA

INTRODUCTION

During the past decade, the large strides made in space science and technology have provided us with new tools with which to undertake spaceborne observations and investigations of the dynamics of the oceans. A better understanding of ocean dynamics, that is, waves, currents, and circulation patterns could lead to more efficient ocean use through improved forecasts of the marine environment.

The ocean plays as fundamental a role in the natural scheme of things as does the atmosphere, although its functions, being considerably more varied and diffuse, are probably neither as well appreciated nor as well understood. The sea profoundly affects the weather and in turn is affected by the atmosphere, acting as both a heat reservoir for storing, distributing, and releasing solar energy, and as the dominant source for atmospheric moisture. Photosynthesis by oceanic phytoplankton is one major process for maintenance of atmospheric oxygen. Geological activity on all time and space scales takes place in the seas and its beds, which serve as the repository for the detritus of man and nature and as an important, practicable source of petroleum and a few useful minerals. Its currents and dilutant powers are called upon to disperse sewage, poisonous and nonpoisonous wastes, solid trash, and excess heat, all the while maintaining a role as the *aqua viva* for an extremely complicated and commercially important food chain, and as a means of recreation and refreshment for people. In the estuaries and the coastal zones, these conflicting demands are especially severe.

However, because of the great length and breadth of the sea,

A. P. Cracknell (ed.), Remote Sensing Applications in Marine Science and Technology, 1–16.

and the harsh environment it presents, the difficulties in obtaining detailed, timely information of sufficient observational density across most of its expanse have prevented an effective monitoring and forecasting system for the oceans. Thus, the prediction of wave heights depends on forecasts of the time and space histories of surface winds--the latter forecasts themselves being fraught with considerable uncertainty, as the loss of ships and oil drilling rigs at sea attests. Similarly, the locations of major ocean currents are known only approximately and the data required for shipping and fishing interests to efficiently exploit currents are lacking. The lack of sufficient wind and pressure data over the oceans has precluded an improved, longer-range weather forecast for continental areas. In order to achieve an effective one-to-two-week forecast, observational data are needed over the oceans with about the same frequency and density as now exist in the continental areas.

One finds a diverse list of parameters entering into oceanic processes; furthermore, many cannot be discerned from spacecraft. By and large, satellite oceanography is confined to surface and near-surface phenomena. This constraint is not as severe as it appears at first glance, because surface data taken from spacecraft can be appended with other, conventionally derived subsurface measurements of certain parameters (e.g., vertical current or temperature profiles) in order to construct a more nearly three-dimensional view of the ocean. In addition, near-surface data are useful in their own right since the interaction between ocean and atmosphere largely takes place in the few tens of meters above and below the sea-air interface. Man's marine activities are mostly limited to that surface as well, so that the kind of two-dimensional oceanography that one can pursue from spacecraft is highly relevant.

The prospect of making planetary observations of the surface of the sea from satellites has been an intriguing one for oceanographers ever since the early spacecraft missions returned color photographs of the ocean that showed surprising amounts of structure. However, it has only been in the past several years that the quality, type, and amount of satellite data have allowed the process of acquiring scientific information from space to resemble "remote measurement". As a result, it is now becoming possible in certain cases to study the dynamics of the ocean surface, insofar as they are represented by changes in surface temperature, color, set-up, or roughness variations.

Remote measurement is certainly not the stand-alone panacea for oceanographers that its most enthusiastic supporters have made it out to be. However, if skillfully used as a new tool in the enlarging tool kit now available, it can be made to yield information of a type and extent that should be of much value to the advance of science. When utilized in conjunction with in-water measurements and an adequate theory, it is often valuable for interpolation between and extrapolation beyond the regions of surface observations, for determining the spatial variability of the surface phenomenon under study, and for deducing scales, speeds, amplitudes, and the like.

USES OF SPACECRAFT DATA

The answer as to who needs what information from spacecraft obviously depends on the type of information that is obtainable. In research areas the disciplines served with some degree of usefulness are marine geodesy and gravity; physical, geological, and biological oceanography; ice dynamics; boundary layer meteorology; and climate dynamics. Various maritime operations, shipping, offshore mining, oil drilling, and fishing, all require an improved and expanded data base and more accurate marine forecasts. The ever-increasing fraction of the population living along the seacoasts needs improved forecasting and warning services for protection of life and property. However, because of the great length and breadth of the sea the difficulties in obtaining timely detailed information of sufficient observational density across its expanse have prevented an effective monitoring and forecasting system for the oceans.

SATELLITES OF UTILITY TO OCEANOGRAPHY

The number of satellites carrying sensors that yield data useful to ocean science is large, and the value of the data from them variable. Table 1 is a listing of the spacecraft that have been or will be sources of data having oceanographic significance. Of the several listed, the most useful are probably the NOAA/Tiros series, Landsat, GEOS 3, GOES, Seasat, and Nimbus-7. The data types are diverse, as is discussed below.

Table 2 lists ocean-related satellites expected to be launched in the next decade, their sponsor, sensors, expected launch dates, and status.

Table 3 lists the generic classes of sensors used in satellites operated by several nations, along with estimates of their typical measurement accuracies. Table 4 gives several sources of oceanographic satellite data, with addresses of the authorities responsible for their operation.

Figures 1-6 give examples of data output from each of these devices that are of particular oceanographic interest.

TABLE 1. U.S. SATELLITES OF UTILITY IN OCEANOGRAPHY

Satellite	Launch Date	Orbit	Utility of Data	Character	Sensors	Oceanic Parameters
Mercury Gemini Apollo Apollo-Soyuz	1962- 1975	Variable	Low to medium	Exploratory	Cameras	Imagery
Nimbus-4 Nimbus-5 Nimbus-6 Nimbus-7	1970 1973 1975 1978	Polar	Medium progressing to high	Experimental	IR and MW radiometers and bolometer; color scanner	Temperature, ice cover, radiation budget, wind, color
ITOS 1-4 ESSA 1-9 NOAA 1-6	1966- 1975 1978-	Polar	Medium and high	Operational	Visible vidicon; IR scanner	Imagery, temperature
ATS 1-3	1966 1967	Synchronous	Medium	Prototype	Visible, IR scanners; data channel	Imagery, temperature, data relay
SMS/GOES 1-5	1974- 1982	Synchronous	High	Operational	Visible, IR scanners; data channel	Imagery, temperature, data relay
GEOS 1-3	1965 1975	Variable	High	Experimental	Laser reflectors; altimeter	Geoid, ocean geoid wave heights
ERTS 1 Landsat 2 Landsat 3	1972 1974 1978	Polar	Medium progressing to high	Prototype	Visible, near-IR scanner; thermal IR scanner	Imagery, temperature
Skylab	1973		Medium progressing to high	Experimental	Cameras; visible IR scanner; spectro radiometer; MW radiometers; altimeter; scatterometer	Imagery, temperature, wave height, wind speed, geoid
Shuttle	1981	Varied	Medium to high	Varied	Synthetic aperture radar	Imagery
Tiros-N	1978	Polar	High	Operational	Visible, IR scanners	Imagery, temperature
Seasat	1978	Near Polar	High	Experimental	Altimeter; synthetic aperture imaging radar; scatometer; MW radiometer; visible/IR scanner	Geoid, wave spectra, heights, wind velocity, surface currents, temperature, atmosphere water, ice cap profiles

TABLE 2. OCEAN-RELATED SPACECRAFT ACTIVITIES: NEXT DECADE

SATELLITE	SPONSOR	SENSOR COMPLEMENT	LAUNCH	STATUS
DMSP	USAF	MR + meteorological sensors	1984	Approved (7-yr coverage
	NASA	MR data processing facility		Proposed planned for MR)
GEOSAT	USN	ALT	1984	Approved (some data to be classified)
MOS-1	JAPAN	CS, IR, MR	1985	Approved
NOSS	USN/NOAA/NASA	ALT, CS, MR, SCAT	1986	Cancelled
ERS-1	ESA	ALT, SAR, SCAT, IR	1987	Approved (essentially)
	NASA	SAR data receiving/processing facility		Proposed
NOAA-H	NOAA	Meteorological sensors (including IR)	1987	Approved (essentially)
	NASA	Contribute piggyback CS		Proposed
SPOT-2	CNES	Terrestrial sensors + piggyback ALT	1987?	Proposed
TOPEX	NASA	ALT + option for SCAT	1988?	Proposed
ROSS	USN	ALT, MR, SCAT	1988?	Proposed
	NOAA	Contribute bus (NOAA-D spacecraft)		Proposed
	NASA	Contribute SCAT		Proposed
MOS-1	JAPAN	SAR	1988?	Proposed
	NASA	Utilize SAR data facility (noted above)		Proposed
GRAVSAT	NASA	Accurate tracking between two satellites	1989?	Proposed
ERS-2	ESA	ALT, SAR, SCAT, ?	1989?	Tentative
RADARSAT	CANADA	SAR	1990?	Proposed
	NASA	Contribute launch & piggyback SCAT		Proposed
MOS-2	JAPAN	ALT, CS, MR, SCAT	1990?	Tentative

TABLE 3. SPACEBORNE SENSORS FOR OCEAN REMOTE MEASUREMENT

Altimeter - a pencil beam microwave radar that measures the vertical distance between the spacecraft and the earth. Measurements yield the topography and roughness of the sea surface from which the ocean geoid, surface current, and average wave height can be estimated.

Vertical height precision:	±10 cm
Wave height precision:	0.5 m or ±0.1 $H_{1/3}$
Horizontal resolution:	1.5 km

Color Scanner - a radiometer that measures the intensity of radiation reflected by the sea in the visible and near-infrared bands in a broad swath beneath the spacecraft. Measurements yield images of ocean color distributions, from which chlorophyll and sediment concentrations can be estimated.

Chlorophyll-a + phaeopigments-a:	±2 X
Sediment concentrations:	±2 X
Horizontal resolution:	1 km

Infrared Scanner - a radiometer that measures the intensity of radiation emitted from the sea in the infrared band in a broad swath beneath the spacecraft. Measurements yield images of sea-surface temperature distributions.

Temperature:	±1.0° C
Horizontal resolution:	1-7 km

Microwave Scanner - a radiometer that measures the intensity of radiation emitted from the sea in the microwave band in a broad swath beneath the spacecraft. Measurements yield microwave brightness temperatures, from which wind speed, water vapor, rain rate, sea-surface temperature, and ice cover can be estimated.

Wind speed:	±2 m/s
Water vapor:	±1 mg/cm^2
Rain rate:	±2 X
Temperature:	±1.5° C
Horizontal resolution:	15-150 km

Scatterometer - a microwave radar that measures the roughness of the sea surface in a broad swath on either side of the spacecraft. Measurements yield the amplitude of short surface waves that are approximately in equilibrium with the local wind and from which the surface wind velocity can be estimated.

Wind speed:	±1.4 m/s
Wind direction:	±18°
Horizontal resolution:	50 km

TABLE 3. SPACEBORNE SENSORS FOR OCEAN REMOTE MEASUREMENT (CONT'D.)

Synthetic Aperture Radar	-	a microwave radar that electronically synthesizes an antenna large enough to achieve a high spatial resolution, data from which are combined to form a high resolution image. Measurements yield information on features (swell, internal waves, rain, current boundaries, and so on) that modulate the amplitude of the short surface waves; they also yield information on the position and character of sea ice from which, with successive views, the velocity of ice floes can be estimated. Swell wavelength: >100 m Swell direction: ±20° Horizontal resolution: 8-25 m

TABLE 4. SOURCES OF OCEANOGRAPHIC SATELLITE DATA

I. EUROPEAN METEOROLOGICAL AND OCEANOGRAPHIC SATELLITE PROGRAM

Execution by the European Space Agency (ESA). The address of the Agency is:

8-10 rue Mario Nikis
75738 Paris, Cedex 15, France

Meteosat
Earth Remote Sensing Satellite (ERS)

II. METEOROLOGICAL SATELLITE PROGRAM OF JAPAN

Name of the satellite program:

Geostationary Meteorological Satellite System (GMSS)
Marine Observational Satellite (MOS)

Authority in charge of preparing for the ground facilities and operating them:

Japan Meteorological Agency (JMA)
1-3-4, Ote-machi
Chiyoda-ku
Tokyo, Japan

Authority in charge of procurement, launching, operating and housekeeping of the spacecraft (GMS):

National Space Development Agency (NASDA)
World Trade Center Building
2-4-1, Hamamatsu-cho
Minato-ku
Tokyo, Japan

TABLE 4. SOURCES OF OCEANOGRAPHIC SATELLITE DATA (CONT'D.)

III. METEOROLOGICAL SATELLITE PROGRAM OF THE UNION OF SOVIET SOCIALIST REPUBLICS

The Soviet system of meteorological satellites in quasi-polar orbits is known as "METEOR". The Hydrometeorological Service of the USSR, 123376, Moscow, Ul, Pavlika Morozova, d. 12, is responsible for the program.

IV. METEOROLOGICAL AND OCEANOGRAPHIC SATELLITE PROGRAMS OF THE UNITED STATES OF AMERICA

Authorities:

Office of Space Science and Applications
National Aeronautics and Space Administration
Washington, D.C. 20546
USA

and

U.S. Department of Commerce
National Oceanic and Atmospheric Administration
National Earth Satellite Service
Washington, D.C. 20233
USA

Authority in charge of relations with users of operational data in real-time, including direct broadcast services:

National Earth Satellite Service
National Oceanic and Atmospheric Administration
Washington, D.C. 20233
USA

Authority in charge of relations with users of archived data:

Satellite Data Services Branch
National Climate Center
Environmental Data Service
National Oceanic and Atmospheric Administration
World Weather Building
Washington, D.C. 20233
USA

TIROS-N Program spacecraft names:

NOAA-1, 2, 3 et seq., TIROS-N

Geostationary Operational Environment Satellite (GOES) spacecraft names:

SMS-1, SMS-2, GOES-1, 2, et seq.

Synchronous Meteorological Satellite (SMS) is the name of the two prototype spacecraft for the GOES system.

Nimbus Program names:

Nimbus-5, 6, 7. Nimbus 7 has considerable emphasis on oceanography.

Ocean Dynamics Satellite (Seasat) - Seasat is a dedicated oceanographic satellite.

GEOS-3 - For geodetic and oceanographic use.

Figure 1 illustrates samples of data from the U.S. spacecraft Seasat on (a) significant wave heights, and (b) satellite-to-surface heights, along a pass starting in the South Pacific, then crossing the Antarctic continent, continuing across the South Atlantic, and ending in Africa. The upper trace shows significant wave heights approaching 6 to 8 m in the "roaring 40's" of south latitude. The altimeter height variability over ocean areas shown in the lower trace is dominantly due to the ellipsoidal shape of the earth; on this scale of display (tens of kilometers) no oceanographic signal is visible. Later papers in this volume will treat the much smaller (10-200 cm) ocean-induced departures from an equipotential surface, which are of great promise for understanding ocean dynamics.

Figure 2 shows a color-coded image obtained from the Coastal Zone Color Scanner, CZCS, a five-channel scanner carried aboard the Nimbus-7 satellite. The area illustrated shows Florida and the Gulf of Mexico, and the quantity represented is the near-surface concentration of chlorophyll-a plus phaeopigments-a, coded according to the scales shown. Horizontal resolution is near 800 m. The range of concentrations is from 0.05 to >1.0 mg/m^3, with the absolute accuracy estimated to be better than a factor of 2. In order to obtain these measurements, relatively complicated algorithms have been applied for correction of atmospheric effects and for calculation of concentrations from functions of the radiances in the various channels. Further discussion of this instrument is found in the chapters ahead.

In Figure 3 is shown an enhanced thermal infrared image of the western North Atlantic Ocean, taken from the Advanced Very High Resolution Radiometer, AVHRR, onboard the NOAA-5 spacecraft. The Gulf Stream and detached mesoscale rings and eddies are visible as warm (dark) features, with cooler slope water (light) seen to the north and east. Horizontal resolution is near 1 km. Under the best of circumstances, absolute surface temperatures $\pm 1.0^{o}C$ are obtainable when atmospheric corrections to the signals in the 10-12 μm band are made.

Data on ocean brightness temperatures from the Scanning Multichannel Microwave Radiometer on Seasat (SMMR) are shown on Figure 4. Because oceanic surface temperatures change only slowly, it is possible to composite observations over several days and still retain a valid representation of surface temperatures over large ocean areas. The absolute accuracy of these measurements is estimated to be better than $\pm 2^{o}C$, but problems remain with data use near continents, in rainfall or in sunlight. Horizontal resolution is about 150 km.

Figure 5 is a composite of data from several sources. The ocean area shown is the Pacific off the west coast of North

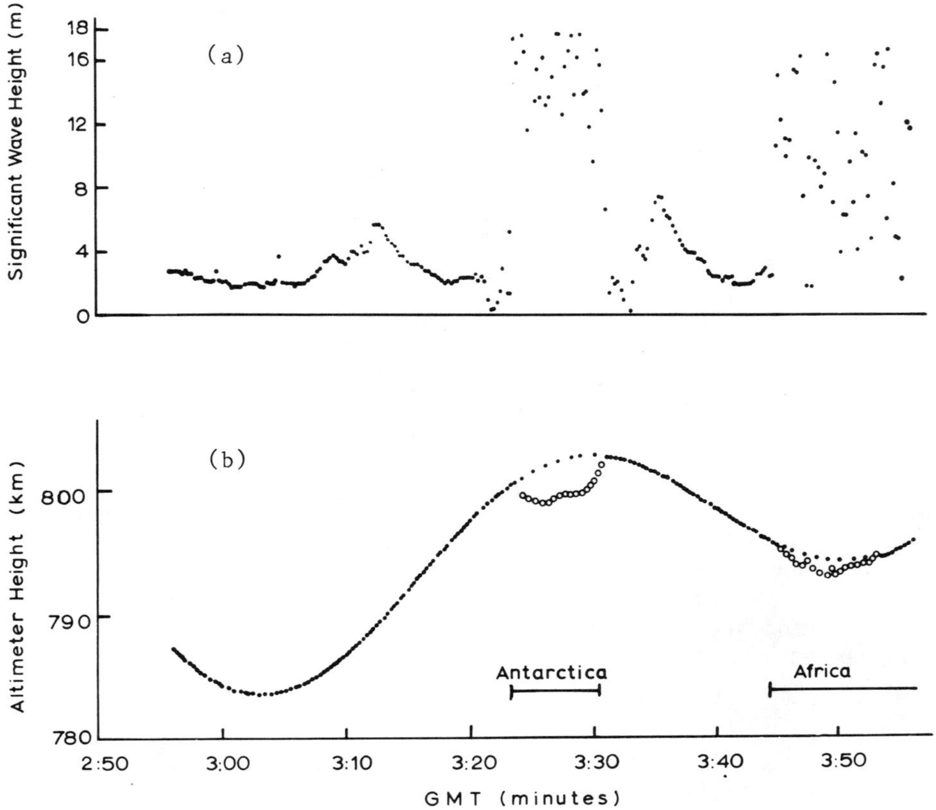

Figure 1. Example of results from the altimeter on SEASAT.

Figure 2 will be found in the Colour Section, p. 449.

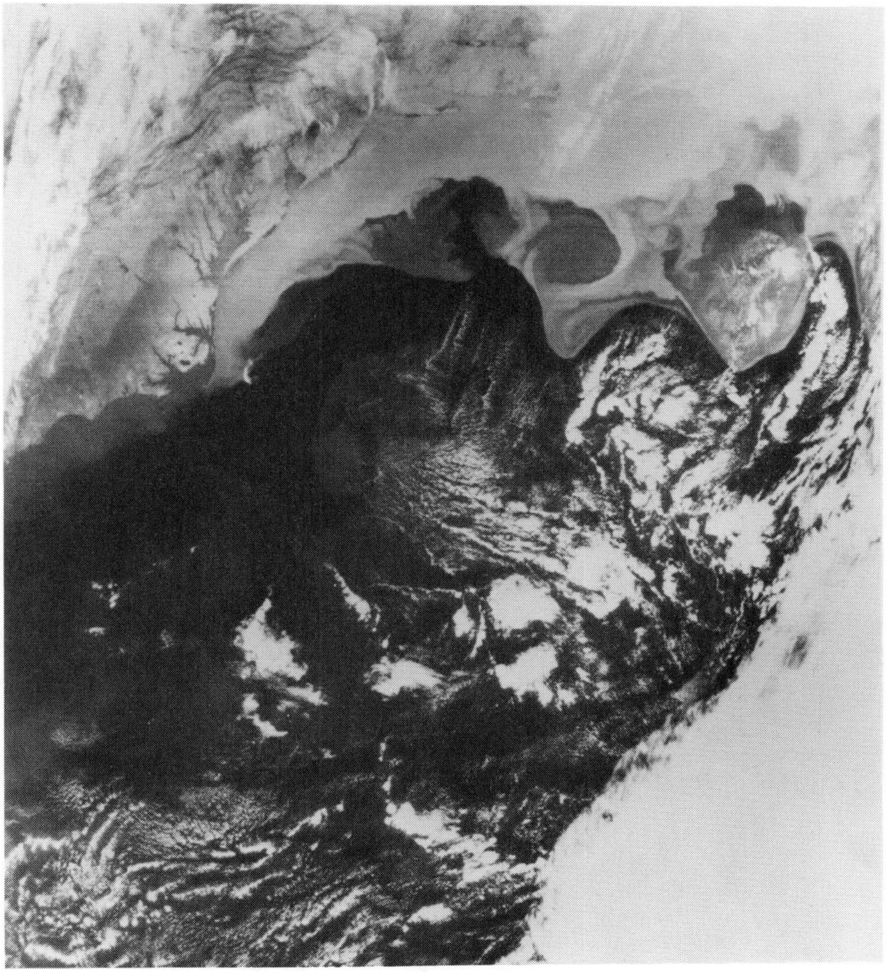

Figure 3. Enhanced thermal infrared image of western North Atlantic Ocean from AVHRR.

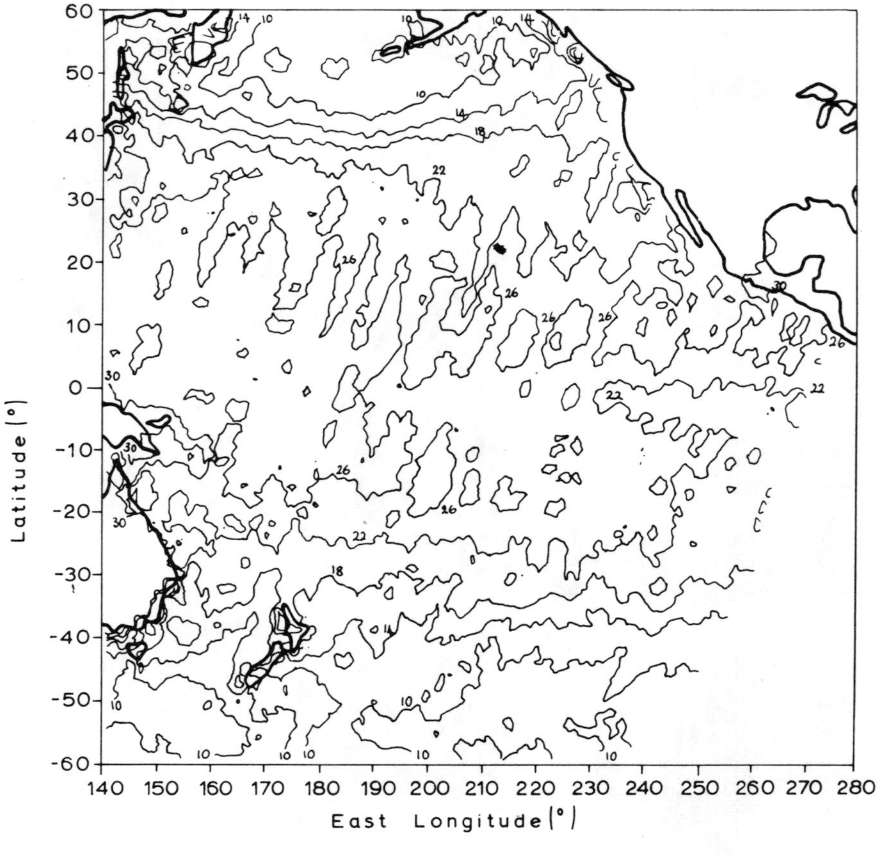

Figure 4. July sea surface temperatures contour map for the Pacific Ocean from the SMMR on SEASAT.

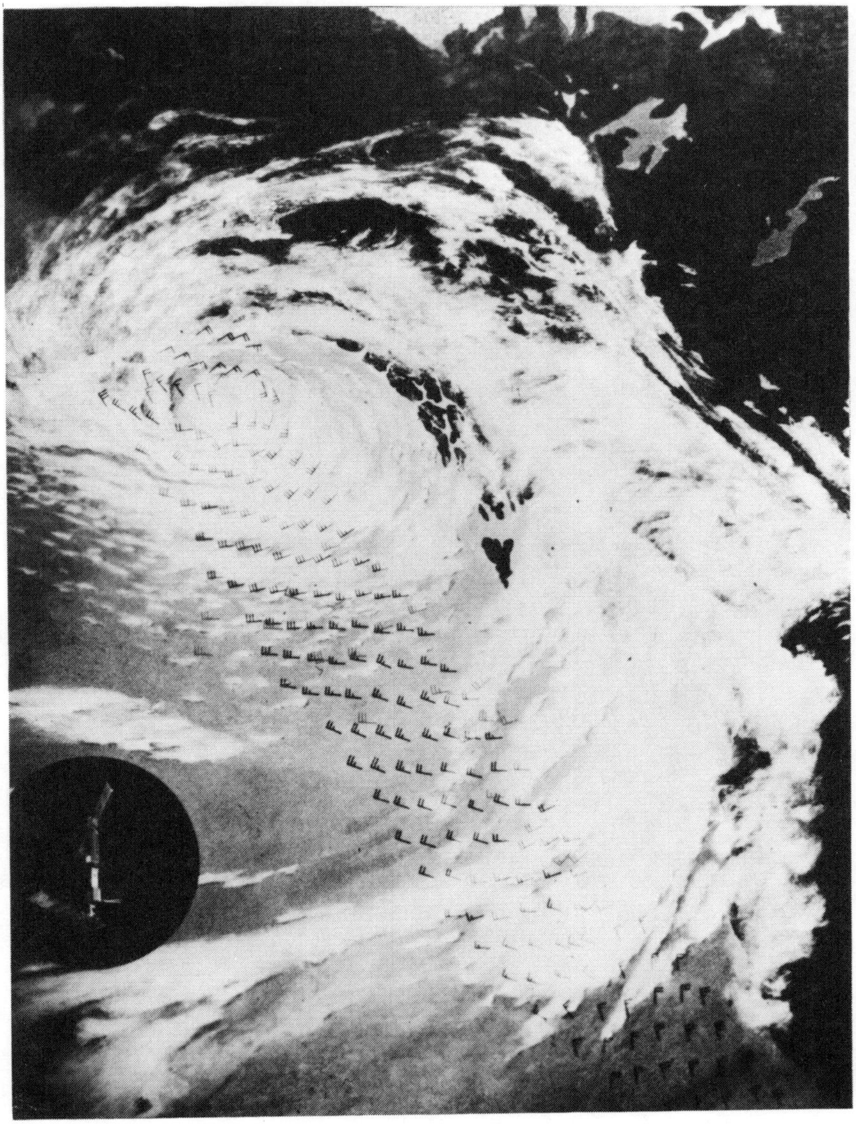

Figure 5. Part of the Pacific Ocean off the west coast of North America.

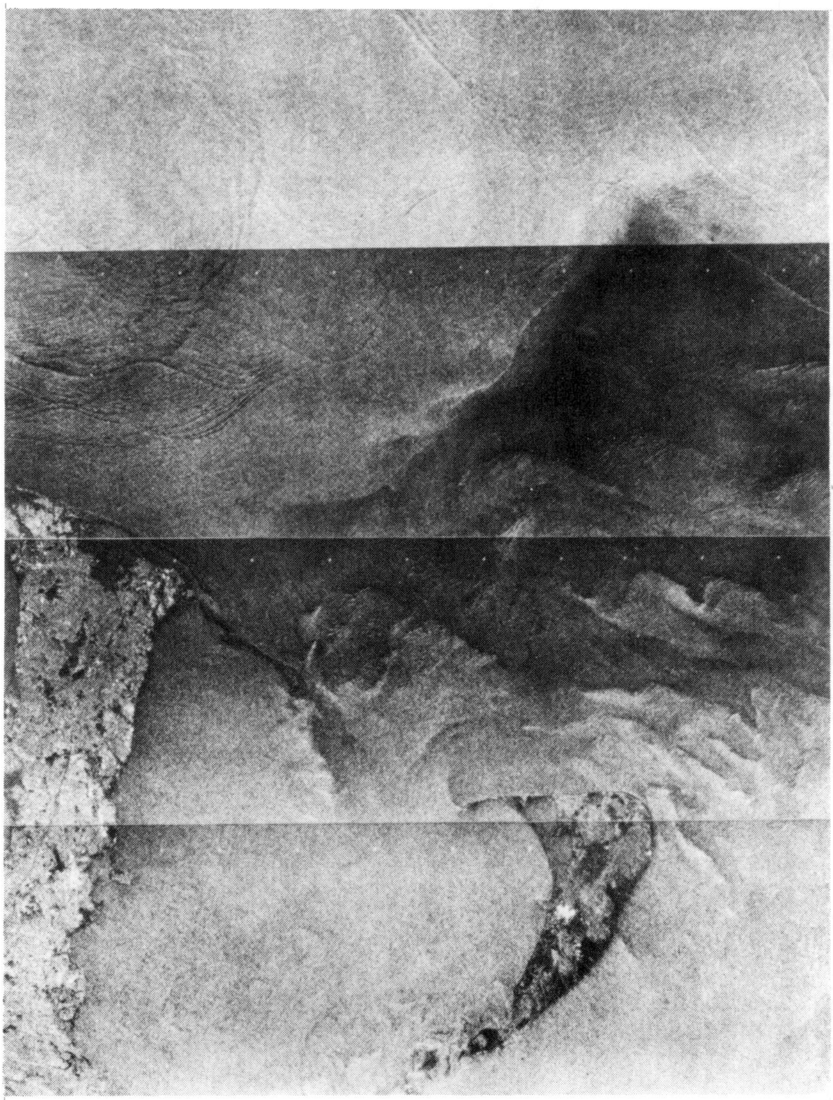

Figure 6(a). Synthetic aperture image from SEASAT.

Figure 6(b). Photograph from Skylab.

America. The visible imagery showing clouds and land mass (dark) comes from the GOES-West geosynchronous satellite and has an intrinsic resolution near 1 km; the darker wind barbs laid out in a diagonal swath are obtained from the Seasat Scatterometer, SASS, while the lighter ones derive from surface ship and buoy measurements. Note how the SASS has resolved the low-pressure system in the Gulf of Alaska and the 90° change in wind direction off California associated with the upper air jet made visible by the cloud structure near the bottom of the image. Spatial resolution is about 25 km, and wind speed accuracies better than ±2 m/s.

Figure 6(a) is a high resolution (40 m) image made by the Seasat Synthetic Aperture Radar, SAR, southeast of Cape Cod, Massachusetts, USA; the island of Nantucket is shown near the bottom. The image dimensions are approximately 75 x 100 km. Visible are regions of surface roughness overlying bottom features that are at depths up to 20 m, as well as surface signatures of internal waves. Tidal flow over variable bathymetry is responsible for the visibility of both of these phenomena. Figure 6(b) is a black-and-white reproduction of an Ektachrome photo of the same area, taken several years earlier from Skylab. The light regions visible in the ocean are sand banks and shoals; the correlation between radar backscatter in Figure 6(a) and upwelling light in Figure 6(b) is quite high.

These samples of satellite data are quite diverse and yield a wide range of geophysical and biological information. It is clear that the synthesis of such data into spatial and temporal fields can yield much greater intelligence than is contained in a single sample. The value of such information is much enhanced by using it in conjunction with in-water data, and with models of the processes being studied. The validity of the results can be tested in part by their predictive skills, which will depend to a considerable degree on the imaginations and talents of the investigators.

SUMMARY

It has almost invariably been the case that the introduction of a significant new instrument technology has yielded for the science to which it was applied a number of unsuspected and often highly significant results. Such serendipitous discoveries can surely be expected from instruments as advanced as those being orbited on ocean-looking satellites. Oceanographers have been hard put to gain the overview of their domain required to understand synoptic or planetary scale events in the sea; for a limited but important group of phenomena, satellites promise to provide the vantage point for this vision.

CHAPTER 2

NOAA'S ACTIVITIES IN THE FIELD OF MARINE REMOTE SENSING

Bruce H. Needham

NATIONAL OCEANIC AND ATMOSPHERIC ADMINISTRATION
SATELLITE DATA SERVICES DIVISION
WASHINGTON, D.C.

ABSTRACT: The U.S. National Oceanic and Atmospheric Administration's (NOAA) Satellite Data Services Division (SDSD), part of the Environmental Data and Information Service's (EDIS), National Climatic Center (NCC), represents an unique archive of data and products from NOAA's operational geostationary and polar orbiting satellites, and several of NASA's experimental satellites. The SDSD archive, containing photographic imagery, paper charts, and digital data tapes, dates from April 1960 to the present. The remainder of this report shall highlight these satellite systems, sensors, data reception and archiving, products, services and future plans.

1. SATELLITE SYSTEMS AND SENSORS

Figure 1 illustrates the U.S. satellites with data on archive at SDSD which are useful for observing oceanographic and meteorological phenomena from space. These satellites, dating back to TIROS-I (Television Infrared Observational Satellite) in 1960 up to the present NOAA-7 satellite, are basically comprised of two types of spacecraft: polar orbiting and geostationary.

1.1 Polar Orbiting Satellites

Polar orbiting satellites are in a relatively low orbit around the Earth (approximately 500 to 900 miles or 800 to 1500 km). This allows them to circle the globe from 12 to 14 times a day and obtain imagery and quantitative digital data along a path on the earth's surface up to 1,550 miles (2,500km) in width during both daytime and nighttime. Table 1 lists pertinant or-

A. P. Cracknell (ed.), Remote Sensing Applications in Marine Science and Technology, 17–28.

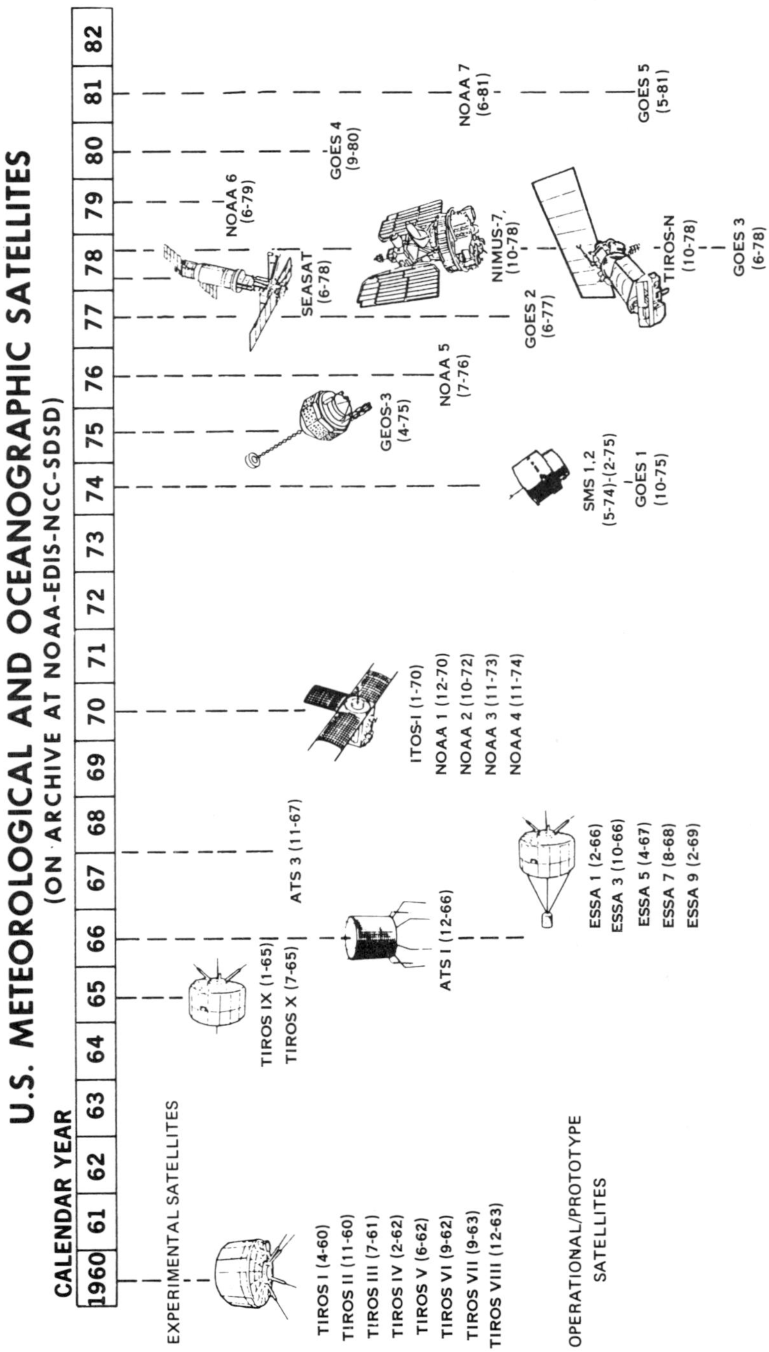
U.S. METEOROLOGICAL AND OCEANOGRAPHIC SATELLITES
(ON ARCHIVE AT NOAA-EDIS-NCC-SDSD)
CALENDAR YEAR
1960 61 62 63 64 65 66 67 68 69 70 71 72 73 74 75 76 77 78 79 80 81 82
EXPERIMENTAL SATELLITES
TIROS I (4-60)
TIROS II (11-60)
TIROS III (7-61)
TIROS IV (2-62)
TIROS V (6-62)
TIROS VI (9-62)
TIROS VII (9-63)
TIROS VIII (12-63)
TIROS IX (1-65)
TIROS X (7-65)
ATS I (12-66)
ATS 3 (11-67)
GEOS-3 (4-75)
SEASAT (6-78)
NOAA 6 (6-79)
NIMUS-7, (10-78)
OPERATIONAL/PROTOTYPE SATELLITES
ESSA 1 (2-66)
ESSA 3 (10-66)
ESSA 5 (4-67)
ESSA 7 (8-68)
ESSA 9 (2-69)
ITOS-I (1-70)
NOAA 1 (12-70)
NOAA 2 (10-72)
NOAA 3 (11-73)
NOAA 4 (11-74)
SMS 1,2 (5-74)-(2-75)
GOES 1 (10-75)
NOAA 5 (7-76)
GOES 2 (6-77)
TIROS-N (10-78)
GOES 3 (6-78)
GOES 4 (9-80)
NOAA 7 (6-81)
GOES 5 (5-81)

Figure 1.

bital characteristics and other information on these satellites. Table 2 lists the sensors that each of these NOAA Satellites carried along with their spectral characteristics, resolution, etc. Table 3 lists similar information for the NASA experimental satellites whose data are archived at the SDSD (GEOS-3, SEASAT, NIMBUS-7).

1.2 Geostationary Satellites

The Geostationary Satellites are essentially "parked" in an orbit about 22,000 miles (36,000km) above the Earth's surface at the Equator and travel at a rate of about 6,820 mph (11,000km/hr). At this altitude and speed, a satellite remains continuously above the same point on the Earth and, thus, is termed geostationary, geo-synchronous, Earth-synchronous, or merely synchronous. The NOAA series of geostationary satellites commenced in 1966 with the launch of the first ATS satellite (Applications Technology Satellite), followed in 1974 with the SMS-Series (Synchronous Meteorological Satellite), and in 1975 with the GOES-series (Geostationary Operational Environmental Satellite). Table 4 lists the pertinent orbital characteristics and other information on these satellites, and Table 5 lists their current locations. Table 6 lists the sensors that each of these satellites carried along with their spectral characteristics, resolutions, etc. The sensors on board acquire data and imagery of the complete Earth disk (about ¼ of the surface) every 30 minutes, 24 hours a day.

At the present, the U.S. has seven functional geostationary satellites in space. Two are currently "active", and the rest are in a "standby" mode. The active U.S. satellites are located above the Equator at 75°W and 135°W as shown in Figure 2.

2. DATA RECEPTION & ARCHIVING

2.1 Polar Orbiting Satellites

Data from the NOAA operational orbiting satellites are archived in several formats including raw (or level Ib) digital tapes, digital tapes of derived products (SST, ICE, etc.), photographic imagery, and paper charts. The signals from the satellites are acquired at two U.S. ground stations (Fairbanks, Alaska and Wallops Island, Virginia), where it is relayed via a domestic communications satellite to the NESS Control Center in Suitland, Maryland. These data include real-time data as it is being acquired and transmitted as the satellite passes over these ground stations, as well as data recorded on board the satellite from elsewhere around the globe. After both of these data types arrive at NESS, they are retained in their "raw" state, processed into "operational" products and photographic imagery, and made

TABLE 1. U.S. POLAR-ORBITING ENVIRONMENTAL SATELLITES AND EXPERIMENTAL SATELLITES WHOSE DATA ARE ARCHIVED AT SATELLITE DATA SERVICES DIVISION

SATELLITE	LAUNCHED	PERIOD (MIN)	PERIGEE (KM)	APOGEE (KM)	INCLINATION (DEGREES)	DATES OF DATA ON ARCHIVE AT SDSD
TIROS-I	4/1/60	99.2	796	867	48.3	4/ 1/60 - 6/14/60
TIROS-II	11/23/60	98.3	717	837	48.5	11/23/60 - 9/27/61
TIROS-III	7/12/61	100.4	854	937	47.8	7/12/61 - 1/23/62
TIROS-IV	2/8/62	100.4	817	972	48.3	2/ 8/62 - 6/18/62
TIROS-V	6/19/62	100.5	680	1119	58.1	6/19/62 - 5/14/63
TIROS-VI	9/18/62	98.7	783	822	58.2	9/18/62 -10/21/63
TIROS-VII	6/19/63	97.4	713	743	58.2	6/19/63 - 2/26/66
TIROS-VIII	12/21/63	99.3	796	878	58.5	12/21/63 - 2/12/66
TIROS-IX	1/22/65	119.2	806	967	96.4	1/23/65 - 9/ 9/66
TIROS-X	7/2/65	100.6	848	957	98.6	7/ 2/65 - 4/ 2/66
ESSA-1	2/3/66	100.2	800	965	97.9	2/ 4/66 -10/ 6/66
ESSA-3	10/2/66	114.5	1593	1709	101.0	10/ 4/66 - 6/ 1/67
ESSA-5	4/20/67	113.5	1556	1635	101.9	6/ 1/67 -12/ 3/68
ESSA-7	8/16/68	114.9	1646	1691	101.7	4/16/68 - 3/31/69
ESSA-9	2/26/69	115.3	1637	1730	101.9	4/ 1/69 -11/15/72
ITOS-1	1/23/70	115.1	1648	1700	102.0	4/28/70 - 6/17/71
NOAA-1	12/11/70	114.8	1422	1472	102.0	4/26/71 - 6/20/71
NOAA-2	10/15/72	114.9	1451	1458	98.6	11/16/72 - 3/19/74
NOAA-3	11/6/73	116.1	1502	1512	101.9	3/26/74 -12/17/74
NOAA-4	11/15/74	101.6	1447	1461	114.7	12/17/74 - 9/15/76
GEOS-3	4/9/75	100.6	840	848	115.7	4/14/75 -12/ 1/78
NOAA-5	7/29/76	116.2	1504	1518	102.1	9/15/76 - 3/16/78
SEASAT	6/27/78	100.8	794	808	108.1	7/ 7/78 -10/ 9/78
TIROS-N	10/13/78	98.92	849	864	102.3	2/ 5/79 -11/ 1/80
NIMBUS-7	10/24/78	99.28	943	955	104.9	10/24/78 - Present
NOAA-6	6/27/79	98.68	805	822	98.7	6/27/79 - Present
NOAA-7	6/23/81	102.04	852	870	98.9	6/23/81 - Present

NOTES: 1. ESSA-2,-4,-6, and -8 were Automatic Picture Transmission (APT) satellites; hence no negatives are archived.
2. NIMBUS-7 - Only data from the Coastal Zone Color Scanner (CZCS) are archived at SDSD.

TABLE 2. POLAR ORBITER
NOAA SATELLITE SENSOR CHARACTERISTICS

SATELLITE (SERIES)	SENSOR NAME	BAND	SPECTRAL RANGE	GROUND RESOULTION	SWATH WIDTH
TIROS-I thru X	VIDICON	VIS	0.45-0.65μm	3.8 km	800 km
ESSA 1,3,5,7,9	ADVANCED VIDICON CAMERA SYSTEMS (AVCS)	VIS	0.45-0.65μm	2.2 km	2300 km
	LOW RESOLUTION INFRARED RADIOMETER (LRIR)	IR	---------	------	VARIABLE

SATELLITE (SERIES)	SENSOR NAME	BAND	SPECTRAL RANGE	GROUND RESOULTION	SWATH WIDTH
ITOS-1/NOAA-1	AVCS	VIS	0.45- 0.65μm	2.2 km	2300 km
	SCANNING RADIOMETER (SR)	VIS	0.50- 1.00μm	4 km	4000 km
		IR	10.50- 0.65μm	8 km	4000 km
NOAA-2 thru 5	SCANNING RADIOMETER (SR)	VIS	0.50-12.50μm	4 km	4000 km
		IR	10.50-12.50μm	8 km	4000 km
	VERY HIGH RESOLUTION RADIOMETER (VHRR)	VIS	0.60- 0.70μm	0.8 km	2580 km
		IR	10.50-12.50μm	0.8 km	2580 km
	VERTICAL TEMPERATURE PROFILE RADIOMETER (VTPR)	1	14.92-14.99μm	68 km	1876 km
		2	14.65-14.87μm	68 km	1876 km
		3	14.29-14.49μm	68 km	1876 km
		4	14.03-14.22μm	68 km	1876 km
		5	13.70-13.88μm	68 km	1876 km
		6	13.30-13.48μm	68 km	1876 km
		7	18.38-19.01μm	68 km	1876 km
		8	11.93-12.08μm	68 km	1876 km

SATELLITE (SERIES)	SENSOR NAME	BAND	SPECTRAL RANGE	GROUND RESOLUTION	SWATH WIDTH
TIROS-N/NOAA-6	ADVANCED VHRR (AVHRR)	1	0.550- 0.90 μm	1.1-4 km	2580/4000 km*
		2	0.725- 1.10 μm	1.1-4 km	2580/4000 km*
		3	3.550- 3.93 μm	1.1-4 km	2580/4000 km*
		4	10.050-11.05 μm	1.1-4 km	2580/4000 km*
	TIROS-N OPERATIONAL VERTICAL SOUNDER (TOVS)				
	HIGH RES IR SPECTROMETER (HIRS)	1-20	.660-14.98 μm	17.4 km	1120 km
	STRATOSPHERIC SOUNDING UNIT (SSU)	1- 3	14.970 μm	147.3 km	1473 km
	MICROWAVE SOUNDING UNIT (MSU)	1- 4	50.300 57.95 GHz	109 km	2320 km
NOAA-7	[SAME AS ABOVE WITH ADDITION OF EXTRA CHANNEL IN AVHRR]	4	10.300-11.30 μm	1.1-4 km	2580/4000 km*
		5	11.500-12.50 μm	1.1-4 km	2580/4000 km*

* Note: AVHRR resolution is 1.1 km in high resolution mode; 4 km in low resolution mode. Swath width is 2580 km in high resolution mode; 4000 km in low resolution mode.

TABLE 3. NASA EXPERIMENTAL SATELLITE SENSOR CHARACTERISTICS

SATELLITE (SERIES)	SENSOR NAME	BAND	SPECTRAL RANGE	GROUND RESOLUTION	SWATH WIDTH
GEOS-3	ALTIMETER (ALT)	1	13.900 GHz (2.16 cm)	3.6 km (HR)	3 km (HR)
				14.3 km (LR)	9 km (LR)
SEASAT	ALTIMETER (ALT)	1	13.900 GHz (2.16 cm)	2.4- 12 km	2.4 km - 12 km
	SCATTEROMETER (SASS)	1	14.599 GHz (2.05 cm)	50 km	1000 km
	SCANNING MICROWAVE RADIOMETER (SMMR)	1	6.600 GHz (4.50 cm)	136 x 89 km	600 km
		2	10.700 GHz (2.80 cm)	87 x 58 km	600 km
		3	18.000 GHz (1.70 cm)	57 x 35 km	600 km
		4	21.000 GHz (1.40 cm)	44 x 29 km	600 km
		5	37.000 GHz (0.80 cm)	28 x 18 km	600 km
	VISIBLE INFRARED RADIOMETER (VIR)	1	0.470 - 0.94 μm	2 km	2280 km
		2	10.500 -12.50 μm	4 km	2280 km
	SYNTHETIC APERTURE RADAR (SAR)	1	1.350 GHz (22 cm)	25 m	100 km

SATELLITE (SERIES)	SENSOR NAME	BAND	SPECTRAL RANGE	GROUND RESOLUTION	SWATH WIDTH
NIMBUS-7	COASTAL ZONE COLOR SCANNER (CZCS)	1	0.433 - 0.453 μm	825 m	1600 km
		2	0.510 - 0.530 μm	825 m	1600 km
		3	0.540 - 0.560 μm	825 m	1600 km
		4	0.660 - 0.680 μm	825 m	1600 km
		5	0.700 - 0.800 μm	825 m	1600 km
		6	10.500 - 12.500 μm	825 m	1600 km

TABLE 4
U.S. GEOSTATIONARY SATELLITES
WHOSE DATA ARE ARCHIVED AT SATELLITE DATA SERVICES DIVISION

SATELLITE	LAUNCHED	PERIOD	PERIGEE (KM)	APOGEE (KM)	INCLINATION (DEGREES)	DATES OF DATA ON ARCHIVE AT SDSD
ATS-I	12/6/66	24 Hr.	41,257	42,447	0.2	1/1/67 - 10/16/72
ATS-III	11/5/67	24 Hr.	41,166	41,222	0.4	3/2/68 - 9/2/74
SMS-1	5/17/74	24 Hr.	35,605	35,975	0.6	6/27/74 - 1/7/76
SMS-2	2/6/75	24 Hr.	35,482	36,103	0.4	3/10/75 - 8/4/81
GOES-1	10/6/75	24 Hr.	35,728	36,847	0.8	1/8/76 - 3/15/80
GOES-2	6/16/77	24 Hr.	35,600	36,200	0.5	8/15/77 - 9/15/80
GOES-3	6/15/78	24 Hr.	35,600	36,200	0.5	7/13/78 - 3/5/81
GOES-4	9/9/80	24 Hr.	35,782	36,200	0.2	3/5/81 - Present
GOES-5	5/15/81	24 Hr.	35,600	35,600	0.2	7/9/81 - Present

TABLE 5
LOCATION OF GEOSTATIONARY SATELLITES
(AS OF JULY 31, 1982)

SATELLITE	OPERATIONAL/STANDBY	LOCATION	DESIGNATION	COMMENTS
SMS-2	Standby	75°W		
GOES-1	Standby	127°W		
GOES-2	Standby	107°W		
GOES-3	Standby	130°W		
GOES-4	Operational	135°W	GOES-WEST	Effective 3/5/81
GOES-5	Operational	75°W	GOES-EAST	Effective 8/4/81

TABLE 6. GEOSTATIONARY
NOAA SATELLITE SENSOR CHARACTERISTICS

SATELLITE (SERIES)	SENSOR NAME	BAND	SPECTRAL RANGE	GROUND RESOLUTION	SWATH WIDTH
ATS-I & III	SPIN SCAN CAMERA (SSC)	VIS	0.550- 0.70μm	4 km	HORIZON TO HORIZON
SMS-1 & 2	VISIBLE INFRARED SPIN SCAN RADIOMETER (VISSR)	VIS	0.550- 0.70μm	14 km	HORIZON TO HORIZON
		IR	10.500-12.60μm	8 km	HORIZON TO HORIZON
GOES 1-5	VISSR	VIS	0.550- 0.70μm	14 km	HORIZON TO HORIZON
		IR	10.500-12.60μm	8 km	HORIZON TO HORIZON
	VISSR ATMOSPHERIC SOUNDER (VAS) [GOES-4 AND 5 ONLY]	V-1	0.550- 0.70μm	14 km	HORIZON TO HORIZON
		IR-1	14.600-14.81μm	16 km	HORIZON TO HORIZON
		IR-2	14.290-14.62μm	16 km	HORIZON TO HORIZON
		IR-3	14.060-14.39μm	16 km	HORIZON TO HORIZON
		IR-4	13.790-14.18μm	16 km	HORIZON TO HORIZON
		IR-5	13.120-13.48μm	16 km	HORIZON TO HORIZON
		IR-6	4.496- 4.537μm	16 km	HORIZON TO HORIZON
		IR-7	12.500-12.82μm	16 km	HORIZON TO HORIZON
		IR-8	10.360-12.12μm	16 km	HORIZON TO HORIZON
		IR-9	7.143- 7.353μm	16 km	HORIZON TO HORIZON
		IR-10	6.390- 7.067μm	16 km	HORIZON TO HORIZON
		IR-11	4.386- 4.484μm	16 km	HORIZON TO HORIZON
		IR-12	3.623- 4.310μm	16 km	HORIZON TO HORIZON

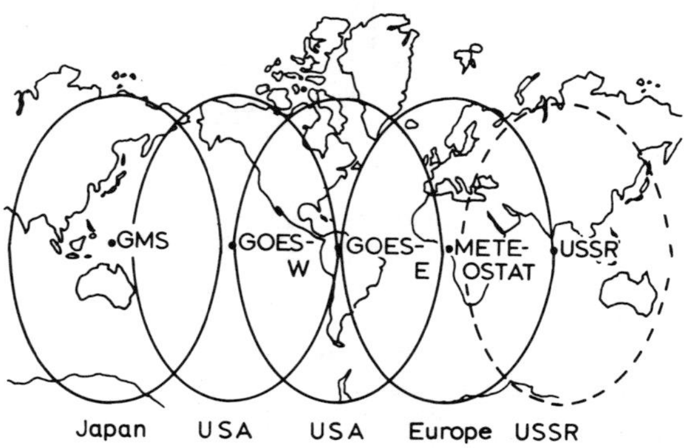

Figure 2. Positions of geostationary satellites.

available to SDSD for archival normally within 24 hours.

2.2 Geostationary Satellites

Data from the two operational geostationary satellites are transmitted to NOAA's Wallops Island ground receiving station every ½ hour, 24 hours a day. The signals are reformatted at Wallops, retransmitted to the satellite and then transmitted directly to the NESS facility in Suitland. Here it is then sent via microwave link to the NESS facility in the World Weather Building. From here several things happen. Selected data are extracted and sent quasi-real-time to several NESS Satellite Field Service Stations throughout the U.S. Both digital and photographic data products are produced at this time. Because of the high data content of the VISSR sensor, only selected, reduced resolution data are retained on digital tapes at SDSD. Full resolution digital data from both GOES-EAST and GOES-WEST are archived and reproduced for SDSD by the University of Wisconsin. The NESS products are turned over to SDSD within 24 hours for archival.

Digital data tapes and products from the NOAA operational polar orbiting satellites, the GOES satellites, and NASA's GEOS-3, SEASAT and NIMBUS-7 are archived at SDSD's computer facility in Suitland, Maryland. Photographic and paper chart products from these systems are archived at SDSD's Photographic Archive in the World Weather Building.

The SDSD Archive currently has data from 36 satellites which carried 25 different instruments from 1960 to the present. Holdings include over 105,000 digital tapes and approximately 10,000, 000 photographic images in various formats and sizes.

3. PRODUCTS & SERVICES

3.1 Geostationary Satellites

Data from the two operational geostationary satellites are obtained every ½ hour, 24 hours a day from each of the satellites. GOES-EAST data are acquired on the hour and ½ hour (i.e., 1800Z, 1830Z), while GOES-WEST data are acquired on the quarter hour (i.e., 1815Z, 1845Z). The GOES data products are available in several different formats.

3.1.1 Magnetic Tape Products

Type A: Reduced Resolution (8km VIS. 8km IR)

Dates: July 26, 1976 to September 5, 1978

Times:	GOES-EAST	GOES-WEST
	0930Z IR	0915Z IR
	1000Z IR	0945Z IR
	1600Z VIS & IR	1515Z IR
	2130Z IR	1545Z IR
	2200Z IR	2145Z VIS & IR

Dates: September 6, 1978 to present (every 3 hours)

Times:	GOES-EAST	GOES-WEST
	0000Z - 2200Z IR	0245Z - 2345Z IR
	1500Z - 2400Z VIS	1445Z - 2345Z VIS

Type B: Full Resolution (8km IR, 1km VIS) (every ½ hour)

Dates; GOES-EAST: February 18, 1978 to present
GOES-WEST: November 20, 1978 to present
GOES-Indian Ocean: December 1, 1978 to November 30, 1979

3.1.2 Photographic Products (VISSR)

Type A: Full Disk & Sectors (25 x 25 cm negatives) VIS & IR (every ½ hour)

Dates: SMS-1 (GATE): 6/27/74 - 9/21/74
GOES-EAST : 9/22/74 - present
GOES-WEST : 3/10/75 - presnet

Type B: Full Disk VIS & IR Microfilm (35mm) 15 days/reel (every 1 hour)

Dates: SMS-1 (GATE): 6/27/74 - 9/21/74
GOES-EAST : 9/1/74 - 12/31/80
GOES-WEST : 3/11/75 - 7/31/80

NOTE: Sample GOES photographic products are shown in Figures 3, 4, 5, 6 & 7 in Satellite Data Users Bulletin, Volume 2, No. 1, 6/80.

3.2 Polar Orbiting Satellites

Data from the NOAA polar orbiting satellites are also maintained as digital tapes, photographic imagery, and paper chart products. (Samples of the photographic products are illustrated in Figures 8-11, in Satellite Data Users Bulletin, Volume 2, No. 1, 11/81).

The TIROS-I through X series are available as 35mm microfilm (2-3 days imagery/reel) from 4/1/60 - 4/20/66, and as computer tapes.

The ESSA 1, 3, 5, 7 & 9 series data are available as 35mm microfilm (2/4/66 - 11/16/71), individual 25 x 25 cm negatives, and Polar and Mercator Mosaics (25 x 25 cm, 35mm and CCTs).

The ITOS-1-NOAA-5, TIROS-N, and NOAA-6 and NOAA-7 series

archive contains a multitude of archived data. Basically, they consist of the individual pass-by-pass photographics (25 x 25 cm, 35mm, and CCT); high resolution frames (25 x 25 cm and CCT); Polar & Mercator Mosaics (25 x 25 cm, 35mm, & CCTs); Sounding Data (CCTs); SST products (charts, 25 x 25 cm, CCTs); Northern Hemisphere Snow & Ice, Gulf Stream, Ice Charts, etc.

NOTE: A complete listing of dates and products are found in the National Holdings of Environmental Satellite Data of the National Oceanic and Atmospheric Administration, August 1981.

3.3 NASA Experimental Satellites

3.3.1 GEOS-3

Data from the Altimeter (ALT) on GEOS-3 are archived in two formats for the period from 4/14/75 - 12/1/78.

"G" - TAPES - 11 CCTs, geophysical data, 1 sample/second.
"I" - TAPES - 131 CCTs, raw data, 10 samples/second.

3.3.2 SEASAT

Data from the sensors on board SEASAT are archived as magnetic tapes and photographic products. Data from the ALT, SASS, and SMMR are available only on CCT as global coverage for the entire mission.

Data from the SAR are available as 70mm photographic film over most areas of the U.S., Canada, Central America, Western Europe, and adjacent waters; and as CCTs and film products for those limited numbers of scenes that were digitally correlated. Illustrations of the SAR imagery are shown in Figures 1 & 2 in Satellite Data Users Bulletin, Volume 2, No. 1, 6/80.

3.3.3 NIMBUS-7 CZCS

Data are archived as photographic imagery (25 x 25 cm negatives) and digital tapes. Coverage is essentially global, but data are normally only processed over cloud free areas. Two levels of data are archived (Level I and Level II). A sample CZCS Level I image is illustrated in Figure 12, Satellite Data Users Bulletin, Volume 2, No. 1, 6/80.

3.4 User Services

The staff at SDSD, in addition to providing a safe archive for all these data, also responds to user requests for duplication of the data. Requests are handled from users throughout the world, and the costs to the user are kept to a reasonable charge to cover the retrieval, reproduction and dissemenation expenses.

The SDSD can provide magnetic tapes in various formats and density sizes, as well as photographic reproductions to various scales. Complete duplication of the archive tape of image is not necessary since the SDSD can extract specific areas, dates, channels, etc., from the tapes, or enlarge small areas from an archive negative. Normal delivery for photographic products is 5 working days and digital data in 1-2 weeks.

4. FUTURE SERVICES & PLANS

NOAA is constantly striving to provide better and quicker products and services to its users. While the prime purpose of NOAA's satellite service is to provide up-to-the-minute satellite data and information to the U.S. National Weather Service, it is also looking for better ways to provide information to the shipping industry for ice conditions and severe storms; the fishing industry for locating prime areas of fishing potential; and also the recreational and scientific research communities.

Recently, NESS has updated the process by which it calculates the Sea Surface Temperature of the world's oceans. Previously, this was accomplished by utilizing only one channel of satellite information. The process now utilized includes inputs from two channels as well as observations from ships throughout the world. The weekly Gulf Stream Analysis Chart is now produced in two sectors, five times a week, and includes SST observations in addition to the visually observed location of the Gulf Stream and eddies.

In response to user demands for the highest resolution data available, NESS commenced in 1979 to save, for an indefinite period of time, the raw (level Ib) digital data from it's polar orbiting satellites which were previously only archived for 90 days. The SDSD, in conjunction with the University of Wisconsin, started in 1978 to archive the full resolution digital data for both of the GOES satellites.

Access to the raw, full resolution digital data allows researchers to enhance the data however they see fit, rather than have to contend with a "product" from NOAA which may have been derived for a different purpose.

In the near future, the SDSD will have the capability, through its own organization, to provide enhanced products to users (either digital tapes or photographic products) to their own specifications.

The SDSD also realizes that the requirement exists to have a central "clearing house" of information for satellite data users throughout the world. An interactive, remote terminal "electronic catalogue" has been envisioned by SDSD, but due to budget constraints only the preliminary design was able to be completed. Within this "catalogue" would have been detailed information on satellites, sensors, and data availability from NOAA's satellites, as well as reference material and points of contact for non-NOAA and non-U.S. satellites. Eventually, users would have been able to query this "catalogue" for data availability, confirm the location of the archive, obtain price quotations on the cost of reproduction, and actually place an order, all electronically.

Since SDSD still believes strongly in this requirement, it is looking forward to possibly incorporating at least some of its information on products and services into similar systems, such as the UNESCO/IOC Marine Environmental Data Information Referral System (MEDI) in Paris, France and the NOAA/EDIS National Environmental Data Referral System (NEDRES) in Washington, D.C. A listing of pertinant names, addresses, and telephone numbers is enclosed as Table 7.

The SDSD, and NOAA in general, will continue to interact with users to provide the best quality service available.

NOAA will continue to operate two geostationary satellites at 75°West and 135°West Longitude into the foreseeable future. Additional satellites to replace those in orbit will be launched as necessary as the current ones fail.

The NOAA polar orbiting satellite program, initially devised to operate two satellites in orbit at any one time, has recently been altered. Current plans are to only operate one polar orbiter at any one time. As NOAA-7 fails, it will be replaced with the next in this series NOAA-8. Due to recent budgetary cutbacks within NOAA, it is planned to attach some experimental sensors (i.e., a new Coastal Zone Color Scanner) on the future polar orbiters, rather than launch an entirely dedicated new experimental satellite.

5. SUMMARY

As explained, NOAA and the SDSD have been able to provide an excellent service to users of satellite remotely sensed data throughout the world. Continuous research is underway to ensure that the best quality data and current state-of-the-art techniques are utilized. The SDSD has, even with its relatively small staff, been able to make the data available known to users throughout the world by issuing its Satellite Data Users Bulletin, participation in conferences and symposia, and through university lectures. The SDSD has seen the numbers of requests for data increase by more than 10% per year, and is pleased to note that currently over 20% of its "business" is from non-U.S. users.

Additional information on these satellites, sensors, and products, including current price lists, brochures, sample imagery, copies of publications in stock, and current documentation on such systems can be obtained by writing to the address in Table 7.

TABLE 7
POINTS OF CONTACT FOR DATA AND INFORMATION

U.S. ENVIRONMENTAL SATELLITES	JAPANESE SATELLITES
Satellite Data Services Division NOAA/EDIS/NCC World Weather Building, Room 100 Washington, D.C. 20233 (Telephone No. 301-763-8111)	Data Processing Department Meteorological Satellite Center 3-235 Nakaiyoto Kiyose Tokyo, 180 JAPAN (Telephone No. 81-03-0424-93-1100)
NASA EXPERIMENTAL SATELLITES	LANDSAT
National Space Science Data Center NASA/Goddard Space Flight Center Code 601 Greenbelt, Maryland 20771 (Telephone No. 301-344-6695)	Users Services U.S. Geological Survey EROS Data Center Sioux Falls, South Dakota 57198 (Telephone No. 605-594-6151)
U.S. DEFENSE METEOROLOGICAL SATELLITE	CLIMATIC DATA
Space Science & Engineering Center DMSP Archive University of Wisconsin 1225 West Dayton Street Madison, Wisconsin 53706 (Telephone No. 608-262-5335)	User Services/NOAA National Climatic Center Federal Building Asheville, North Carolina 28801 (Telephone No. 704-258-2850)
EUROPEAN (METEOSAT) SATELLITE DATA	OCEANOGRAPHIC DATA
Meteorological Data Management Department European Space Operations Center Robert-Boschstrasse #5 6100 Darmstadt, Germany (Telephone No. 49-6151-8861)	NOAA/EDIS National Oceanographic Data Page Building #1 2001 Wisconsin Avenue N.W. Washington, D.C. 20233 (Telephone No. 202-634-7500)

CHAPTER 3

UNIVERSITY OF DUNDEE SATELLITE DATA RECEPTION AND ARCHIVING FACILITY

P.E. Baylis

Department of Electrical Engineering and Electronics,
University of Dundee, Dundee DD1 4HN

1. INTRODUCTION

The first images received at Dundee were from Automatic Picture Transmission TV vidicons on ESSA-2 and NIMBUS-2 in 1966. During 1975 construction of an ITOS/NOAA Very High Resolution Radiometer receiving system was completed. With the aid of a grant from the Natural Environment Research Council, daily archiving of VHRR data commenced on 23rd August 1976. A total of 2500 VHRR recordings were taken from NOAA-5 during the period up to 28th February 1979 when the mission was terminated.

A METEOSAT Primary Data User Station for the reception of Visible and Infra-red Spin Scan Radiometer digital data was constructed prior to the launch of METEOSAT 1 in November 1977. A receiving system for Advanced Very High Resolution Radiometer digital data reception from TIROS-N was commissioned in time for the launch of TIROS-N on 13th October 1978. Archiving commenced soon after launch and has continued, first with NOAA-6 and now with NOAA-7.

The Dundee station was upgraded to receive NIMBUS-7 Coastal Zone Colour Scanner data during the first half of 1979. Daily archiving of CZCS data commenced on 3rd August 1979.

Prime funding is now in the form of a contract between the University of Dundee and NERC Scientific Services.

A. P. Cracknell (ed.), Remote Sensing Applications in Marine Science and Technology, 29–34.

2. STATION OPERATION

The station is operational seven days a week throughout the year with two overlapping shifts per day, covering the period approximately 0800 to 2200 GMT. Work is divided between tracking selected orbits and archive retrieval. All archived data are stored on high density instrumentation tape unprocessed, as received from the spacecraft and will be kept indefinitely.

2.1 VHRR and AVHRR

Basically, at least two satellite passes per day closest to overhead the U.K., have been recorded since archiving commenced in August 1976. With NOAA-5, a N-S pass was recorded each morning and a S-N each evening. With TIROS-N, recordings were taken from two successive S-N passes each afternoon up until AVHRR failure on 1st November 1980. Since the launch of NOAA-6 (June 1979) recordings have been taken each morning from the closest to overhead U.K. N-S pass. However, after the failure of TIROS-N AVHRR recordings were taken each evening from two successive S-N NOAA-6 passes instead of the TIROS-N passes, in addition to the NOAA-6 morning passes. However, since the launch of NOAA-7 (June 1981) afternoon recordings have been resumed and the evening NOAA-6 recordings discontinued. From August 1982 all NOAA-6 reception has been discontinued and all AVHRR data currently received is from NOAA-7.

In addition to the above, extra passes are recorded on occasion, by special request to support oceanographic research activities. Typical examples are JASIN, MARSEN, EURASEP, KONTUR etc. In particular, early a.m. recordings were taken from NOAA-7 during the N. Sea KONTUR project (14th September - 22nd October 1981) in order to enable groups interested in sea surface temperature measurement to validate the new AVHRR five channel instrument. Night-time data from NOAA-7 are now received and recorded automatically while the station is unmanned.

2.2 CZCS

The CZCS is usually, but not always, switched on over European waters during the S-N daytime passes of NIMBUS-7. Normally it is switched off at about latitude 60°N, but sometimes is left on up to the northern horizon in polar regions. Since 3rd August 1979 the Dundee station has tracked NIMBUS-7 every day for 2-3 successive passes, recording CZCS data whenever it is switched on.

2.3 VISSR

Recordings from METEOSAT are taken only by prior request. Routine archiving is not performed at Dundee since such an archive would be enormous and would duplicate material in the archive maintained by the spacecraft managers at Darmstadt in the Federal Republic of Germany.

3. PRODUCTS AVAILABLE FROM THE DUNDEE ARCHIVE

Primary output from the VHRR/AVHRR/CZCS archive is hard copy photofacsimile imagery, from any chosen channel, linearised to correct Earth curvature distortion, but not Earth rotation de-skewed. Linearisation means that the scale of the images is very nearly (but not exactly) constant in all directions and at all points in the image. However, in the case of CZCS further distortion in the image is caused by tilt of the scan mirror to avoid sun glint.

Images are basically in one of two categories:

(a) Full pass (x1) with or without latitude/longitude and land outline grids. These images are contact printed from full size negatives. They can also be supplied as Xerox copies, of reduced quality, which are quite readable in terms of major meteorological features and are much less costly.

(b) Sectorised enlargements. These are first generation products derived directly from the archive tape and are selected sectors of the full pass image. Sectorisation is by electronic stretching and the resulting image represent the best quality available from the archive. Gridding of sectors is not at present available, but they are linearised.

A summary of VHRR/AVHRR/CZCS sizes and areas covered is given in Table 1. VHRR/AVHRR and CZCS images can be grey scale stretched to any desired extent to bring out such features as sea surface temperature gradients in IR images.

A complete set of VHRR/AVHRR/CZCS x1 contact prints is kept on the station to indicate the contents of the archive. However, not all scanner channels are included. For CZCS channel 5 is used and for VHRR/AVHRR VIS. and IR from the closest to overhead U.K. passes are included. For other passes either VIS. or IR is selected. Periodically the AVHRR browse file prints are photographed on to 35 mm film and then contact printed 36 to a page, either as positive prints or positive transparencies as required. Both full size prints and miniatures are distributed to users on the mailing list. Back numbers dating back to August 1976 are available. Full size x1 contact prints are also distributed on a weekly batch basis.

The station browse file may be consulted by visitors,

alternatively browse file searches can be undertaken on behalf of data users.

TABLE 1

Magnification	Lines/cm	Size cm	Total Lines	Grid	Area Coverage H x W Km
VHRR x 1	246	25 x 21	6250	Optional	5625 x 4400
VHRR x 2	118	25 x 25	3000	No	2800 x 2800
VHRR x 4½	59	25 x 25	1500	No	1350 x 1350
VHRR x 6¼	39	25 x 25	1000	No	900 x 900
AVHRR x 1	177	24 x 16	4275	Optional	4700 x 3000
AVHRR x 1½	118	25 x 25	3000	No	3300 x 3300
AVHRR x 3	59	25 x 25	1500	No	1650 x 1650
AVHRR x 4½	39	25 x 25	1000	No	1100 x 1100
AVHRR x 9	20	25 x 25	500	No	550 x 550
AVHRR x 18	10	25 x 25	250	No	275 x 275
CZCS x 1	118	25 x 17½	3000	No	2400 x 1682
CZCS x 2	59	25 x 25	1500	No	1200 x 1200
CZCS x 3	39	25 x 25	1000	No	800 x 800
CZCS x 6	20	25 x 25	500	No	400 x 400
CZCS x 12	10	25 x 25	250	No	200 x 200

CZCS or AVHRR scenes can be transcribed from the archive tapes on to CCT against specific requests. Data on CCT are unprocessed, as received from the spacecraft, and therefore contain the full scan width and all channels. The format used is 1600 BPI/PE and approximately 30 cm of CCT is required for each scan line of CZCS or AVHRR. VHRR CCTs are not available.

CCTs of any video data, provided they are correctly formatted, may be sent to Dundee for hard copying in black and white photo-fascimile with a maximum line density of 118 lines/cm. (300 LPI) up to size 25 x 25 cm.

4. CONCLUSION

The continuing requests, not only from Britain but from throughout Europe and even from farther afield, for data from the Dundee archive indicate a requirement for the facility. The greatest use of the archive data is in oceanography where the requirement is for sea surface temperature and water colour, but there are also a number of users from the meteorological community as well. Although the station is not primarily geared up to provide real-time data, images can be produced and dispatched to users very quickly in support of particular projects.

A university electrical engineering department provides a good environment for the station because of the excellent facilities for in-house design and construction of specialised electronic equipment. Much of the equipment and software are the results of student effort at both undergraduate and post-graduate level.

BIBLIOGRAPHY

The following references may be useful to anyone requiring more detailed information about the satellite systems and station equipment discussed in this paper.

Baylis, P.E., "Guide to the Design and Specification of a Primary User Receiving Station for Meteorological and Oceanographic Satellite Data" in "Remote Sensing in Meteorology, Oceanography and Hydrology" ed. A.P. Cracknell (Ellis Horwood, Chichester, 1981) p. 81.

Baylis, P.E., "University of Dundee Satellite Image Data Acquisition and Archiving Facility" in "Matching Remote Sensing Technologies and their Applications 16-18 December 1981" ed. D. Lynn and J.A. Allan (Remote Sensing Society, Reading) p. 517.

Baylis, P.E. and Brush, R.J.H., "NOAA SR/VHRR Receiving and Archiving Facility", U.K. Remote Sensing Society, 1977, Conference Proceedings.

Baylis, P.E., Mather, J.R. and Brush, R.J.H., "A METEOSAT Primary Data User Station" ESA/ESOC, Darmstadt, Germany, 1980.

Bentley, W.E., "Squeeze More Data onto Mag. Tape", Electronic Design, 21 October, 1975.

Brush, R.J.H., "An Algorithmic State Machine Decommutator for METEOSAT Primary Data User Stations", The Radio and Electronic Engineer.

EMI Engineering Handbook, "Modern Instrumentation Tape Recording", EMI Technology Inc., 1978.

Mather, J.R., "Study and Specification of a Low Cost/High Performance METEOSAT Primary Data User Station", Dept. of Electrical Engineering and Electronics, University of Dundee, 1981.

"METEOSAT Ground Segment Handbook", ESA/ESOC Darmstadt, Germany, 1977.

"NIMBUS G Data Systems Description" NASA Goddard Space Flight Centre.

NOAA Technical Memorandum NESS 95 "The TIROS-N/NOAA A-G Satellite Series", NOAA/NESS Washington, D.C.

NOAA Technical Report NESS 75 "Guide for Designing RF Ground Receiving Stations for TIROS-N", NOAA/NESS Washington, D.C.

Spilker, J.J., "Digital Communications by Satellite", Prentice-Hall, 1977.

CHAPTER 4

DATA ACQUISITION AND QUALITY

A.P. Cracknell

Carnegie Laboratory of Physics, University of Dundee,
Dundee DD1 4HN, Scotland, U.K.

ABSTRACT

The object of these lectures was to consider the problem of data acquisition from remote sensing satellites, principally from the point of view of a physicist. By this I mean that we consider some of the physical principles involved in satellite remote sensing programmes. This should enable us to gain an appreciation of the nature of the information that can be obtained about the marine environment from this data; it should also enable us to obtain a better appreciation of the errors in the data and of the limitations to the information that can be obtained from the data. However, material that has been covered in the proceedings of our 1980 summer school and in other chapters in the present proceedings is, by and large, not duplicated in this written version of the lectures.

1. INTRODUCTION

At the beginning, some of the simpler physical ideas associated with electromagnetic radiation, which is the vehicle by which information is carried from the surface of the Earth to a remote sensing satellite, were reviewed. These include a discussion of the electromagnetic spectrum, Planck's radiation formula, Stefan's law of radiation, the Wein displacement law, the Doppler effect, and the transmission, reflection and detection of electromagnetic radiation. This material will be found in numerous textbooks on physics and there is no need to repeat it here; the particular form in which it was presented followed quite closely the form given in sections 1.5 and 1.6 of the proceedings of our 1980

A. P. Cracknell (ed.), Remote Sensing Applications in Marine Science and Technology, 35–66.

summer school (1).

2. SATELLITE SYSTEMS RELEVANT TO MARINE ENVIRONMENTAL REMOTE SENSING

I shall not attempt to consider all those satellite systems that were listed in tables 1 and 2 of chapter 1; I shall concentrate on those which, in my view, are reasonably widely used in practice by marine scientists and engineers who actually make use of remote sensing data collected by satellites. This should provide a useful background to what follows in a number of later chapters. A comprehensive review of remote sensing satellite programmes is given by Cornillon (2).

Let us first dispose of manned satellite programmes. The USA and USSR have for many years conducted manned satellite programs including cameras, rather than scanners, in their payloads. Some details of some cameras flown on such missions are shown in Table 1.1 of Ref. 1. I think it is fair to say that while manned missions may be more spectacular than unmanned missions they are necessarily of rather short duration and the amount of useful information obtained from them is quite small compared with the total of useful information obtained from unmanned satellites.

Let us now recall the important difference between (polar) orbiting and geostationary satellites. Suppose that a satellite of mass m travels in a circular orbit of radius r around the Earth, of mass M, then it will experience a gravitational force of GMm/r^2 (G = gravitational constant), which is responsible for causing the acceleration $r\omega^2$ of the satellite in its orbit, where ω is the angular velocity. Thus, using Newton's second law of motion, $GMm/r^2 = mr\omega^2$ or $\omega^2 = GM/r^3$. The period of revolution, T, of the satellite is then given by $T = 2\pi/\omega = 2\pi(r^3/GM)^{\frac{1}{2}}$; i.e. the period of revolution of the satellite depends on the radius of the orbit but, other than that, on nothing other than the constants G and M.

A satellite has to be high enough above the surface of the Earth that air resistance is negligible. However, the higher it flies then, for a given sensor, the lower will be the spatial resolution. The usual arrangement is to put a satellite into a near-polar orbit, at about 800-900 km above the surface of the Earth and in consequence having a period of about 90 minutes. But if you increase the radius of the orbit then the period increases; for a radius of 384,400 km the period is about 28 days and we do indeed have a satellite permanently out there at that radius, i.e. the Moon. Somewhere in between these two radii there will be one value of the radius (about 42,250 km, corresponding to a height of about 35,900 km above the surface of the Earth) for which the period is exactly equal to 24 hours, or one day. For a polar orbit that would not be particularly interesting, but for an equatorial orbit and going round in the same sense as the sense of the rotation of the Earth it means that the satellite would remain vertically over the same point on the surface of the Earth. That is, we have a satellite in a geostationary orbit.

2.1 GOES-E, GOES-W, METEOSAT

In principle you can park a geostationary satellite over any desired point on the equator; in fact what is done at the moment is to try to have a geostationary satellite every 60° or 70° around the equator. Given the horizon you can see from the geostationary height this gives global coverage of the Earth with the exception of the polar region, see Figure 2 of chapter 2. METEOSAT, a European satellite is parked over the Greenwich meridian. The ones at 75° W (GOES-E) and at 135° W (GOES-W) are launched and operated by the USA. The addresses of the points of contact from which data from these satellites can be obtained are given in table 7 of chapter 2.

METEOSAT-1 was launched in November 1977 and functioned normally for about 2 years (3). METEOSAT-2 was launched recently. The diameter of the satellite is 2.1 m, its length 3.2 m and its mass is 293 kg at the beginning of its life in orbit, see Fig. 1. It spins at 100 rpm about an axis almost parallel to the N-S axis of the Earth. Changes in inclination, spin rate and longitudinal position can be made using a series of thruster motors which are controlled from the ground. The main instrument onboard the satellite is a scanning radiometer which has three spectral bands

(i)	Visible (VIS)	0.4 - 1.1 μm
(ii)	Thermal infrared (IR)	10.5 - 12.5 μm
(iii)	Infrared water vapour absorption (WV)	5.7 - 7.1 μm.

The first two wavelength ranges are not particularly unusual but the third one was a bit unusual and the pictures in this band are correspondingly unusual. The instantaneous field of view, IFOV, is 2.5 km for the visible and 5 km for the IR and WV channels. An IR or WV image consists of 2500 lines x 2500 pixels. A visible image consists of 2500 lines x 5000 pixels; there are 2 visible detectors and they are offset by one scan line so that a full visible image using both detectors would give 5000 lines x 5000 pixels. But the WV channels share the sampling electronics of one visible channel and therefore each ½-hour one obtains either (i) IR + both VIS or (ii) IR + one VIS + WV. METEOSAT is also used for communications purposes. First, there is a data collection system associated with METEOSAT, see section 5. Secondly, the satellite is used as a communications satellite to

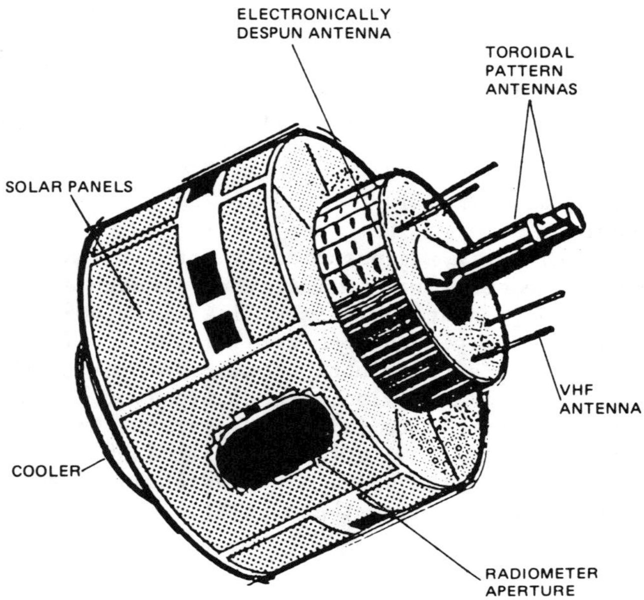

Figure 1. METEOSAT satellite.

transmit processed METEOSAT data from the European Space Operations Centre in Darmstadt to users of the data and also to transmit GOES-E data to users who are outside the direct line of sight of the GOES-E satellite.

The GOES-E and GOES-W satellites operated by the USA are very similar to METEOSAT, but (see table 6 of chapter 2) they only have two, rather than three, spectral channels and the IFOV is rather larger than for METEOSAT.

Let us now consider several important polar-orbiting satellites.

2.2 LANDSAT

This satellite is in a near-polar sun-synchronous orbit at a height of about 918 km above the surface of the Earth; it travels in a direction slightly west of south and passes overhead at about 10 am local sun time. In a single day there will be 14 southbound (daytime) passes (the northbound passes are at night). The distance between successive paths is much greater than the swath width and so not all the Earth is scanned each day. The following day all the orbits will be moved slightly to the west and it is only after 18 days, i.e. on the 19th day, that the orbits repeat themselves exactly. There is some overlap of orbits and in northerly latitudes this overlap becomes quite large. Each of the first three LANDSAT satellites carried a multispectral scanner, MSS, and a return beam vidicon, RBV, camera. The MSS on LANDSAT-3 had a thermal infrared band, but relatively little useful data was actually generated in this band. Otherwise these instruments all involve the visible or near infrared part of the electromagnetic spectrum, see table 1. At these wavelengths the surface of the Earth will be obscured if there is cloud present. Thus adding the facts of only one or two passes in 18 days and the proportion of days which are clear of cloud, one can see that the number of useful LANDSAT passes per annum over a given area is probably fewer than half a dozen. Nevertheless, data from the multispectral scanners on the LANDSAT series of satellites have been used very extensively in a large number of remote sensing programmes. As their name suggests, these satellites were designed primarily for remote sensing of the land, but in certain circumstances useful data are also obtained for marine applications (4).

2.3 TIROS-N

The satellites in this series, TIROS-N, NOAA-6, NOAA-7 ...., have several instruments on board; of these the most famous is the Advanced Very High Resolution Radiometer or AVHRR. This is a multispectral scanner, like those of the LANDSAT series, but with a very much poorer spatial resolution, namely about 1 km. The swath width is correspondingly wider, about 2000 km or more instead of about 185 km; passes that produce images of any given area are obtained much more frequently than with LANDSAT - daily in fact. It is these satellites that are used to give the satellite pictures used by weather forecasters on TV. The AVHRR has four spectral bands though the actual wavelengths are quite different from the LANDSAT MSS wavelengths, see table 1. In the latest version of the instrument the thermal infrared fourth band has been replaced by two bands at 10.3-11.3 μm and 11.5-12.5 μm. The idea behind having several infrared bands is to try to correct "brightness" sea surface temperatures by using the data from several bands to eliminate atmospheric effects (see chapter 12).

The AVHRR is enormously successful for meteorological and deep sea oceanographic purposes where the spatial resolution is perfectly adequate for the purpose. For coastal and estuarine work, however, the spatial resolution is often not adequate. The AVHRR can also, of course, be used for some purposes over land areas, provided its spatial and spectral resolution are appropriate. For instance in the proceedings of our 1980 summer school there is a contribution by Professor Østrem (5) describing its use in connection with the determination of snow cover, and

Table 1. Comparison of characteristics of some multispectral scanners

LANDSAT MSS[1]		TIROS-N AVHRR[2]		NIMBUS-7 CZCS[3]		METEOSAT	
Channel	Wavelength (μm)	Channel	Wavelength (μm)	Channel	Wavelength (μm)	Channel	Wavelength (μm)
4	0.5-0.6	1	0.55-0.90	1	0.433-0.453	1	0.4-11.1
5	0.6-0.7	2	0.725-1.0	2	0.510-0.530	2	10.5-12.5
6	0.7-0.8	3	10.5-11.5	3	0.540-0.560	3	5.7-7.1
7	0.8-1.1	4	3.55-3.93	4	0.660-0.680	Resolution	
Resolution	∿ 80 m	Resolution	∿ 1 km	5	0.700-0.800	Channel 1	∿ 2.4 km
				6	10.5 -12.5	Channels	
				Resolution	∿ 825 m	2 and 3	∿ 5 km

1 Multi-Spectral Scanner

2 Advanced Very High Resolution Radiometer

3 Coastal Zone Colour Scanner

hence melt-water run-off forecasting, for the Norwegian hydro-electric power board.

However the AVHRR is not the only instrument on the TIROS-N series of satellites; there are several others which are often neglected, possibly because they do not generate images. These instruments include the Tiros Operation Vertical Sounder, TOVS, (which itself is a three-sensor atmospheric sounding system), the Space Environment Monitor, SEM, which measures solar particle flux at the spacecraft and the ARGOS data collection system (see section 5 below).

2.4 NIMBUS-7, SEASAT

I have put these two satellites together. They were both launched in 1978 though they have had rather different survival times; SEASAT only lasted 3 months but NIMBUS-7 has already lasted for more than 3 years. Each carried several instruments; mostly they were not the same but one instrument, the Scanning Multichannel Microwave Radiometer, SMMR, was carried by both of them. On SEASAT there were

(i) Altimeter
(ii) Scatterometer
(iii) Synthetic Aperture Radar, SAR
(iv) Scanning Multichannel Microwave Radiometer, SMMR

and these are all relevant to marine work. On NIMBUS-7 the instruments relevant to marine work are

(i) Scanning Multichannel Microwave Radiometer, SMMR
(ii) Coastal Zone Colour Scanner, CZCS.

All the SEASAT sensors and the SMMR on NIMBUS-7 are microwave sensors, see section 3; the CZCS is an optical and infrared multispectral scanner.

The CZCS is a multispectral scanner working in the visible and infrared and it is therefore similar in many ways to the LANDSAT MSS and to the AVHRR. Its IFOV, instantaneous field of view, is comparable with that of AVHRR but it has 6 spectral channels, including some very narrow channels in the visible and a thermal infrared channel, see table 1. It has been claimed that the CZCS spectral bands in the visible are particularly appropriate for marine and coastal work. The use of CZCS data is described in several other chapters in this volume (see chapters 1, 7, 13, 14 and 16). One might argue that for coastal work the IFOV is rather large. The frequency of coverage is more like that of the AVHRR than that of the LANDSAT MSS. But the CZCS has a very high power consumption and so it is only switched on for relatively short periods that fall very far short of the full 90 or so minutes of the complete orbit. Thus even if one knows, from orbit considerations, that it was possible to image a certain area at a certain time with the CZCS there is no guarantee that the instrument was actually switched on at that time by NASA.

One other practical point should, perhaps, be made about the CZCS. It might be supposed, from what has been said, that the thermal infrared band of the CZCS is much the same as that of the AVHRR; however, in fact, this band on the CZCS is much noisier, see Figure 2.

One can think of the SMMR as being a microwave analogue of an optical and infrared multispectral scanner, although in its construction and details of operation it is rather different (see, for example, Ref. 6). It is a 5-channel instrument operating at 6.6, 10.7, 18, 21 and 37 GHz, each with both vertical and horizontal polarization which gives 10 channels in all. The spatial resolution is, not surprisingly, enormously poorer than that of a typical optical or infrared scanner. The IFOV is elliptical, varying in size according to the wavelength of the channel:

	Along Track	Cross Track
at 6.6 GHz	150 km	100 km
at 37 GHz	27 km	17 km

This is to be compared with an IFOV with edge of about 1 km for AVHRR and it can be seen that we are now concerned with global coverage rather than with small seas or lakes. The scanning is achieved by the rotating offset reflector at the top of the SMMR and this produces a swath width of about 800 km.

2.5 Other Satellite Systems

There are too many other systems for us to consider them all (see chapter 1). We should perhaps just mention the HCMM, Heat Capacity Mapping Mission; this has a multispectral scanner with emphasis on the thermal infrared and with a spatial resolution rather similar to that of CZCS or of AVHRR.

We should also, perhaps, mention briefly some future systems. These include the new satellites in the LANDSAT series, LANDSAT-4 and LANDSAT-D', and there is the French SPOT programme. These new satellites are to have improved multispectral scanners such as the thematic mapper on LANDSAT-4 and LANDSAT-D'. This

(b)

(a)

Figure 2. Thermal infrared images of northern Scotland 17 May 1980 from (a) AVHRR (b) CZCS (data received and processed by Dundee University).

instrument has more spectral bands than the scanners on previous satellites in this series, see table 2; the spatial resolution

Table 2. Thematic Mapper Characteristics

Spectral Bands	Wavelength (μm)	Resolution (m)
Blue	0.45 - 0.52	30
Green	0.52 - 0.60	30
Red	0.63 - 0.69	30
Near infrared	0.76 - 0.90	30
Mid-infrared	1.55 - 1.75	30
Infrared	2.10 - 2.35	30
Thermal Infrared	10.4 - 12.5	120

is very much better than previously, with an IFOV of only 30 m in the visible bands and an IFOV of only 120 m in the thermal infra-red band. The scanner on the new French SPOT series of satellites will have fewer spectral bands than the thematic mapper but it will have even better spatial resolution, with an IFOV of only 20 m. Finally, there is the ESA remote sensing satellite, ERS-1, that is due for launch in about 1987. ERS-1 is to be relevant to marine applications of remote sensing. Originally it was intended that it should have a multi-spectral scanner, the Ocean Colour Monitor (OCM) working in the visible and infrared; but that instrument was removed from the payload for financial reasons. The main part of the payload will be a set of microwave instruments which will probably be rather like those that were flown on SEASAT. There is going to be another experiment on board and this is expected to be an infrared instrument, the Along Track Scanning Radiometer (ATSR) proposed by a team from the Rutherford Appleton Laboratory, for accurate sea-surface temperature determination.

3. ACTIVE AND PASSIVE SENSORS

Under the heading of passive instruments we include radiometers or scanning radiometers, known as scanners for short, working at a wide range of frequencies. These instruments are usually multispectral, that is they record the intensity of the radiation received in each of several spectral bands from the instantaneous field of view (IFOV). In the marine context such scanners working at optical wavelengths are used to study the quality of the top layer of the water; by the word quality we mean suspended sediment concentration, chlorophyll concentration, etc. Infrared scanner data are used to study water surface temperatures while passive microwave scanners are used to study sea surface temperatures (with much lower spatial resolution than in the infrared part of the spectrum), near surface wind speed and (in principle at least) salinity. Microwave radiometers have also been used in delineating sea ice and in monitoring oil pollution. In some of these applications, such as salinity and oil pollution monitoring, it is only aircraft-flown, rather than satellite-flown, instruments that have been used so far. The spatial resolution achieved with a microwave scanner, such as the SMMR flown on SEASAT and NIMBUS-7, is very much poorer than that achieved with an optical or infrared scanner (see section 2 above). Certain information about the atmosphere can also be extracted from scanner data in a variety of ways.

In describing these instruments as "passive" we mean that the instruments simply receive the electromagnetic radiation, in a given wavelength range, that happens to fall on the receiving aperture of the instrument; the radiation originates either from the Sun, by reflection, or from the body itself below the scanner, by emission. These instruments, of course, record the intensity of the radiation received by the instrument and, since the instrument is high above the surface of the Earth, this will not be the same as the radiation leaving the surface of the Earth. As the radiation that has left the surface of the Earth passes upwards through the atmosphere, there will be a variety of processes that occur and some of these processes will augment the intensity of the radiation while others of these processes will reduce its intensity. One can, therefore, use the data coming from a scanner to obtain information about the atmosphere itself, or one may regard the atmosphere as introducing errors, which have to be eliminated if one is trying to extract data about the surface of the land or the sea. If one is trying to observe the surface of the land or of the sea using optical or infrared scanners the atmospheric effects are often catastrophic. Thus, if there is cloud cover the surface of the land or of the sea is completely obscured and the scanner is simply observing the tops of the clouds. This is, of course, very useful for meteorological purposes and, indeed, meteorological remote sensing satellites are primarily concerned with the observation of cloud systems and their motions. In effect the opacity of the clouds means that remote sensing of the land or sea at optical and infrared wavelengths is an activity that is restricted to fine weather. An all-weather capability to observe the surface of the Earth can be achieved by using a microwave scanner instead of optical or infrared instruments. However, as indicated already, the spatial resolution of a microwave scanner is very poor; in fact the IFOV may be 50 km or even 100 km square. To achieve a spatial resolution comparable with those of optical or infrared scanners it is necessary to use an active instrument, i.e. a radar. In this case, the radiation that is received by the instrument was generated by the instrument itself and reflected by a target. Although we introduced the idea of using microwaves in connection with overcoming problems associated with cloudy conditions, we should not think of microwave remote sensing solely in this context. Even in clear weather the information that can be gained from the microwave sensors should be regarded as complementary to that obtained from optical and infrared instruments. As far as marine applications are concerned, we can regard the signal as giving information about the quality of the water, i.e. suspended sediment, chlorophyll and, in the infrared, temperature. The wavelength is very much smaller than typical distortions of the shape of the surface. On the other hand, radio waves will not penetrate the body of the water and are just reflected by the surface; their wavelengths are comparable with the size of surface distortions. Therefore, loosely speaking, the optical instruments tell you about the quality of the water while radio waves tell you about the shape of the surface. We shall devote some attention to radars.

The original purpose of radar, as you may recall from the etymology of the word, was detection and ranging. The detection was achieved by the fact that one got an echo from the target one was trying to detect, such as an enemy ship or aircraft. The range was determined, essentially, from the time for the double journey to the target and back. In the early days of radars one was looking for a target that one could expect to stand out fairly well against a dull background. But in our remote sensing applications for environmental science we are unlikely to be concerned with that sort of situation but rather with some field, such as a wave field on the surface of the water, in which there are relatively small variations from point to point in the field.

One can see the origins of the ground wave and sky wave radar systems, which will be described in chapter 9, in conventional navigation or military radars or in conventional radio techniques. Also if one is operating from a low-flying aircraft a fairly conventional radar system may again be used and one commonly arranges that the radar system is viewing a swath on the ground that is not vertically below the aircraft but is off to one side, hence the name Side-Looking Airborne Radar, SLAR, which is commonly used in this context. When it comes to satellites the instruments may look rather different and, although they may still be called radars as in the case of synthetic aperture radar, quite often the term radar has disappeared from their titles and they acquire such names as scatterometers or altimeters, see also chapter 8. The synthetic aperture radar can be used to produce images but the scatterometer and altimeter are not image-producing instruments. If one were to attempt to use an imaging radar at the height involved if one is using a satellite, then one would find that to achieve a useful spatial resolution in an image of the ground one would need to use an antenna with a length of several km. It is clearly not feasible to have an antenna of such a size on a space-craft and it is to overcome this problem that the alternative of synthetic aperture radar, SAR, was developed for imaging uses. Though this can be used on an aircraft the SAR really comes into its own on a spacecraft.

The idea of a synthetic aperture radar is not entirely trivial. What is involved in synthetic aperture radars is to take successive positions of a quite small real antenna and use the signals transmitted and received at a succession of positions of this real antenna to synthesise, build up, or reconstruct what you would have got if you had had a long real antenna occupying all those positions. This is an ingenious idea, but there are many problems: (i) the amount of data involved for a single image is enormous, (ii) there is no simple relationship between the raw data and an image and a great deal of (optical or digital) processing goes into the production of an image from the raw data, and (iii) care is sometimes needed in the interpretation of an image. One example of (iii) is concerned with objects that are moving on the ground. The theory used in the reconstruction of an image from SAR data involves using the Doppler effect. If there is a moving object in the field of view then reflections from this object will have an additional Doppler shift. The processing will fail to distinguish between Doppler shifts of different physical origins. The net effect is that a moving object has a displacement in the image relative to stationary objects; the displacement depends on the magnitude and orientation of the velocity of the object, relative to the SAR. Examples of SEASAT SAR imagery showing a train that appears to be displaced from the railway line on which it is running or a ship that appears to be separated from its wake are, I think, well known. Another example of curiosities attached to SAR images arises with metal wires such as, for example, high voltage electricity transmission lines. If these wires are parallel to, or very close to parallel to, the direction of flight of the platform carrying the SAR they show up very clearly on the SAR image, but for other directions relative to the line of flight these transmission lines are invisible.

As already mentioned, the altimeter and scatterometer are examples of active microwave instruments that do not produce images. In the altimeter one is essentially making use of the ranging facility; we suppose that we are looking directly down from the satellite in a vertical direction and the altimeter enables us to measure the vertical distance from the (mean) sea surface to the satellite. From the shape of the return pulse one can also extract data about the roughness of the sea surface and hence about the wind speed, but not wind direction. With the scatterometer one can study wind direction as well as wind speed.

4. RESOLUTION, ETC.

There are three important quantities that need to be considered (7), namely (a) Spectral resolution, (b) Spatial resolution, (or instantaneous field of view, IFOV), and (c) Frequency of coverage. We shall consider each of these three important quantities briefly, in turn, with particular reference to multispectral scanners, see table 1; however, many of the ideas involved apply to other imaging systems such as radars and even to some non-imaging systems too. (a) Is determined by the construction of the sensor system itself, while (b) and (c) are determined both by the construction of the sensor system and by the conditions under which it is flown. To some extent there has to be a trade-off between spatial resolution and frequency of coverage; good spatial resolution, i.e. small IFOV, tends to be associated with low frequency of coverage. Thus we have

Satellite/Sensor	Frequency of coverage
LANDSAT MSS	Once or twice in 18 days
AVHRR	Two or three times per day
CZCS	Once or twice per day
METEOSAT	Every half hour.

4.1 Spectral resolution

For some applications one wants as many spectral channels as possible, for example for the determination of suspended sediment loads and chlorophyll concentrations. For this the LANDSAT MSS is poor but CZCS is tolerable. For other applications, such as sandbank mapping and sea-surface temperature determinations, a multitude of spectral channels in the visible is un-necessary. All one needs is the appropriate infrared channel which, by and large, has been available on existing scanners; in work on sea-surface temperatures, additional infra-red channels may be valuable in trying to eliminate atmospheric effects. For detecting oil pollution the panchromatic band of the AVHRR or the visible bands of LANDSAT MSS would seem to be adequate from the spectral point of view.

4.2 Spatial resolution

In coastal and estuarine work the spatial resolution of the LANDSAT MSS is adequate for many purposes. The spatial resolution of other instruments is not; even a large river like the Tay is quickly crossed in half a dozen, or fewer, pixels for AVHRR, CZCS or HCMM.

For deep-sea oceanographic work the spatial resolution of

AVHRR, CZCS or the scanners on the geostationary satellites is generally adequate. For the AVHRR or the CZCS the IFOV is of the order of 1 km square. For the METEOSAT radiometer the spatial resolution is considerably poorer because the satellite is very much higher above the surface of the Earth; the IFOV is about 5 km square for the thermal infrared channel of METEOSAT. At the other extreme, of course, for a thermal infrared scanner flown in a light aircraft at a rather low altitude the IFOV may be only about one metre square or so; aerial surveys using such scanners are now widely used to monitor heat losses from the roofs of large buildings (8). In areas of open ocean the spatial resolution of the AVHRR, CZCS or METEOSAT radiometer is perfectly adequate - it actually enables the oceanographer to have synoptic maps of sea surface temperatures over enormous areas; these could not previously be obtained on such a scale in any other way before the use of satellites. Such maps are also beginning to find uses in marine exploitation and management, e.g. in assisting in the location of fish (9).

However, with the AVHRR, CZCS and METEOSAT radiometer, their spatial resolution makes it rather difficult to use data from these instruments for smaller areas, such as rivers or estuaries to study, for example, the thermal pollution by the cooling water from a power station or factory. The width of the river would only correspond to two or three times the linear dimension of the IFOV. The scanners on the early satellites in the LANDSAT programme did not have a thermal infrared channel. For good thermal maps of estuaries and coastal regions we await data from new instruments and satellite programmes, such as the thematic mapper on LANDSAT-4 which was launched very recently.

4.3 Frequency of coverage

We have already mentioned that there is a trade-off between the frequency of coverage and the spatial resolution. For the LANDSAT MSS the spatial resolution is rather less than 80 m but coverage is only once or twice in 18 days. For the AVHRR or CZCS the spatial resolution is reduced by about an order of magnitude to about 1 km, but the frequency of coverage is - from orbit considerations - about an order of magnitude better, of the order of once or twice a day. However, it is not just a matter of instrument specifications and orbit considerations but there are also other considerations associated with both power requirements and the recovery of the data. The CZCS, for example, requires too much power to be left switched on for all the orbit and so as well as needing an overpass you also need to have the instrument switched on as well before one can obtain data. A similar problem associated with power requirements applied to the synthetic aperture radar, SAR, that was flown on SEASAT. There are essentially two ways in which the data generated by a sensor on an unmanned spacecraft may be recovered on the ground. First, the data may be transmitted on a carrier radio wave immediately after the data is generated by the sensor. In this case the data can be received by any suitably-equipped ground station, provided that the ground station is in the direct line of sight from the satellite. A signal coming via any other path than the direct one would be far too weak to be detected. Looked at from the point of view of the ground receiving station, the satellite must be above the horizon. Secondly, the data gathered in one part of the orbit may be recorded on a tape recorder on board the satellite; the tape can then be played back and the data transmitted to a chosen ground receiving station at an appropriate later stage in the orbit. In this manner one can, in principle, obtain global coverage with a single ground receiving station. However, this may not be possible in practice for various reasons. For example, the on-board tape recorder may not have sufficient storage capacity to hold all the data generated when the satellite is below the horizon from the ground-receiving station. Thus in the case of the AVHRR, although global coverage is acquired by NOAA by using the on-board tape recorder, it is only possible to record part of the data from each orbit. Thus the AVHRR archive built up in Dundee (see chapter 3) from direct read-out transmissions contains much data that has not been acquired by NOAA. Another reason why it may not be possible to obtain global coverage by using on-board tape recorders is that the data rate of the output from the sensor may be too high to enable it to be recorded on board at all. This was the case with the synthetic aperture radar on SEASAT; hence SEASAT SAR data could only be recovered for areas near to the ground receiving stations situated in the USA, Canada and Great Britain that were able to receive the data directly. On the other hand, data from the low data-rate instruments on SEASAT were recovered on a global basis.

This discussion of resolution has been primarily in terms of optical and infrared scanners, though with some discussion of synthetic aperture radar. We should just consider microwave scanners here too. The Scanning Multichannel Microwave Radiometer, SMMR, flown on SEASAT and on NIMBUS-7 can be used to extract sea surface temperatures. As in the case of the thermal infrared, the basic problem is the inversion of the Planck distribution function in order to get at the sea-surface temperatures; we are, as with the thermal infrared, dealing with emitted radiation and not reflected solar radiation. There is, as always, a trade-off between spatial resolution and frequency of coverage. We have seen in section 2 that the spatial resolution for the SMMR is very much poorer than for the thermal infrared bands of the AVHRR or the CZCS. However, this is compensated by a greater frequency of coverage; the data rate of the SMMR is very low and so data acquired during the complete orbit can be recovered whereas, as we have seen, data can only be obtained from part of the orbit for AVHRR and for CZCS.

4.4 Relevant and irrelevant data

It is important, before ordering satellite data, to ensure that it will actually be relevant to the problem you wish to tackle. In the case of sea surface state (winds, waves and currents) there is, in fact, the choice between satellite microwave data and the use of ground-based radar systems. The ground-based systems, whether using ground-wave or sky-wave techniques, will only gather data for a relatively small area of the surface of the sea whereas, in principle at least, a satellite will give global coverage. However, one may not need global coverage and the restricted coverage obtained with a ground-wave or sky-wave radar may be quite adequate. There is little choice among microwave instruments in remote sensing satellite systems. There is the passive SMMR flown on NIMBUS-7 and on SEASAT. There were the various active microwave instruments flown on SEASAT, which only lasted about three months, and the next active microwave system is not due to be flown until the launch of ERS-1 in 1987. In the case of water quality parameters there is more choice of satellites and sensors. The constraints of spectral resolution, spatial resolution and frequency of coverage will assume varying importance according to the marine parameter that is being studied. Water-quality parameters that are commonly measured in situ include the temperature and salinity and the concentrations of chlorophyll and of suspended sediment. The suspended sediment may be of natural origin, as it is for instance in the Tay, or it may result from the activities of mankind. Pollutants are also of great importance; these include heavy metals dissolved in the water, oil floating on the surface of the water, and oil-water emulsions which may be on the surface or below the surface.

There are two stages involved in studying the problem of extracting water-quality information from remote-sensing data. The first stage is to ascertain whether the parameter under investigation can, in principle, be extracted from the remotely-sensed data. The second stage, if a positive answer is obtained at the first stage, is to determine how much useful information can actually be obtained at present about the parameter in question, given the current status of the space and ground segment technologies of remote sensing satellite programmes.

It is important to realize that there are some water-quality parameters about which remote sensing can provide virtually no information. For example, most heavy metal ions which may be present in sufficient concentration to have quite serious toxic effects on marine life are unlikely, at such concentrations, to make a significant change to the optical properties of the water and are therefore unlikely to be detected using optical scanners. The position with regard to salinity is somewhat similar. There is no doubt that variations in the salinity cause variations in the optical properties of sea water and we are, indeed, doing some theoretical investigations of this ourselves. Nevertheless, the changes in the optical properties are relatively small and are almost certain to be masked by other effects as far as remote sensing from satellites is concerned. On the other hand, oil pollution floating on the surface of the sea should, in principle, be clearly visible in a conventional aerial photograph or in an image from an optical scanner. Although chlorophyll and suspended sediment are not so clearly visible as a floating oil film would be, their presence can often be seen in imagery from optical scanners. In principle it should be possible to extract estimates of the concentrations of these materials from remote sensing data, either with the use of simultaneous in situ data for calibration within a training area on the scene or, preferably, without the use of any simultaneous in situ data.

For oil pollution monitoring spatial resolution is often a serious problem, although under favourable conditions it is possible to observe and monitor oil slicks using satellite data. For example, in the case of the IXTOC-1 blowout in the Gulf of México evidence of the oil slicks resulting from the spill can be seen in many images from scanners on several satellites, namely LANDSAT MSS, AVHRR, CZCS and the scanner on the geostationary satellite GOES-E. Some examples of these images are shown in Fig. 3. The United States Coast Guard did indeed make some use of satellite remote sensing data in monitoring the movement of the oil originating from this spill. But this was an enormous spill which lasted for several months. Most other spills are much smaller and the oil involved disappears much more quickly. We have tried to use the Oil Spill Intelligence Report in conjunction with AVHRR, CZCS and LANDSAT MSS imagery to find evidence in satellite imagery of any of the major spills in European waters in recent years. The problems are greater than with IXTOC-1, both because the slicks produced are smaller and more short-lived and because the cloud-cover is generally greater than in the Gulf of México. We have failed to find any evidence of these smaller spills in the satellite imagery.

The conclusions drawn (10) from the experiences of the United States Coast Guard with IXTOC-1 seem to us to be very fair and worth summarizing here. The meteorological satellites were useful in the early stages, in the absence of all other sources of information, in providing a general sense of the magnitude of the spill and of the general direction of drift of the spilt oil. However, the low resolution and small scale and high distortion of the standard image product were felt to limit these remote sensing systems as reconnaissance tools. In practice the oil

Figure 3. Images of IXTOC-1 blowout, Gulf of México, June-August 1979 from TIROS-N series of satellites and from GOES-E (data from NOAA SDSD archives, Washington, D.C.).

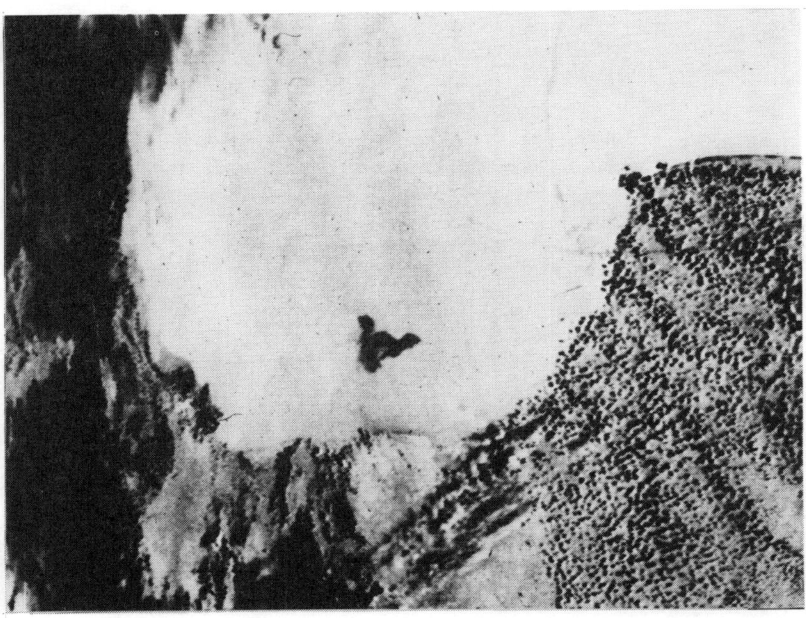

Figure 3(a). IXTOC-1: AVHRR image, 6th June 1979.

Figure 3(b). IXTOC-1: GOES-E image, 18th June 1979.

Figure 3(c). IXTOC-1: GOES-E image, 19th June 1979.

Figure 3(d). IXTOC-1: GOES-E image, 20th June 1979.

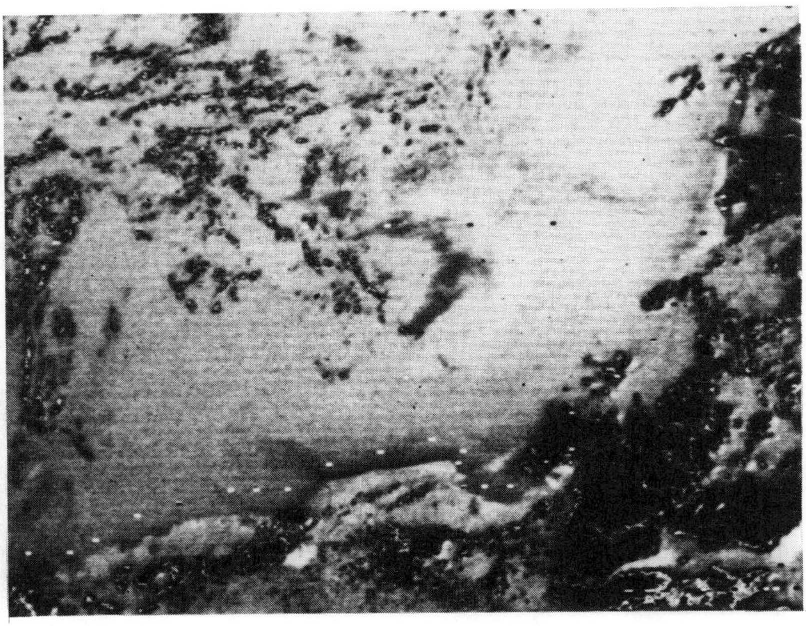

Figure 3(e). IXTOC-1: GOES-E image, 21st June 1979.

Figure 3(f). IXTOC-1: GOES-E image, 5th July 1979.

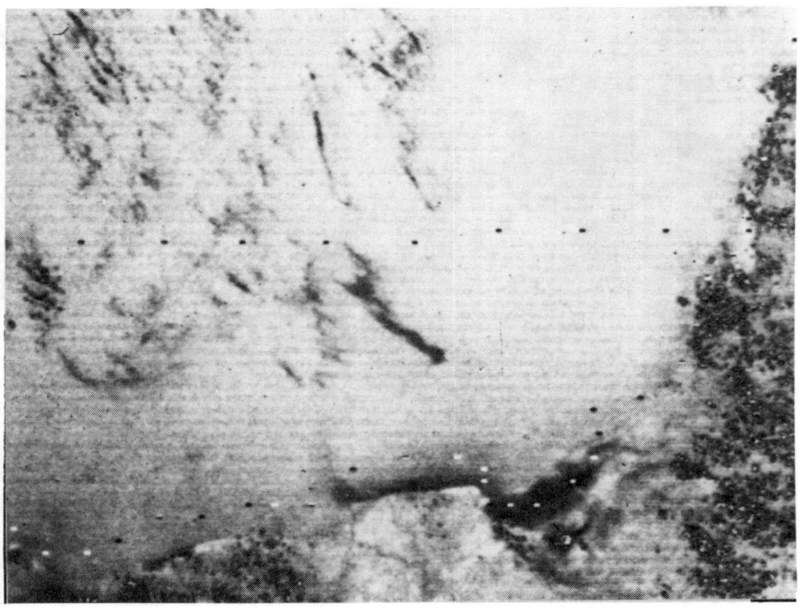

Figure 3(g). IXTOC-1: GOES-E image, 14th July 1979.

Figure 3(h). IXTOC-1: GOES-E image, 15th July 1979.

Figure 3(i). IXTOC-1: GOES-E image, 20th July 1979.

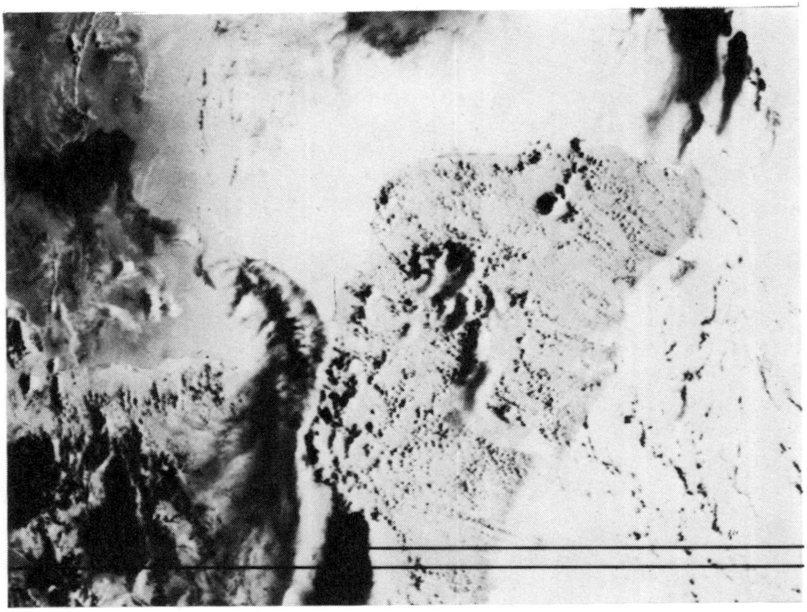

Figure 3(j). IXTOC-1: AVHRR image, 30th August 1979.

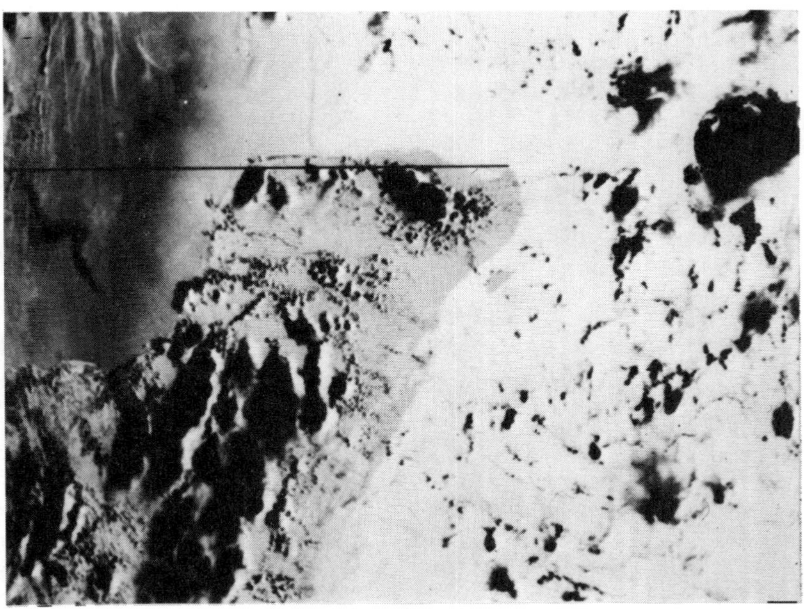

Figure 3(k). IXTOC-1: AVHRR image, 31st August 1979.

only seemed to be detectable with the visible sensors and not by the infra-red sensors. During the time of the blowout both the LANDSAT-2 and LANDSAT-3 satellites were operating and a considerable amount of relevant imagery was obtained from them; it was found to be useful in corroborating the results from aircraft reconnaissance and in helping to establish the overall picture of the situation. But the infrequency of satellite overpasses, the problem of cloud cover, the rather low spatial resolution, and the lack of real-time processing and delivery of imagery prevented the use of LANDSAT for tactical response by the on-scene coordinator of the U.S. Coast Guard. In the event the use of "low altitude visual aircraft reconnaissance backed up by an all-weather airborne remote sensing system (where weather and visibility required it) proved to be the most effective tactical and, along with accurate oil drift computer model forecasting, strategic aid for environmental pollution response planning for the Bay of Campeche oil spill. This will probably be true for some time into the future. The need for frequent, accurate, comprehensive, rapidly processed and analyzed information as input to the on-scene coordinator's decision making processes cannot be overemphasised. Until satellite remote sensing systems can deliver useful products in fine enough detail and observational frequency .... in sufficient time to respond to any likely impact such systems will necessarily occupy a subordinate role in the oil spill support strategy. Developing an all-weather operational satellite oil spill detection and mapping capability is a logical solution to monitor such events which are so unpredictably and widely distributed in time, space and environmental conditions. Orbiting spacecraft could alleviate some of the often immense logistical burdens placed upon conventional reconnaissance methods by the remoteness of the site or hostility of the environment. Whether or not this route shall be chosen depends most upon the development of capable remote sensors from orbital altitude and the ability to find sufficient multiple applications of the system beyond marine environmental protection to satisfy the requirement for sufficient economic and social benefit to justify the cost."

One might think, for example, about trying to detect oil pollution using radars. Indeed the Swedish Coast Guard, for example, does look for oil dumped at sea using light aircraft carrying radars as well as optical instruments. The oil is presumably detected with radars principally from the fact that it alters the shapes of the waves; one remembers the old saying about producing calm and tranquility by "pouring oil on troubled waters". However, in operating a pollution monitoring system to detect oil spills from tankers or oil-well blowouts it is necessary to recall that one may have to be able to distinguish between oil from different ships and also to be able to distinguish man-made spillages from natural seepage of mineral oil and from fish oil slicks.

The problems of spatial resolution and of temporal resolution apply also to the problems of studying suspended sediment, of mineral or human origin, and of biological material of non-human origin. For some relatively large-scale phenomena LANDSAT MSS, NIMBUS-7, CZCS and AVHRR data may provide adequate spatial resolution. For projects involving relatively small areas in estuarine or coastal regions there may still be considerable advantages in the use of aircraft rather than satellites (11). However, the question of spectral resolution is rather different. In the case of monitoring oil pollution the panchromatic view of the AVHRR visible channel or the rather small number of visible wavelength channels of the LANDSAT MSS constituted much less of a problem than some of the other difficulties that we noted in the use of data from these instruments. In the case of sea surface temperatures, the optical channels were more or less irrelevant and only the data from the thermal infrared channel, or channels if more than one were available, were relevant. In other words, in those cases, spectral resolution was not a serious problem because the extraction of the required parameters was quite straightforward. For the oil slick monitoring this was principally based on visual interpretation, coupled with geometrical rectification if desired, while for sea surface temperature determination a well-established algorithm, the Planck radiation formula, is available. For suspended sediment and chlorophyll monitoring the extraction of the required parameters from the remotely sensed data is substantially more difficult.

Quantitative information about the concentration of suspended sediment or of biological material is not obtained from the intensity recorded in a single spectral channel of a multispectral scanner, irrespective of whether the channel in question is restricted to a narrow range of visible wavelengths or is panchromatic, i.e. covering the whole range of visible wavelengths. This information has to be gained by using some algorithm for extracting the concentration from data in many spectral channels. There is no single algorithm that is universally accepted for the recovery of suspended sediment or chlorophyll from remote sensing data. However, what is fairly generally accepted is that one needs optical multispectral scanning data from a scanner with quite a large number of spectral channels in the visible region of the electromagnetic spectrum. Our own experience (12,13) has been that the number of visible channels in the LANDSAT MSS is rather too few; essentially only two of the bands (4 and 5) carry any useful data about the contents of the water. With the larger number of channels on the CZCS it is possible to do some reasonable work on algorithm development and testing (14) and about four of the six CZCS channels are useful for this purpose, see also chapter 16.

5. DATA COLLECTION SYSTEMS

So far we have discussed satellites primarily in the context of using them as platforms on which remote sensing instruments are mounted and from which the remote sensed data are transmitted back to Earth. Satellites may also be used in a communications role. We leave aside the general field of broadcasting and telecommunications, but recall that some remote sensing satellites are also used in a communications role. We have already noted this in connection with METEOSAT in section 2. Also, remote sensing data, having once been received from a satellite at a ground station may then be re-transmitted via a communications satellite.

There is another aspect of the use of satellites for communications purposes that is, perhaps, of relevance to us if we are involved in scientific or engineering environmental work. This is that we can use them to collect data from platforms in difficult or inaccessible locations on the surface of the land or sea. You may have instruments out at sea on a moored or drifting buoy; you may have instruments on a weather station in a hostile environment on land, up in the hills, in the Arctic circle, in the desert, and so on. Let us assume that the data gathering station is unmanned. There are then several options for recording and retrieving the data (i) cassette tape recorders with occasional visits to change tapes, (ii) direct radio link to ground, and (iii) radio link via satellite. (i) may be satisfactory if the amount of data is relatively small, but if there is a lot of data and one is only able to visit the site occasionally this may not be very suitable. (ii) may be satisfactory over short distances but become progressively more difficult over longer distances. (iii) has some attractions and is worth mentioning briefly here. There are two satellite-based data collection systems which are of importance. One involves using a geostationary satellite such as METEOSAT, while the other system is the ARGOS system which involves the polar-orbiting satellites of the TIROS-N, NOAA-6, NOAA-7 series. It may be wondered why one bothers with the satellite and does not just use direct radio transmission and reception. The answer is, essentially, that it is a matter of convenience. It saves on the cost of reception equipment and simplifies problems of radio frequency allocations. Moreover, there may be problems of distance. If the satellite is orbiting it can store the messages on board and play them back later, when the satellite may be round the other side of the Earth. One can sit in one's laboratory in Britain and receive data from buoys drifting in the Pacific Ocean, for example, using the ARGOS system.

To some extent the METEOSAT and ARGOS data collection systems are complementary. A data collection system based on using a geostationary satellite has the advantage that the satellite is always overhead and thus always able, in principle, to receive data. With a data collection system based on using a polar-orbiting satellite the satellite will be out of sight of the data collection platform for a great deal of the time. One cannot use a data collection system on a geostationary satellite if the data collection platform is situated in extreme polar regions, see Fig. 2 of chapter 2. On the other hand a data collection system that uses a polar orbiting satellite will perform better in polar regions because the satellite will be in sight of a platform that is near one of the poles much more frequently than for a platform near the equator. A sketch illustrating the data flow in the ARGOS system is given in Fig. 4. For the METEOSAT system buoys or stationary platforms on land can be equipped with transmitters to send records of environmental parameter measurements up to the satellite and the data are recovered by the European Space Operations Centre in Darmstadt and then relayed to the user.

In addition to gathering data one can use the satellite data collection system, or at least the ARGOS one, to locate the platform as well as simply to gather the data. This is not particularly interesting for a land-based platform because you know where it is - though if someone steals it but leaves the transmitter functioning it could be a useful feature! At sea, however, location is very valuable; it enables you to gather data from drifting buoys and ensures that you know where the buoys are when the data is gathered. It also enables you to check that a buoy that is meant to be moored has not broken free from its moorings; if it has broken free and you want to recover it then the location facility helps you to know where to start looking for it. On a slightly more frivolous note, the yachts competing in the "Observer" transatlantic yacht race are equipped with ARGOS system transmitters to locate the yachts; to be more correct the transmitters are fitted to the yacht's life-boat or life-raft for fairly obvious reasons. The location of a platform is determined by making use of the Doppler effect in the carrier frequency of the transmission from the platform; the transmitting frequency is fixed. Each measurement made by the satellite corresponds to a field of possible positions of the platform under consideration. This field takes the form of a cone with the satellite at its apex and the velocity vector along the axis of symmetry. Since the altitude of the satellite is assumed to be known, the intersection of several of these cones (each corresponding to a separate measurement) with the altitude sphere yields the solution. This actually yields two symmetrical solutions relative to the ground track. One of these points is the required solution, the other is its "image". The ambiguity cannot be resolved with data from a single orbit without some additional information, such as the previous position and likely speed; the ambiguity can be resolved by using data from two successive orbits.

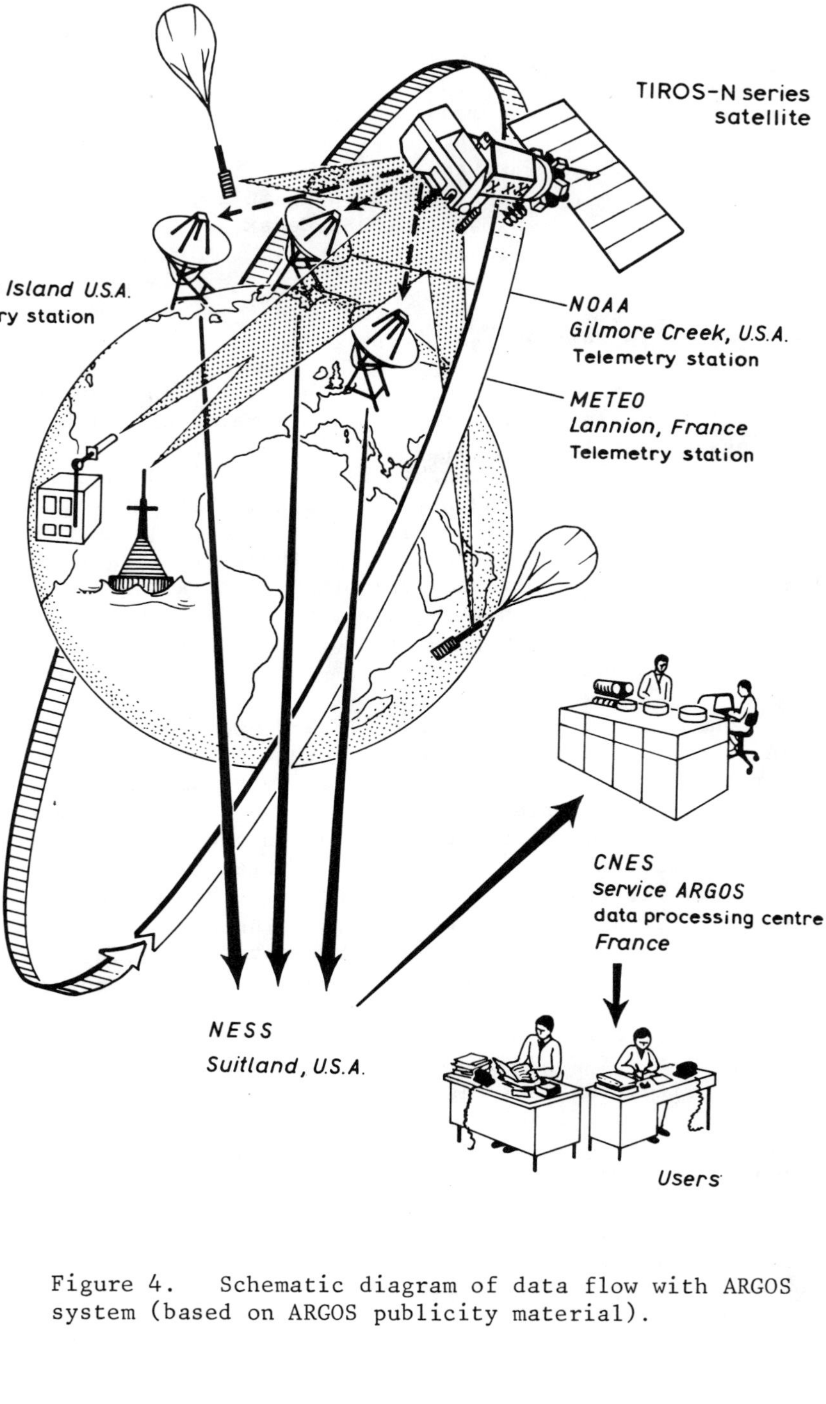

Figure 4. Schematic diagram of data flow with ARGOS system (based on ARGOS publicity material).

Further details of these two data collection systems can be obtained from

METEOSAT: European Space Operations Centre,
Robert-Bosche-Strasse 5,
6100 DARMSTADT,
Germany.

ARGOS: Service ARGOS,
18 Avenue Edouard Belin,
31055 TOULOUSE Cedex,
France.

6. SATELLITE DATA SOURCES

Some of the principles involved in the design and specification of a satellite data receiving station were described in the proceedings of our 1980 summer school (15) and some further discussion of the reception and archiving facility in Dundee will be found in chapter 3 of the present volume. If one sets up a receiving station one can receive for oneself the current data transmissions from a satellite while that satellite is above the horizon. For current data from other areas and for historic data one has to rely on the reception and distribution facilities provided by the body responsible for the operation of the satellite system in question.

Consider, for example, the TIROS-N, NOAA-6, NOAA-7 satellites. NOAA in the USA acquires global coverage by playing back tape-recorded data as the satellite passes over their receiving station. The scheduling and playback is controlled from the NOAA control room. Further details will be found in chapter 2. The Dundee reception facility, see chapter 3, only receives data from these satellites relevant to the area centred on Dundee and going as far as Greenland, North Africa, etc. The data are archived on magnetic tapes and distributed as requested by users. Anyone, anywhere, is free to set up their own station to receive data from this series of satellites. "Off-the-shelf" systems are now available for sale to enable one to do this.

The situation with regard to LANDSAT is quite different. The hardware is a little more sophisticated; moreover, one is only allowed to receive the data if one pays a quite large annual licence fee. LANDSAT carries tape recorders and global coverage is held by the EROS Data Centre. LANDSAT stations are springing up in various parts of the world outside the USA, in Europe, South Africa, Canada, India, Japan and Australia, while others are on the way (Africa, China, etc.). European LANDSAT data is distributed by EARTHNET, a part of ESA. Originally all the

European data was received at Fucino in Italy, but more recently data for northern Europe has been transferred to Kiruna in Sweden for its reception and distribution. Distribution is via the National Points of Contact in the member states of ESA; addresses of the sources of LANDSAT data are given in tables 1.3 and 2.2 of ref. 1.

Data from the geostationary satellites GOES-E and GOES-W are distributed by NOAA in Washington, see chapter 2. Data from METEOSAT are handled and distributed by ESOC in Darmstadt, see the previous section. Data from other satellites such as SEASAT, HCMM and NIMBUS-7 are distributed by NOAA and EARTHNET.

7. CONCLUSION

In this chapter I have tried to comment on the physical nature of the data obtainable from remote sensing satellites and on the quality and accuracy of the information about marine parameters that can be extracted from this data.

8. REFERENCES

1. Cracknell, A.P.: 1981, "Remote Sensing in Meteorology, Oceanography and Hydrology", Ellis Horwood, Chichester.

2. Cornillon, P.: 1982, "A Guide to Environmental Satellite Data", Marine Advisory Service, University of Rhode Island, Narragansett.

3. Mason, B.D.: 1981 in "Remote Sensing in Meteorology, Oceanography and Hydrology", (ed. A.P. Cracknell), Ellis Horwood, Chichester, pp. 56-65.

4. Cracknell, A.P., MacFarlane, N., McMillan, K., Charlton, J.A., McManus, J. and Ulbricht, K.A.: 1982, Int. J. Remote Sensing, 3, pp. 113-37.

5. Østrem, G.: 1981 in "Remote Sensing in Meteorology, Oceanography and Hydrology", (ed. A.P. Cracknell), Ellis Horwood, Chichester, pp. 258-69.

6. Thomas, D.P.: 1981 in "Remote Sensing in Meteorology, Oceanography and Hydrology", (ed. A.P. Cracknell), Ellis Horwood, Chichester, pp. 357-69.

7. Cracknell, A.P. in "Matching Remote Sensing Technologies and Their Applications. Proceedings of the Annual Conference of the Remote Sensing Society, London 16-18 December 1981", (ed. D. Lynn and J.A. Allan), Remote Sensing Society, Reading, pp. 285-97.

8. Anderson, J.M. and Callison, R.D.: 1981 in "Remote Sensing in Meteorology, Oceanography and Hydrology", (ed. A.P. Cracknell), Ellis Horwood, Chichester, pp. 508-12.

9. Domain, F., Citeau, J. and Noel, J.: 1980 in "Coastal and Marine Applications of Remote Sensing", (ed. A.P. Cracknell), Remote Sensing Society, Reading, pp. 59-67.

10. Hayes, R.M.: 1980 in "Proceedings of the 14th International Symposium on Remote Sensing of Environment, Costa Rica, April 1980", Environmental Research Institute of Michigan, Ann Arbor, pp. 1187-96.

11. Valerio, C.: 1981 in "Remote Sensing in Meteorology, Oceanography and Hydrology", (ed. A.P. Cracknell), Ellis Horwood, Chichester, pp. 218-42.

12. Singh, S.M. and Cracknell, A.P.: 1979 in "Proceedings of Workshop on the EURASEP Ocean Colour Scanner Experiments 1977", (ed. B.M. Sørensen), Joint Research Centre, Ispra, pp. 171-9.

13. Cracknell, A.P. and Singh, S.M.: 1980 in "Proceedings of the 14th Congress of the International Society of Photogrammetry, Hamburg, July 1980", International Archives of Photogrammetry 23, B7, pp. 225-34.

14. Cracknell, A.P. and Singh, S.M.: 1981 in "Proceedings of EARSeL-ESA Symposium on Application of Remote Sensing Data on the Continental Shelf, Voss, May 1981", E.S.A., Paris, SP-167, pp. 157-68.

15. Baylis, P.E.: 1981 in "Remote Sensing in Meteorology, Oceanography and Hydrology", (ed. A.P. Cracknell), Ellis Horwood, Chichester, pp. 81-96.

CHAPTER 5

STATUS AND FUTURE PLANS FOR THE FIRST EUROPEAN REMOTE SENSING SATELLITE ERS-1

Guido Paci
presented by Alain de Leffe

European Space Agency, Toulouse, France

1. INTRODUCTION

On the 28 October 1981, the Member States of the European Space Agency, joined by Norway and Canada, decided to initiate the first ESA Remote Sensing Satellite (known as ERS-1) Programme. This first mission is oriented towards ice and ocean monitoring, the main mission objectives being of both scientific and economic nature and aiming at:

- increasing the scientific understanding of coastal zones and global ocean processes which, together with the monitoring of polar regions, will provide a major contribution to the World Climate Research Programme. ERS data used alone, or more commonly used in conjunction with complementary data from buoys, radio-sondes, research vessels, other near-surface platforms and other satellites in pre-arranged global or regional experiments, will enable significant advances to be made in physical oceanography, glaciology and climatology;

- developing and promoting economic/commercial applications related to a better knwoledge of ocean parameters and sea-state conditions.

This is of importance in view of the increasing development of coastal and offshore activities and the adoption by countries of the 200 nautical mile economic zone. In addition, monitoring of sea-ice and icebergs will be of importance for industrial activities performed at high latitudes.

In the following sections an overview is given of the present status and future plans for ERS-1 due to be launched in 1987 and preparing the way for a fully operational multi-satellite system in the 1990's.

A. P. Cracknell (ed.), Remote Sensing Applications in Marine Science and Technology, 67–81.

2. ERS-1 STATUS

Since 1978 a number of studies have been conducted within the European Industry aiming at the preliminary definition of the ERS-1 Programme including overall mission concept, preliminary instruments definition and accommodation and data acquisition/processing/distribution strategy. A brief description is provided of the present overall ERS-1 system including the payload configuration, the launcher and satellite platform, the orbit, the ground segment and the development plan.

2.1 Payload Description

The priority in the payload has been given to a comprehensive set of active microwave instruments able to observe as completely as possible the surface wind and wave structure over the oceans. The set of instruments consists of a Wind Scatterometer to provide wind speeds to an accuracy of 10-20%, and wind directions to within about 20°, and of a Radar Altimeter to provide significant wave heights to about 10% accuracy. Further, the SAR on SEASAT was successful in imaging waves yielding, in most cases, the directions and wave-lengths of the dominant waves and potentially the full two-dimensional wave energy spectrum.

In the selected ERS-1 configuration, SAR small-scene wave images will be interleaved with scatterometer data of the same microwave wavelength in a global sampling scheme, thereby enabling the development of algorithms including the important coupling between the wind, the short back-scatterometer ripples and the longer modulating waves. The simultaneous global sampling of wind and wave fields will furthermore enable the application of mutually supportive objective analysis schemes in which the wind and wave fields are estimated jointly through the application of dynamic wave model.

For ERS-1, the Wind Scatterometer, Wave Scatterometer, and SAR will be combined as an Active Microwave Instrumentation (AMI). This approach leads to a reduction of the required mass, volume and cost by sharing common hardware.

To summarise, therefore, the ERS-1 baseline payload will consist of the following instruments:

- an Active Microwave Instrumentation (AMI) operating in the C-band and combining the functions of a Synthetic Aperture Radar (SAR), a Wave Scatterometer and a Wind Scatterometer, with the aim of measuring wind fields and the wave image spectrum and of taking all-weather high resolution images of coastal zones, open oceans and ice areas and (on an experimental basis) over land;

- a Ku-band Radar Altimeter (RA), with the aim of measuring significant wave-height, wind speed, and of providing measurements over ice and major ocean currents;

- Laser Retroreflectors for accurate tracking from the ground;

- an additional package* resulting from an Announcement of Opportunity to the scientific community, consisting of:

  . the Along-Track Scanning Radiometer with a Microwave sounder (ATSR-M) to be provided and funded by the UK and France. This is a 3-channel infra-red radiometer for accurate sea-surface temperature determination from measurements at two different angles, plus a 2-channel nadir microwave radiometer for the accurate determination of the total vertical column water vapour amount. Two additional experimental channels are also proposed (90 and 150 GHz) in this sounder.

  . the Precise Range and Range Rate Experiment (PRARE), to be provided and funded by Germany. This is a two-way X-band RF tracking system using pseudo-noise coding techniques. S-band one-way is also included to correct for the ionospheric error.

A short description of the main performances of the instruments is given in Annex 1.

The data from the payload instruments will be collected and formatted within the Instrument Data Handling and Transmission Subsystem (IDHTS) before transmission at X-band to the ground.

Because of the global character of the ERS-1 mission, and in order to enable data acquisition in areas where the satellite is not within visibility of a ground-station, the payload will include a data storage capability which will be read-out when the satellite passes over a dedicated ground station. Thus instrument data and necessary auxiliary data will be:

---

*Final decision to embark the additional package will be taken at the end of Phase B.

- either transmitted to the ground in real-time,
- stored in the on-board data storage facility, with the exception of data generated by the SAR in its imaging mode which will be transmitted in real-time only,
- or both transmitted in real-time to one station while recording for later play-back to a dedicated ground station.

Most representative sensor combinations to be considered include:

- AMI in its high-resolution imaging mode, either alone, or in combination with the Radar Altimeter. This combination will be mostly used over land, ice and coastal areas within visibility of ground station visibility;
- AMI in its wave and wind measurement interleaved mode, in combination with the Radar Altimeter. This combination will be used over open ocean areas. The data are recorded on-board and/or transmitted directly when in ground station visibility;
- ATSR for which it is assumed that it shall function continuously and in combination with all other instruments.

2.2 Launcher and Satellite Platform

ERS-1 will be launched from the Kourou Space Centre (French Guiana) by the ARIANE-2 (or 3) launcher as a single passenger due to its dimensions.

It will make use of the Multi-Mission Platform (PFM) developed in the framework of the French SPOT Programme. Main characteristics of this Multi-Mission Platform are given in Annex 2.

2.3 Orbit Configuration

Considering the conflicting requirements:

- to monitor appropriately fast-changing features such as surface wind fields or sea-state;
- to achieve a global coverage of the Earth's surface with the sensors, the polar caps being accepted exceptions;
- to operate the sensors under constant geometrical conditions, and in some particular cases, under constant local time conditions to ease the data interpretation (AMI imaging mode for important applications);

to simplify the technical requirements for the satellite thermal control and power generation,

and the diverse characteristics of the sensors selected for ERS-1 in terms of field of view and modes of operation, and the variety of their potential applications, it is difficult to achieve a compromise between coverage and repetition rate which is fully adequate for all sensors and for all potential applications; this difficulty will be overcome by the orbit manoeuvre capability to change the orbit repetition pattern and to test various coverage/repetition rate profiles on an experimental basis.

The nominal orbit is sun-synchronous and circular at an altitude of 777 km, providing a 3-day repetitive ground track pattern. The stability of the ground tracks will be better than ± 1 km. The local time of Equator crossing is not frozen and its optimisation is still the subject of further consideration by the users. However, a local time around 10.30 (descending node) is a realistic assumption. This local time will be maintained within ± 15 minutes.

Main characteristics of the ERS-1 nominal orbit are given in Annex 3.

2.4 Ground Segment Concept

2.4.1 Mission Requirements. Benefits to be expected from a mission such as ERS-1 will result essentially from the possibility of generating improved short-term and medium-term forecasts of weather and ocean conditions on a local or global basis. All-weather high-resolution imaging from the AMI in the imaging mode will be used to monitor ice coverage in polar or near-polar regions or for surface pollution monitoring in combination with optical data from other sources e.g. LANDSAT, SPOT. As concerns ground segment design, it appears that some products will have to be generated within a few hours in operational systems.

Outside these operational users, there are many scientific activities for which longer delivery times are acceptable. This leads to a natural separation of the users'community into two classes: a "real (or quasi-real) time" class for which the ERS-1 data will have to be processed and delivered within 3 to 6 hours, and a "non real-time" class for which delivery times from one day to several weeks are adequate.

2.4.2 Instrument Duty Cycles. The following duty cycles have been used for the preliminary definition of the ground segment concept:

- High bit rate mode (AMI imaging mode)
  Because of the high bit rate (100 Mbps) on-board recording cannot take place and, therefore, data acquisition is only possible from areas in the coverage of available stations. The SAR will operate with a duty cycle of about 10%.

- Low bit rate mode
  During the low bit rate mode, the following sensors operate with a duty cycle of 40 to 100%:

Altimeter:	about 10 Kbps
Wind Scatterometer:	" 1 Kbps
AMI wave mode:	" 600 Kbps
ATSR:	" 100 Kbps.

2.4.3 Coverage. ERS-1 is intended to allow world-wide coverage, compatible with the duty cycles of its payload instruments. The Agency will accordingly ensure real-time transmission from the satellite and effect any necessary coordination with national facilities for data acquisition.

Furthermore, the on-board tape recorders will give access to data from any part of the world via the play-back data acquisition facility provided that the global amount of data recorded per recording period will have an upper limit of around 5 Gbits.

2.4.4 Deliverable Products & their Quality

LEVEL A Raw Data
Raw data as received from the satellite and recorded on HDDT in satellite format.

LEVEL B RT/DT Quick Annotated Raw Data/System-Corrected Data
Raw sensor data, time ordered, time-tagged Earth located. Internal calibration tables are attached but not applied. Geometric auxiliary information will include Earth location data, making use of the latest available orbits and attitude information, i.e. the corrections defined in the auxiliary products can be applied in order to generate system-corrected data.

LEVEL C RT/DT — Quick-Look Data
Data, in digital or photographic form, for payload monitoring, browse facility support and, if requested, fast delivery to users. Depending on the sensor, the spatial resolution and radiometric quality could be lower than that of the final product.

LEVEL D RT/DT — Quick Thematic Products
Geophysical data, from low bit rate sensors, after application of instrument transfer functions and removal of environmental effects. They will be extracted mainly from Level B data. They include, amongst others, wave spectra from SAR when this instrument is operated in its reduced mode.

RT = real-time (normally hours)
DT = delayed time (one day to weeks).

2.4.5 Ground Segment Concept Description

The ERS-1 ground segment will include:

- the Mission Management and Control Facilities located at ESOC, and combined in the MMCC (Mission Management and Control Centre),
- one station located in Kiruna for coverage reasons, for:
  - acquisition of all payload data (DAF: Data Acquisition Facility)
  - telemetry, telecommand and tracking

and for provision of near real-time products (RTPF: Real-Time Processing Facility) of a global nature.

For the distribution of the near real-time data to the users, two possibilities will be investigated:

- transmission via land lines to regional hubs (WMO solution),
- transmission via satellites such as SIRIO-2/MDD or ECS/ESS.

The other facilities of regional interest, RDAF(s) Real-Time Data Acquisition Facilities, and PAF(s) Processing and Archiving Facilities, which are not necessarily ERS-1 dedicated, but participate in the end-to-end data system, should be the elements of the Earthnet Programme or national facilities coordinated by the Earthnet Programme Office.

2.5 Development Plan

On 15 May 1982, contributions from participating States* reached the level of the financial envelope laid down for starting the detailed system definition phase (Phase B) of the ERS-1 Programme within Industry.

The final decision to proceed with Phase C/D (hardware development) will be taken at the end of 1983 at the end of Phase B.

The development schedule for the ERS-1 is expected to be as follows:

i) Carrying out of ERS-1 Phase B studies (detailed system and subsystem design) by Industry from mid-1982 to end 1983.

ii) Approval and start of ERS-1 Phases C/D at end 1983 (development and manufacture of the spacecraft/ instruments, plus the setting up of an adequate ground segment).

iii) Launch of ERS-1 by ARIANE at the end of 1987. Lifetime of ERS-1 is expected to be 3 years.

It is worthwhile mentioning that in order to prepare the necessary support technology, the ESA Member States approved in early 1979 the Remote Sensing Preparatory Programme (RSPP) aimed at:

- initiating the development of critical technologies both for optical and microwave payload elements and on-board and ground data management,
- and performing experiment campaigns with sensors embarked on aircraft e.g. campaign SAR-580.

*Belgium/France/Germany/Italy/Spain/Sweden/Switzerland/United-Kingdom/Norway/Canada. (In addition Denmark & Netherlands participation is soon to be made official).

3. FUTURE PLANS FOR ERS-1

3.1 Experimental/Pre-Operational System

It is essential when designing the ERS-1 mission to involve the users'community in the development of the requirements, the development of the system and the evaluation of the performance of the system. Basically two main user communities will make use of ERS-1 data, one scientifically oriented and the other applications-oriented, each having different requirements and characteristics. The Agency has put a special effort to involve these communities in the definition of the mission requirements and intends to pursue this effort through the development and exploitation phases of ERS-1, considered as both an experimental and pre-operational system.

It will be experimental since, as the first ESA mission, it will have to demonstrate that the concept and technology for both the space and ground segments are right for the applications envisaged, and that the users are ready and able to use the data generated. This will require a number of activities before and after launch which are very important for the success of the mission, such as:

- simulation and optimisation of sensor performance (airborne testing),
- development and testing of algorithms and models,
- setting up and testing of data products,
- definition of distribution networks to meet user requirements,
- development of pilot-projects (small scale) and demonstration missions (large scale),
- continuation of research and development to optimise the use and value of the data provided by the satellite system.

On the other hand, ERS-1 will have to demonstrate, for some appropriate applications and on a limited scale, an operational capability. This will, of course, require that data or products be delivered in quasi real-time to corresponding existing operational services or end users.

To this effect, the Agency has set up or is setting up teams of experts, called Instruments and Data Teams, to advise ESA on all user-related/scientific/technical aspects of each instrument and, for the overall system, on the operation aspects, data handling/dissemination and the promotion aspects of the Programme.

Scientific and application-oriented experts from the Participating States will contribute to these teams throughout the development and exploitation phases of the ERS-1 Programme.

3.2 Preparation of Follow-on Operational Missions

It is expected that, with ERS-1, a gradual transfer of applications from experimental to operational users will take place, preparing the future users to later operational satellite systems. As a first step, although not yet approved, it is planned to propose to the Participating states to ERS-1 to consider the launch of a second flight unit nearly identical to ERS-1 2 or 3 years after the launch of ERS-1 i.e., in the 1989-1990 time-frame, therefore providing the users'community with 5 to 6 years of continuous data.

In parallel, the Agency is participating to meetings of remote sensing satellite operators and has recently hosted the second meeting of these operators including at present Canada, Japan, India, USA, France and ESA. This offers the possibility for coordination of missions of the same nature (i.e. land-oriented or ocean-oriented) with the aim of providing services to users in terms of measurement frequency, standardisation of products etc.

The ultimate goal would be to set up or to contribute to an operational multi-satellite system for global rather than regional monitoring.

4. CONCLUSION

Within Europe there is a particular interest in a satellite monitoring programme over such important economic areas as the oceans, the coastal zones and ice caps. Equally, there exists an active European scientific community who recognises the potential value of modern spaceborne sensors to longer-term studies of ocean processes and their relationships to climatological problems.

In conclusion, it is clear that if Europe wants to be present on the world scene with operational remote sensing satellite systems in the 1990s, it is necessary to develop now an experimental/pre-operational satellite programme with a first launch around 1987. All necessary capabilities, whether industrial, scientific or economic, exist in Europe and the positive decision which has been recently taken by ESA Member States plus Canada and Norway, to develop the ERS-1 Programme should allow Europe not only to be present in the world competition but also to provide a contribution to future worldwide systems aiming at global monitoring of coastal zones and oceans.

ANNEX 1

PRELIMINARY INSTRUMENT PERFORMANCES

AMI Imaging Mode (SAR)

- Spatial resolution
  100 m x 100 m or 30 m x 30 m
- Swath
  80 km minimum
- Radiometric Resolution
  1 dB for 100 x 100 m at σ_0 = - 18 dB
  2.5 dB for 30 x 30 m at σ_0 = - 18 dB
- Incidence angle (on the ground) ≃ 23° at mid-swath
- Frequency
  5.3 GHz in HH polarisation
- Mean RF Power
  300 W
- Data Rate
  ≃ 100 Mb/s
- Antenna Size
  ≃ 10 m x 1 m

AMI Wave Mode

- Spectral Samples
  from < 100 m to 1000 m in 12 log. steps,
  resolution equal to step size
- Angular Samples
  over 180° in azimuth in steps < 30°,
  resolution equal to step size
- Spatial Samples
  each 100 km along track looking at 5 km square
- Sample Accuracy
  20° spectral energy density of spectrum (target figures)
- Incidence Angle
  ≃ 23° side-looking
- Frequency
  5.3 GHz in HH polarisation
- Mean RF Power
  < 100 W
- Data Rate
  ≃ 600 Kbps

AMI Wind Mode

- Spatial Resolution
  50 km
- Wind Speed
  4 to 24 m s^{-1}
- Accuracy
  2 m s^{-1} or 10%
- Wind Direction Accuracy
  < 20° (1σ); acceptable level of ambiguities TBD*
  (design goal: 0)
- Swath
  400 km one-sided (25° to 55° incident)
- Polarisation
  VV, 3-antenna beams
- Frequency
  5.3 GHz
- Antenna Size
  3.6 m long, 0,3 m high, 2 off
  2.5 m long, 0.3 m high, 1 off
- Data Rate
  ≃ 1 Kbps

Radar Altimeter (RA)

- Altitude Measurement
  < 10 cm (goal 5 cm) (1σ, 1 sec)
- Significant Wave Height
  1 ... 20 m (10% or .5 m)
- Backscatter Coefficient Measurement
  ± 1 dB (1σ)
- Frequency
  13.5 GHz
- Bandwidth
  300 MHz
- Peak Power
  500 W
- Antenna Size
  1 m Ø
- Data Rate
  8.5 Kbps

*Under investigation with simulated data products.

ATSR-M

- Along-Track Scanning Radiometer (ATSR)

- Spectral Channels
  3.7, 11 and 12 μ
- Spatial Resolution
  1 km x 1 km square
- Radiometric Resolution
  0.1° K
- Absolute Accuracy Predicted
  0.5° K over 50 km x 50 km square in 80% cloud cover conditions
- Swath Width
  500 km
- Power Consumption
  48 W
- Data Rate
  110 Kbps (after data reduction)
- Mass
  30 kg

- Microwave Sounder

- 2 channels: 23.8 & 36.5 GHz
- Pixel Size: 22 km for both
- Power Consumption: 30 W
- Mass: 21.5 kg
- Predicted Accuracy: 2 cm

Precise Range & Range Rate (PRARE)

- Up-link: 10 MHz in the band 7.19 - 7.235 GHz
- Down-link: 10 MHz in the band 8.35 - 8.50 GHz
- S-band down-link: 1 MHz in the band 2.20 - 2.29 GHz
- Power Consumption: 35 W
- Predicted Ranging Accuracy: 5 - 10 cm

ANNEX 2

MAIN CHARACTERISTICS OF THE MULTI-MISSION PLATFORM (PFM)

- Compatibility with sun-synchronous, circular orbits of altitudes between 600 km and 1200 km
- Compatibility with local time of satellite passes (ascending and descending node: 0800 to 1600 hours)
- Available Power: 1.9 KW (beginning of life) provided by a solar generator
- Power Storage Capacity: 4 x 23 Ampere hours
- Attitude Control Performance:
  pointing towards Earth centre with yaw steering capability
  stability on yaw: $1.1\ 10^{-3°}$/sec
  pitch and roll: 7.10^{-4}/sec
  accuracy: 0.15° for all axes
- Attitude Measurement Accuracy:
  better than 0.15°
  use of Fine Attitude Measurement System (FAMS) with star sensor could improve above performance
- Orbit Control:
  parallel and perpendicular to the orbit plane
  maximum capability 580,000 Nsec (300 kg of hydrazine)
- Satellite Management:
  by on-board computer with 20 Kwords (16 bits) memory available for the payload management
- Communications with Ground:
  telemetry and telecommand via an S-band transponder compatible with ESA and NASA networks
  transmitted power: up to 200 mW
  data rate: 2 Kbits/sec.

ANNEX 3

MAIN CHARACTERISTICS OF THE ERS-1 NOMINAL ORBIT

Semi-major axis a	7153 km
Inclination i	98.52°
Eccentricity e	1.105×10^{-3}
Perigee Argument	90°
Nodal Period T	100.465 minutes
Repeat Cycle	3 days
Orbital Periods per Day	14 + 1/3

CHAPTER 6

DIGITAL IMAGE PROCESSING

Peter Haberäcker

Messerschmitt Bölkow Blohm GmbH
P.O.Box 80 11 49, 8000 München 80, FRG

This paper is intended to be a short overview of Digital Image Processing. In section 1 fundamental principles in Digital Image Processing are discussed. Digitization of images and their pictorial reproduction are explained. After the definition of some statistical properties of digital images, data structures applied in Digital Image Processing are summarized. Section 1 finishes with simple image enhancement techniques. In section 2 some hard- and software aspects of a general purpose image processing system are discussed. Section 3 gives an overview of advanced techniques such as local operations, Fourier transform and digital filters, orthonormal expansions and geometrical transformations. Finally in section 4 scene analysis which can be devided in the steps preprocessing, segmentation and image understanding is explained. Thresholding, multivariate classification, contour following and region growing are illustrated with several applications. A summary of different features for segmentation is given.

A. P. Cracknell (ed.), Remote Sensing Applications in Marine Science and Technology, 83–135.

0. Introduction

The relationship between Digital Image Processing, Pattern Recognition and Artificial Intelligence may be compared with the fundamental mathematical operations of adding two numbers, multiplying two numbers and computing the power of two numbers: if one wants to understand the multiplication one first has to study the addition and if one wants to understand how to compute the power of a certain number one first has to study the multiplication. Digital Image Processing is restricted to the investigation and evaluation of two or more dimensional digital arrays. In Pattern Recognition which is not restricted to pictorial information, e.g. the processing of acoustic speech signals is an important application, one tries to understand the human perception and to apply the results to various data sources. In many cases digital image preprocessing is necessary in order to achieve good results in Pattern Recognition. In Artificial Intelligence systems are aspired which operate more or less independently in a well defined model of the "world". An example would be robots used in the fabrication of cars. A lot of Image Processing and Pattern Recognition techniques are necessary in order to achieve this independence.

This paper is intended to be a short overview of Digital Image Processing. Since it is impossible to describe all the procedures used in Digital Image Processing within this paper, only the most important techniques will be discussed briefly. In the first chapter the fundamental principles in Digital Image Processing will be described. Section two is a summary of some hard- and software aspects within this field. In section three some advanced techniques such as digital filtering, geometrical rectification and multispectral evaluation will be discussed. Finally section four is titled "Scene Analysis and Segmentation". This chapter is close to Pattern Recognition on the basis of two dimensional digital arrays.

1. FUNDAMENTAL PRINCIPLES IN DIGITAL IMAGE PROCESSING

1.1. Digitization of Images

In order to be able to process images with modern computer systems they have to be transformed into information which is compatible with computers. This transformation is called the digitization of the image. For photographs, paper or transparencies, or line drawings, drum scanners or flying spot scanners may be used for digitization. In many applications, especially in remote sensing, the digitization is performed during the recording time of the data using multispectral line scanners in aircraft or in satellites.

First the digitization of black and white images will be considered. Such an image, e.g. a line drawing, has two different graytones, usually black and white. In the first step the image is overlayed with a regular grid (two dimensional array) of squares. In the second step all squares are checked whether they contain more black or more white information of the original image. A square is then defined to be wholly black or white respectively. If the black and white squares, called pixels (from picture elements), are coded with 0's (black) and 1's (white) we have a binary, digital representation of the original image (figure 1.1.).

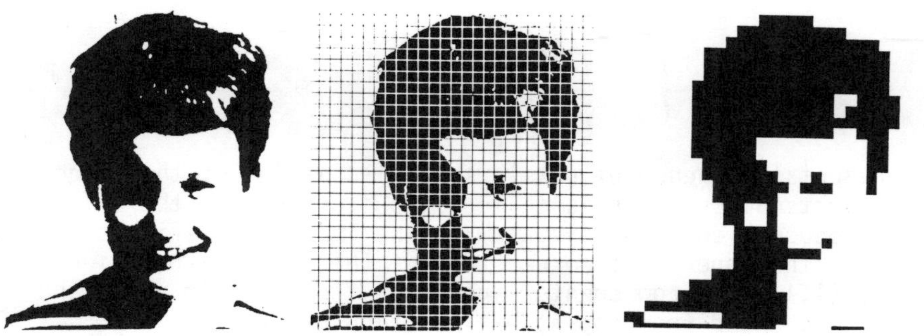

(a) (b) (c)

Figure 1.1. Digitization of a black and white image. (a) original image, (b) overlay with a regular grid, (c) digitized image represented with black and white pixels.

The quality of the digitization of the black and white image depends on the size of the squares. The sampling theorem requires that

$$f_s > 2f_{max},$$

where f_s is the spatial sampling frequency and f_{max} is the maximum spatial frequency of a bandlimited signal. If Δx is the length of the square we have

$$f_s = 1/\Delta x > 2f_{max} \text{ or } \Delta x < 1/(2f_{max}).$$

In other words, if we have an image line with n pixels of length Δx the maximum number of lines which we can resolve within the scanning interval is n/2 corresponding to a maximum spatial frequency of

$$f_{max} = n/2 : (n\Delta x) = 1/(2\Delta x).$$

With a mathematical notation we may characterize a binary image as follows:

$G = (0,1)$	binary gray scale with the two gray values 0 and 1;
$s = s(x,y)$	binary matrix of the digitized image;
x,y	spatial reference coordinates of a pixel in position (line,row) = (x,y)
$s(x,y)$	binary value of the pixel in position (x,y)

$$s(x,y) = \begin{cases} 0 \rightarrow \text{black} \\ 1 \rightarrow \text{white} \end{cases}$$

A horizontal sequence of pixels is called a line of the image and a vertical sequence is called a row (column) on the image. We introduce a coordinate system in the image and define the pixel in the upper left corner to have the coordinates (line, row) = (1,1). The row counter increases from left to right and the line counter from top to bottom (figure 1.2.).

Let (x,y) be a pixel of a given image. The 4-neighbors are the two horizontal and the two vertical neighbors (x,y-1), (x,y+1), (x-1,y) (x+1,y). In addition (x,y) has four diagonal neighbors in the positions (x-1,y-1), (x-1,y+1), (x+1,y-1), (x+1,y+1). The 4-neighbors together with the diagonal neighbors are called the 8-neighbors of the pixel (x,y) (figure 1.2.).

The distance between two pixels with the coordinates (x_1,y_1) and (x_2,y_2) may be computed with the Euclidian distance function

$$d_E = \sqrt{(x_1-x_2)^2+(y_1-y_2)^2}.$$

In this case the distance from a pixel (x,y) to his 4-neighbors is 1 pixel and the distance to his diagonal neighbors is $\sqrt{2}$ pixels. With another metric, the chessboard distance,

$$d_M = \max \left\{ |x_1-x_2|, \quad |y_1-y_2| \right\}$$

all 8 neighbors have the same distance 1 pixel from the central pixel (x,y).

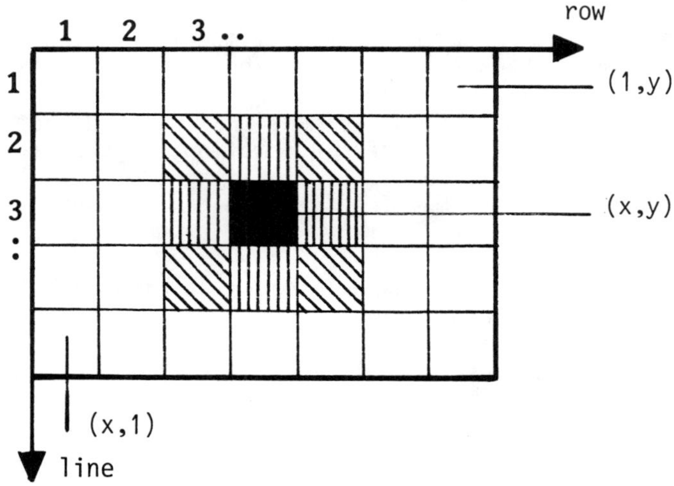

Figure 1.2. Coordinate system of a digitized image with 4- and 8-neighbors

A gray tone image has more than two gray tones. If we want to digitize a gray tone image we have to perform the same procedure as in the case of the binary image. In contrast to above we define a gray scale G with more than two gray values, usually G=(0,1,..., 255). For computer processing 256 gray values may be coded with 8 bits (1 byte). For every square of the sampling grid we have to find the average brightness which is converted (by means of an analog/digital converter) into the gray scale G (quantization). A pixel s(x,y) of a digitized gray tone image may have one of 256 different values. The gray value 0 represents black, 128 medium gray and 255 white. If the gray scale G with 256 values is not sufficient we can use more than 8 bit/pixel, e.g. 16 bit/pixel corresponding to 65 536 gray values. If we need less than 256 gray values the gray scale may be reduced. Techniques in this field will be discussed in section 1.4.

Digitizing a color image a color separation using adequate blue, green and red filters has to be performed. Each scan is equivalent to the gray tone image procedure with the exception that we use a blue, green and red filter. A pixel of a digitized color image is a three dimensional vector $s(x,y,n)=(g_b,g_g,g_r)^T$, where g_b is the average intensity in blue, g_g in green and g_r in red. For example, the pixel (127,0,0) represents a pure blue color of medium intensity, (0,0,0) represents black and (255,255,255) white.

A multispectral scanner with N spectral bands produces N-dimensional pixels $(g_1,g_2,\ldots,g_N)^T$, where g_i is the coded intensity in the spectral band i. In our mathematical model a multispectral image will be written as a three dimensional matrix $S=s(x,y,n)$, where x,y are the spatial coordinates and n the spectral band. In some cases the components of a N-dimensional pixel do not represent spectral but arbitrary other information, e.g. coded elevation of a digital terrain model. Digital images of this type we call multichannel images. Multispectral images are special cases of multichannel images.

Finally we may receive multichannel images at different times t_1, t_2, ... A multichannel time sequence of images we denote as a four dimensional matrix $S=s(x,y,n,t)$, where x,y are the spatial coordinates, n counts the channels and t the time. In this case a pixel would be a two dimensional matrix.

For the clarity of the paper we omit indicies which are not necessary in the given circumstances. Thus we denote a pixel as $s(x,y)=g$ or $s(x,y,n)=\underline{g}$ or $s(x,y,n,t)=(\underline{g}_t)$, respectively.

1.2. Pictorial Reproduction of Digitized Images

Sometimes it is necessary to inspect the gray value $s(x,y)$ of a given image S. In this case a hexadecimal, decimal or octal print out (line by line) of the image is useful. In general such a printout which is very paper consuming is not a good reproduction of the original image.

With line printers we can reproduce digital images if we overprint different symbols in order to achieve different gray tones. E.g. the symbol "blank" could reproduce white pixels and the overprint of the symbols * 0 M W L and T (→ ■) which yields a rather dark rectangle could be used as black pixels. With additional combinations one can achieve about 16 different gray tones. If one regards such images from some distance they look quite good. The problems with this method are the low resolution (enlargement of the original, paper consuming) and the geometrical distortion due to the rectangular shape of the pixels.

Figure 1.3. Part of a digitized topographical map reproduced with an electrostatic printer plotter

Electrostatic printer plotters may be used to produce images of acceptable quality. In the case of a binary digital image we output the image line by line and write a dark dot at every position of a black pixel (figure 1.3.).

With a multithreshold technique, similar to the technique in printing, or techniques with random number generators halftone images of good quality may be produced (figure 1.4.).

Important devices for the pictorial reproduction of digital images are raster image displays on TV-monitors. They have a high resolution of e.g. 512*512 or 1024*1024 pixels with 256 different gray values. In an image refresh memory several images may be stored simultaneously and displayed on the screen. If the device has a color screen truecolor representation may be achieved by combining the blue, green and red channels of the digitized color image. In a falsecolor representation other spectral channels, e.g. near or thermal infrared, are combined in order to obtain a color image. If the different gray values of a gray tone image are coded on the screen with different colors we have a pseudocolor representation. The raster image devices may be combined with special output devices which produce hardcopies (black and white or color) from the image on the screen on film or polaroid material.

Figure 1.4. Part of an aerial CCD-scanner image reproduced as halftone image by a multithreshold technique on an electrostatic printer plotter

Finally digital film recorders, laser beam recorders and ink jet plotters should be mentioned which yield a high quality output (black and white or color) of digital images of large size. In this case the computer reads the image data directly from the disk files and transfers them to the recorder.

1.3. Statistical Properties of Digital Images

Let $S=s(x,y)$ be an image with gray values $G=(0, 1, \ldots, 255)$. The histogram of S is

$$p_S(i) = N_i/M, \quad i=0, 1, \ldots, 255, \text{ where}$$

N_i is the frequency of occurrence of gray value i in S and

$M=L \cdot R$ the number of pixels in S with L lines and R rows.

The graphical representation of $p_S(i)$ is a bar graph with 256 bars (figure 1.5.).

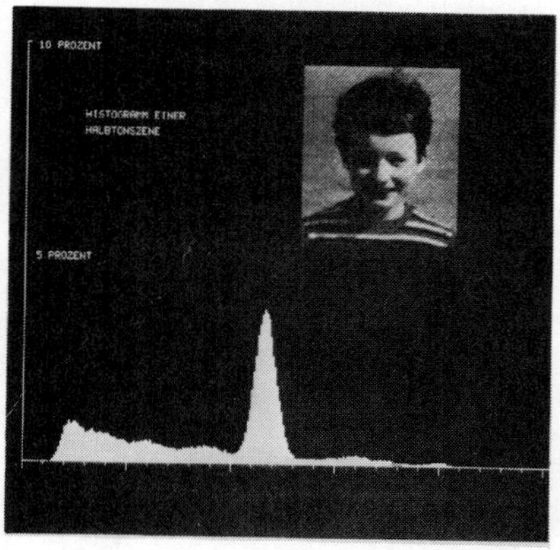

Figure 1.5. Histogram as bar graph of an image. The maximum corresponds to the gray values of the background, dark gray values are on the left, bright on the right.

A histogram is an approximation of the probability density of the gray values of the image. An approximation of the probability distribution function of S is

$$h_S(i) = \sum_{k=0}^{i} p_S(k), \quad i=0, 1, \ldots, 255.$$

If the bars of the histogram of S are located close to i=0 the image is dark, if they are close to i=255 the image is bright and if the histogram is narrow the corresponding image has low contrast. Techniques of image enhancement using histograms will be discussed in section 1.5.

The mean value of S is

$$m_S = 1/M \sum_{M} s(x,y) = 1/M \sum_{i=0}^{255} i \; p_S(i) \text{ and the}$$

approximation for the standard deviation is

$$\sigma_S = \sqrt{1/(M-1) \sum_{M} (s(x,y)-m_S)^2}$$

$$= \sqrt{1/(M-1) \sum_{i=0}^{255} (i-m_S)^2 p_S(i)}.$$

If we have the histogram of a given image we may compute the entropy

$$H = - \sum_{i=0}^{255} p_S(i) \log_2 p_S(i).$$

The entropy is a measure useful for image coding, because it is not possible to code without distortions an image with fewer than H bits.

If the gray values of an image tend to "cluster" around two distinct values the histogram is said to be bimodal. The unisotropy coefficient

$$\alpha = \left(- \sum_{i=1}^{na} p_S(i) \log_2 p_S(i) \right) / H$$

where na is determined by the condition

$$\sum_{i=1}^{na} p_S(i) \geqq 1/2$$

is a measure of the bimodality and may be used in thresholding. Bimodal histograms are important in image segmentation.

For an image $S=s(x,y,n)$ with N channels we may compute the covariance of channel n_1 and n_2:

$$c_{n_1 n_2} = 1/(M-1) \sum_{M} \left((s(x,y,n_1)-m_{n_1}) \cdot (s(x,y,n_2)-m_{n_2}) \right)$$

where

m_{n_1}, m_{n_2} are the mean values of channels n_1 and n_2.

The covariance matrix C of S is the symmetric matrix $C=(c_{n_1 n_2})$ of the covariances of pairs of channels n_1 and n_2. C is used in the computation of the principal components and in classification.

The N-dimensional histogram of S=s(x,y,n) is

$$p_S(i_1, \ldots, i_N) = N_{i_1 i_2 \ldots i_N}/M, \text{ where}$$

$N_{i_1 i_2 \ldots i_N}$ is the frequency of occurrence of the combination of gray value i_1 in channel 1, i_2 in channel 2, ... and i_N in channel N.

Similar to the 1-dimensional case the N-dimensional histogram is important in image segmentation (clustering).

1.4. Data Structures, Data Reduction

After digitization an image is stored in the computer system in disk files which usually are record oriented. In a line interleaved image data structure one line of the image is one record of the file. The lines are stored sequentially from the first line 1 to the last line L. Direct access to lines may be possible. In the multichannel case the file starts with line 1, channel 1, line 1, channel 2, ..., line 1, channel N, line 2, channel 1, etc. A data structure with separate files for the different channels is said to be image interleaved. The third case is pixel interleaved, where the pixels are stored as sequential vectors.

If the digitized image is very large the necessity for special data structures and data reduction may arise. The computation of principal components which reduces the number of (spectral) channels will be discussed in section 3. At the moment we will investigate some techniques with 1-channel images as input.

Reducing the number of gray levels (the number of bits per pixel) is an important example for data reduction. The histogram gives us the minimum and the maximum gray value and with this information we can requantise the image. A lot of other techniques, e.g. the entropy are applied in this field.

In natural images the gray values usually change very slowly from pixel to pixel. Thus we can achieve data reduction if we store instead of the values of the pixels those of the difference of two adjacent pixels. To code these differences we may need only four bits instead of 8 bits.

The run length coding will be described briefly with binary images. Suppose we have an image line which starts with the sequence "0000111111100000011111000...". In the run length code only the starting position and the length of a sequence of 1's is stored: (5,7), (16,5), ... The efficiency of the run length code is image dependent: an image with large homogeneous areas will produce a more compressed code than a image with many frequent changes between black and white.

A quadtree is something like a two dimensional run length code for a binary image. Let S be an image of the size 2^k*2^k. These 2^{2k} pixels are represented with a node of the highest level k of the quadtree. For every node at level l, l=k-1, k-2, ..., 0, we inspect whether all pixels which represent the node are "black", all pixels are "white" or some pixels are black and some pixels are white ("gray"). In the cases "black" and "white" nothing more is to be done. In the case "gray" four new nodes at level (l-1) are created which correspond to the four quadrants of the node at level l.

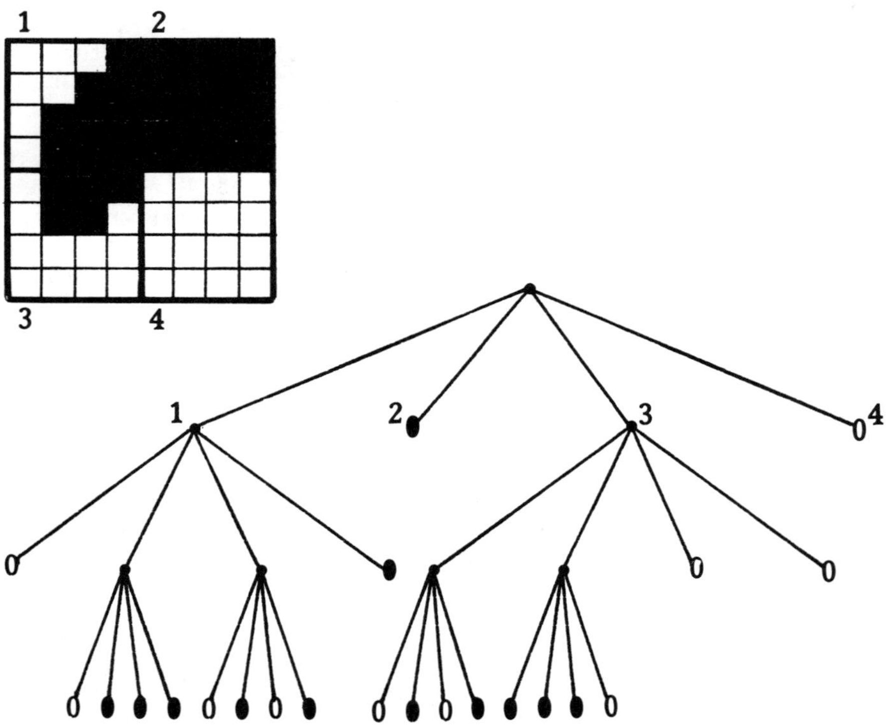

Figure 1.6. Example of a quadtree structure for a simple image (**0** ≙ white, ● ≙ black).

To each of the four nodes which represents an image section of size $2^{(l-1)} * 2^{(l-1)}$ we apply the same procedure as before. This dividing algorithm ends at latest when we reach the pixel level ($l=0$). Figure 1.6. is an example of a simple quadtree ($k=3$).

In the realization every node has a link to his "father" and to his four "sons". Thus the quadtree may be evaluated at random. In applications where no random access to the nodes is necessary the links may be omitted in order to achieve a higher compression ratio. One advantage of the quadtrees is that a lot of operations such as the logical operations AND, OR and NOT between two quadtrees may be computed directly in the quadtree structure as the intersection, the union or the negation of the quadtrees. The run length and the quadtree coding schemes may be generalized to images with more than two gray values. The quadtrees have important applications in image segmentation.

Another coding scheme is chain coding. The basic idea is that every pixel has 8-neighbors. If we label these 8 neighboring pixels from 0 to 7 in a counterclockwise sense, we can represent a digitized line structure simply by a sequence of octal digits together with one coordinate pair of the starting point. Figure 1.7. is an example of a chain code of a simple line structure.

Many operations such as inverse, length, width and height of a chain, as well as its integration, moments, rotation, etc. can be performed on this data structure. The chain code scheme can be generalised to images with more than two gray levels and to 3 dimensional image data.

The last compression technique which will be mentioned is coding with general polygonal line structures. With this technique a line structure is represented by means of polygons (sequences of coordinates). The most important application is in computer aided design (CAD).

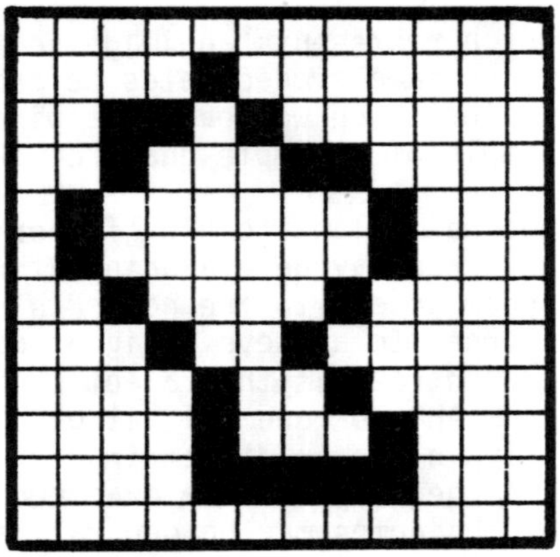

3	2	1
4		0
5	6	7

labels of the 8-neighbors of the pixel

Starting point: (2,5)
chain: 770765577644442233321201

Figure 1.7. Example of a chain code of a simple line structure. The structure is coded as a sequence of labels for neighboring pixels together with the coordinates of the starting point.

1.5. Simple Image Enhancement

In linear scaling we apply a transformation $S \rightarrow S'$: $s'(x,y,)=(s(x,y)+a)*b$. The constant a causes a change in the brightness (e.g. $a < 0$ is a shift of the histogram to the left) and b causes a change of the contrast (e.g. $b > 0$ is an expansion of the histogram). In practice the above transformation must be implemented as piecewise linear scaling in order to avoid producing gray values less than 0 or greater than 255. The parameters a and b may be obtained from the original histogram of S or may be found interactively. Obviously other than linear transformations may be used (general gray scale transformations) if we take care that the output pixels are confined to the range (0,255).

With piecewise constant transformation functions we have the technique of gray level slicing which has many simple applications in digital image processing. It should be mentioned that all gray scale transformations can be implemented as table-look-up procedures which are very effective with regard to computation time.

Another enhancement technique is the gray scale correction where we try to eliminate effects such as vignetting caused by a lense or stripes in the image caused by imperfections by CCD-arrays.

In histogram modification an image S with histogram $p_S(i)$ is transformed into an image S' with histogram $p_{S'}(i)$ which is similar to the histogram $p_R(i)$ of a given reference image R. These techniques are applied e.g. if we want to "compare" two images which are taken from the same object and the same direction at different times.

References for section 1 (basic text books and tutorial papers are underlined):

2, 4, <u>5</u>, <u>9</u>, 25, 26, 28, 36, 38, 39, <u>41</u>, <u>43</u>, 44, 47, <u>48</u>, <u>50</u>, <u>53</u>.

2. HARD- AND SOFTWARE ASPECTS IN DIGITAL IMAGE PROCESSING

2.1. Image Processing Systems

The applications of Digital Image Processing and Pattern Recognition are manifold. In a research oriented environment the objectives are the development of new algorithms and their application in various fields of science such as physics, medicine, biology, geoscience, etc. Image processing systems in this field should be "general purpose image processing systems" which have a highly flexible data and program structure that may be used without placing restrictions on the different data sources.

In applications where the image processing systems are used in special environments the software and the hardware will be optimized with regards to the actual requirements. In an image processing system which allows to digitize, to preserve, to modify and to output line drawings e.g. of a technical documentation handbook,some of the above mentioned flexibility may be abandoned in order to achieve better image thruput and comfort in system handling. Instead of a general raster image display one would use e.g. a special screen which allows both raster images and vector images. The system should work interactively which means that the response time of the computer is generally less than 15 or 20 seconds.

The next step would be an environment with strong real time conditions. If a vidicon has to trace bottles on a conveyor-belt and the system has to detect dirty or broken bottles, the system has a well defined time intervall to compute the decision. Hard- and software is highly specialised in this case and there is no possibility to process e.g. medical data with such a system. Special hardware such as array processors or two-dimensional fourier analyzers may be the core of the hardware.

A system of the highest integration degree will have image processing components suitable for tasks such as navigation, tracking, etc. Here it is often necessary to develope new special hardware which has the capability to solve the real time problems.

In this section we will describe the components, hard- and software, of a general purpose image processing system, because the objectives of this paper is to give a more research oriented overview than a description of operational systems.

2.2. Hardware Configuration of a General Purpose Image Processing System

Figure 2.1. shows the hardware configuration of a fictitious general purpose image processing system. The heart of the system

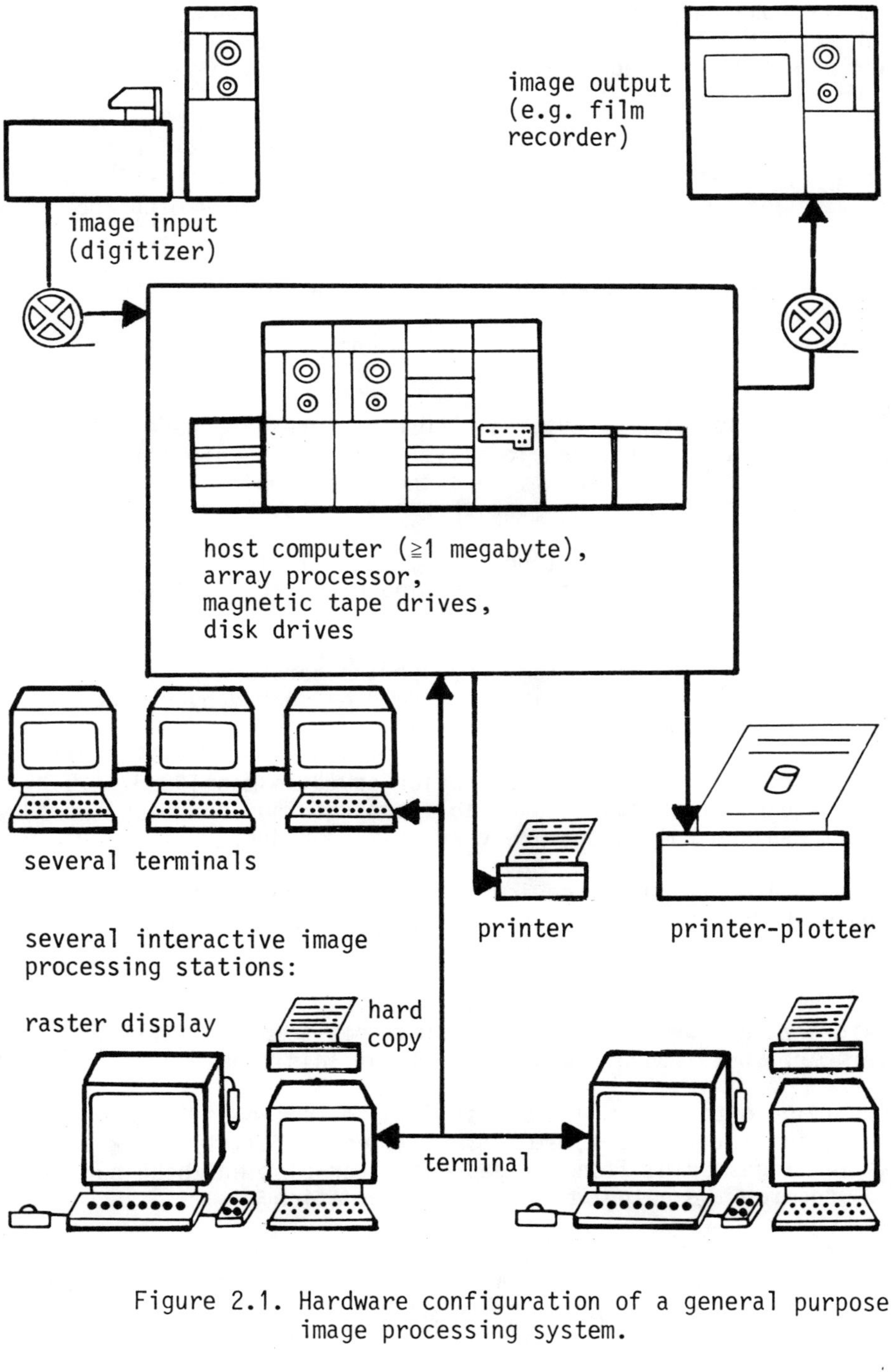

Figure 2.1. Hardware configuration of a general purpose image processing system.

is the host computer. Most of the available computers may be used in digital image processing but if the host has some features which will be listed briefly fulfilling the image processing requirements will be easier:

- byte oriented machine. Mostly 256 gray values are used which can be stored in 1 byte. The possibility to address the bytes simplifies the programming and helps to save storage place.

- enough memory. Depending on the number of users and the main applications the memory should be not too small. Memory should be planned to be greater then 1 megabyte.

- real time features. These features are necessary if the system has to solve application with time constraints. In any case the host should allow easy integration of new peripheral devices.

- virtual operating system. A virtual operating system would simplify programming but is not absolutely necessary.

- multiuser system. Because the hardware of such a system is very expensive it should be shared by a group of users.

There are a lot of other facts which could be mentioned, e.g. floating point hardware, available programming languages, etc. But these considerations are very task dependent.

As standard peripheral devices our system has two tape drives, two disk drives (each about 300 megabyte), a system console, a line printer and several terminals.

For the input of digitized images a raster scanner is necessary. This may be a drum scanner, a flying spot scanner, a flatbed scanner, a vidicon or something else of this type.

For image output we may use raster image recorders (drum recorders, ink jet plotters, etc.) or for medium quality black and white images or halftone images a printer-plotter which can also be used for the output of graphic information.

It is obvious that the tapes are also important media for image input and output, e.g. Landsat data on computer compatible tapes (CCT-s).

In many applications it may be necessary to integrate an array processor into the system in order to have more computing power.

An essential part of the system are the interactive user stations. A user station should have a terminal for user input and system output, a raster image display and perhaps a hardcopy device which is connected with the raster image display. Some important features of the raster image display system include:

- a minimum of three image refresh memories (512*512 pixels, 8 bit deep);

- the possibility of reconfigurating the refresh memories, e.g. to obtain two images of 1024*1024 pixels, each 4 bit deep;

- the functions scroll image, roam image and zoom image;

- look-up-tables (gray scaling and pseudocolor coding);

- a trackball, joy stick, data tablet or similar equipment;

- the capability of graphic information.

More comfortable raster image display systems have an integrated processor, sometimes an array processor which allows to do a lot of processing in very short time without the host computer. If there is enough money available the system should have more than one user station.

When the future development of personal computers will also offer not too costly image processing it will be possible to work without a host computer. An additional link-up to a host computer would then afford faster input and output of images and the occasional use of greater computing power.

2.3. Description of the Software of a General Purpose Image Processing System

The software of an image processing system may be devided into two major sections: The image processing programs on the one hand and the software which controls these programs as well as the images, the user interactions, etc.

First the features of the control software, which is something like an image processing operating system will be discussed.

If the host computer runs on a multiuser system the image processing system may also be a multiuser system. This means that several users may work with the system simultaneously. Peripheral devices which exist only once in the system (e.g. a printer-plotter or a raster image input device) must be controled and shared by the system.

Normally the user works with the system in an interactive mode: He enters commands to the system which cause a reaction, e.g. the execution of a certain program, and he receives information from the system, e.g. that a certain program has finished executing. Generally, during the interactive mode the user terminal is assigned to the current process and no further process can be activated from it. Thus in the case of programs which require long computation times the system should provide processing such as subtasks or (and) batch. A subtask which has been created by the user runs simultaneously with the interactively controled foreground tasks. The system should provide capabilities for the control of the background tasks (status information, commands to kill background processes, etc.).

The structure of the image data should be highly flexible. As an example the image data structure of the digital image processing system IGARIS (Interactive Graphic And Raster Image Analysis System) of Messerschmitt-Bölkow-Blohm (MBB) will be described (figure 2.2.). An IGARIS-image always consists of two files. One of these two files contains the image data, the second contains image description data. The image data may be of different types: ordinary raster scan image data stored in the usual way, special compressed raster image data (run length or quad tree) or vector images. At the moment the following types of raster image data are implemented:

- type 1: 1 byte/pixel, integer,
- type 2: 2 byte/pixel, integer,
- type 4: 4 byte/pixel, real,
- type 8: 8 byte/pixel, complex,
- type 9: extended image data structure (e.g. quad trees, vectors, etc.).

Additional data types, e.g. type 0: 1 bit/pixel, may be implemented if necessary. A raster image may consist of several image channels. The maximum number of channels is 32. The raster image data are stored line interleaved. If necessary several modes of compressed stored images are available. At the moment the data compression technique of quad trees is implemented for binary and logical images. Other techniques such as run length coding or even vector image data would be possible.

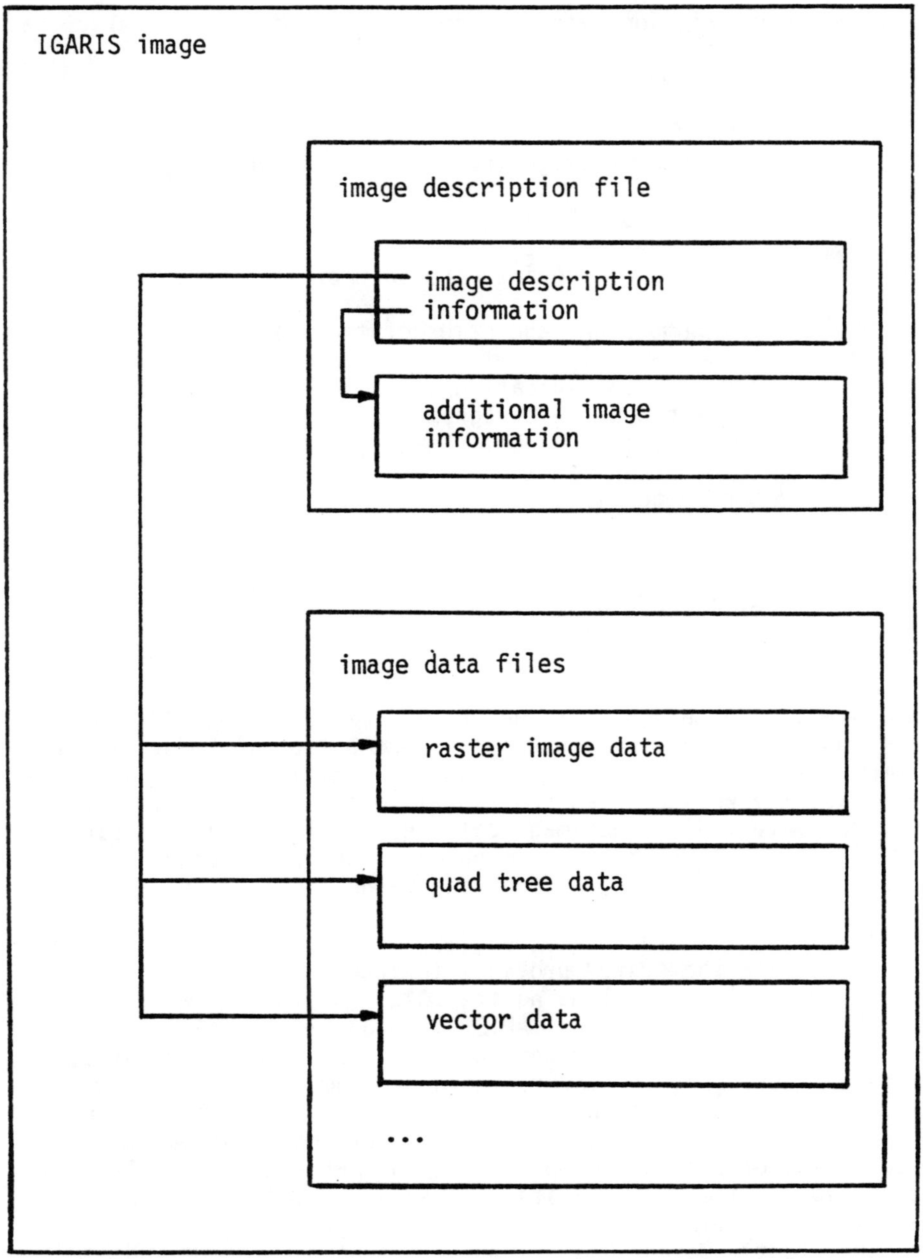

Figure 2.2. Image data structure of the image processing system IGARIS (Interactive Graphic And Raster Image Analysis System) of Messerschmitt-Bölkow-Blohm (MBB).

The image description file contains:

- the name of the image and the name of the image file which may be different,
- the date of image creation and of the last change,
- the type of the image,
- the size of the image (number of lines and rows),
- the number and the names of the image channels,
- a window-specification of an image section which an image processing program should use as input data (default is the whole image),
- the image history,
- a description of additional image information types such as look-up-tables, pseudocolor tables (x,y)-coordinates, etc,
- the specific data of the additional information types.

We give a brief example of the use of the additional image information. If interactive contrast enhancement of an image is performed using a special raster scan image device, the final look-up-table may be stored as additional image information. It is not necessary to modify the image data because the look-up-table is available together with the image data. Only in case of a hardcopy the pixels have to be modified using the look-up-table.

Another essential feature of an image processing system is the comfort which the system provides in interactive handling and in programming. In the following the IGARIS philosophy will be described as an example. In IGARIS all programs and even the images communicate with the user through an interactive input/output monitor. For every IGARIS program a program description has to be made. This consists of a description of the algorithm and the function of the program, a description of the input parameters and a list of possible help, prompt and error messages. The description of the input parameters has a special syntax which enables them to be of several types (integer, real, logical, vector, text). Default values of the input parameters and plausibility control information (e.g. control information of the range of parameter values) are provided. In addition the input parameters may be structured hierarchically using sub- and subsubparameters. The input/output monitor uses this program description to communicate with the user. Every input parameter may be called with a keyword and the default value may be changed within the boundaries of the plausibility control information. Help functions may be called at any moment the user is allowed to enter data. Figure 2.3. is an example of a program description of a simple program which performs gray level slicing. Using this scheme all programs of IGARIS present themselves to the user with the same "face": if the user is familiar with the syntax and the communication mode of the input/output monitor, he can handle all available programs. It should be mentioned that the input/output monitor can be used also in batch mode.

The IGARIS programs, system software and image processing software, are devided into thematic program sets. For example one set are the system programs, a second set are image conversion programs (input and output of images), a third set the fourier analysis programs, etc. All program descriptions of one thematic program set are included in a thematic program file which may be edited by the editor of the host computer system. If translations of this thematic program files into other languages are available, IGARIS may be used in different languages simultaneously. To obtain hardcopy descriptions of the programs, the possibility of printing out a short overview of the IGARIS programs, a list of one thematic program set or a whole handbook is provided.

Since the structure of our image processing system is highly modular it is very easy to integrate new programs or even new thematic program sets into it. It is also possible to maintain private thematic program sets which are accessible only to authorized user groups. This may be useful during development of new programs or for the protection of privacy. The program development can be standarized and simplified if a "form" of a typical image processing program is provided, from which new programs can be derived.

After the description of the structure of the control software we are going to give a short summary of the image processing programs which should be available in a general purpose image processing system. We only list the headlines of different areas of application, some of the algorithms behind them will be discussed in sections 3 and 4 of this paper.

- Image input conversion programs. Conversion from the external into the internal format, e.g. input from tapes or from raster scan devices.

- Image output conversion programs. Conversion from the internal into an external data format, e.g. output on tapes, printer-plotter, raster image recorder or raster image display.

```
***************************************************************
*PROGRAM-NAME: GSLICE
*T
*T      Gray Level Slicing
*T      ==================
*T      Author:
*T      Date:
*T
*T      With the program GSLICE a gray tone input image can be
*T      transformed into an output image with sliced gray
*T      levels. There are three modes available:
*T
*T         - mode 1: the gray level slices are coded in the
*T                   output image with new gray values;
*T
*T         - mode 2: only edges between the different gray
*T                   level regions are coded as dark lines;
*T
*T         - mode 3: mode 1 and mode 2 integrated in one image.
*T
*T      Description of the input parameters:
*T
*T      INPUT:       Input image, type 1
*T      OUTPUT:      Output image, type 1
*T
*T      MODE:        Mode 1, 2 or 3 (see above)
*T
*T      FILE:        Input medium for the gray level slices
*T
*PARAMETERS:
*
*1      INPUT  = ' '          <LNG=8>
*1      OUTPUT = ' '          <LNG=8>
*
*1      MODE   =  3           <MIN=1, MAX=3>
*
*       Input medium for gray level slices:
*       ------------------------------------
*1      FILE   = 'TERMINAL'   <LNG=30>
*
*PEND
+++ERROR: Input image file does not exist
+++ERROR: Error during parameter input
+++ERROR: Wrong input image type
+++ERROR: Output image: no storage place on the disk
...............................................................
```

Figure 2.3. Example of a (compressed) program description of a program which performs gray level slicing

- Interactive programs for the handling of the raster image display, e.g. linear gray scaling with the trackball, definition of pseudocolor tables, output of (x,y)-coordinates together with the gray value.

- Fundamental image processing programs, e.g. statistical programs, image scaling, gray level slicing, smoothing, edge enhancement, principal components,
- calibration, image accumulation, differences, ratios.

- Data structures and data compression, e.g. run length, quad trees, chain code.

- Digital Filtering, e.g. 2D-fourier transform, convolution, filter construction.

- Geometrical rectification, e.g. enlargement, reduction in size, rectification with control points, high precision correction, 3D-processing.

- Image segmentation, e.g. thresholding,region growing, unsupervised and supervised classification, edge oriented segmentation, interpretation of segmented images.

With this summary we finish the discussion of our general purpose image processing system. In the next two sections we will look behind some catchwords listed above.

References for section 2 (basic text books and tutorial papers are underlined):

1, 12, 13, 18, 23, 37.

3. ADVANCED TECHNIQUES IN DIGITAL IMAGE PROCESSING

3.1. Local Operations

Smoothing of an image is an operation which is often applied prior to further processing steps, e.g. gray level slicing. In image smoothing we regard a square (n*n)-environment of every pixel of the input image. The gray values of the pixels in the environment are accumulated and the pixel of the smoothed output image is assigned the average gray value:

$$s'(x,y)=1/n^2\left(\sum_{i=x-m}^{x+m}\ \sum_{j=y-m}^{y+m} s(i,j)\right)$$

for all pixels in position (x,y), where m=(n-1)/2, n=3, 5, 7, ... Compared with the original the smoothed image is less sharp, the contrast is lower and the histogram is softer. Figure 3.1. is a graphical representation of this procedure.

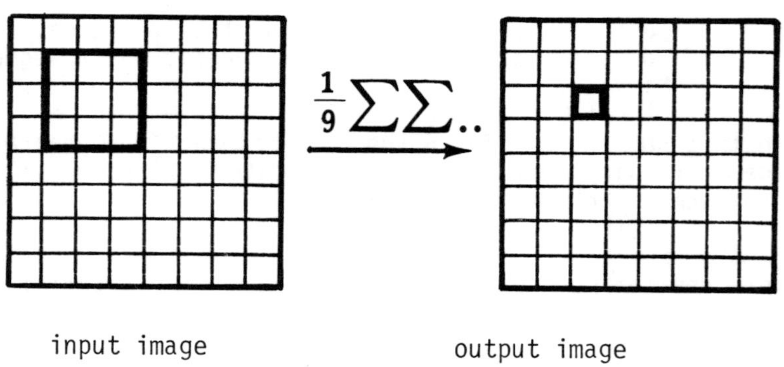

mask: $\begin{pmatrix}1 & 1 & 1\\ 1 & 1 & 1\\ 1 & 1 & 1\end{pmatrix}$

Figure 3.1. Image smoothing (moving average) with a (3*3)-environment

This smoothing operation may be interpreted as the convolution of a square mask, which describes the weights of the pixels in the (n*n)-environment, with the original image. In case of the described "moving average" the mask has only the value 1 in all positions (Figure 3.1.). From the contents of the mask we can derive the effect of the application of the convolution. Because one pixel of the output image depends only on the environment defined by the mask these operations are called "local operations".
The moving average which is something like a discrete integration

of the image is equivalent to a low pass filter.

Image sharpening (differentiation, high pass filter) we can implement with similar operations. Examples of masks for image sharpening are the Laplacian operator or the vector gradient from which we can compute the magnitude and the direction of the gray value gradient (Figure 3.2.).

$$\begin{pmatrix} 0 & -1 & 0 \\ -1 & 4 & -1 \\ 0 & -1 & 0 \end{pmatrix} \quad \text{or} \quad \begin{pmatrix} -1 & -1 & -1 \\ -1 & 8 & -1 \\ -1 & -1 & -1 \end{pmatrix} \quad \text{Laplacian}$$

$$\begin{pmatrix} 1 & 2 & 1 \\ 0 & 0 & 0 \\ -1 & -2 & -1 \end{pmatrix} \quad \text{and} \quad \begin{pmatrix} 1 & 0 & -1 \\ 2 & 0 & -2 \\ 1 & 0 & -1 \end{pmatrix} \quad \text{vector gradient}$$

horizontal vertical

Figure 3.2. Masks of the Laplacian operator and the vector gradient.

In the sharpened image we have more contrast than in the original, but it is very likely that we also enhance the noise in the image.

No we will discuss some techniques for edge enhancement. These techniques are used if we are only interested in the gray value edges of an image and not in the distribution of the gray values. An example: if we have to produce a documentation of a certain technical machine (e.g. a car), this documentation will contain, together with text, many line drawings. Usually the line drawings are designed on the basis of photographs. Instead, we could try to enhance the edges from the original digitized photo and use the resulting representation as input for a computer aided design system. Figure 3.3. is an example of such an application. The image processing procedure is to apply a local operator such as the Laplacian or the vector gradient and threshold the results there by producing a binary image. The threshold may be static, that means the same threshold value holds for the whole image, or dynamic. In this case the threshold is computed for every pixel depending on its environment. More sophisticated techniques for edge enhancement would be relaxation or dynamic programming. Here we use the output of the vector gradient as a first iteration. A certain quality function is derived and we iteratively try to improve this function.

Further local operations are blow up and shrink operations which we will explain with binary images. If we blow up all black regions at their border by one pixel and then shrink the regions by one pixel we have eliminated isolated small white holes in the black regions. But this operations should be used with care because in the case of images with a fine structure the quality of the output image may be less than that of the original.

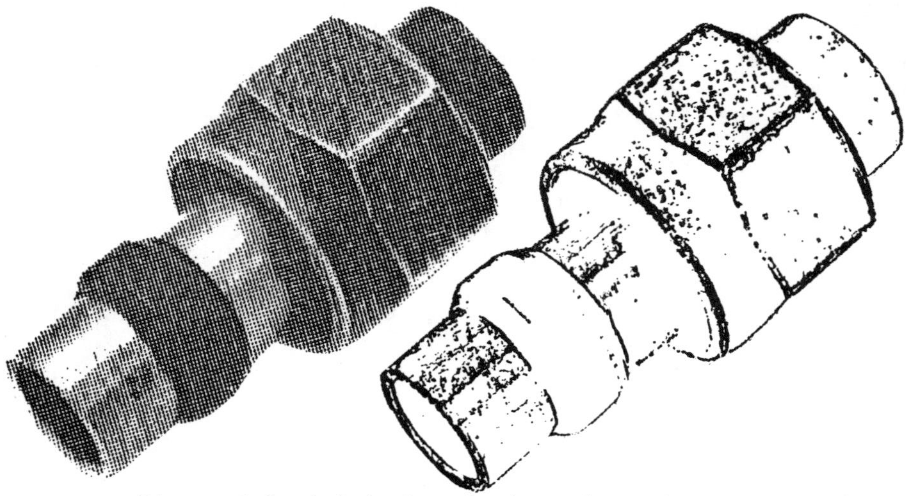

Figure 3.3. Original and edge enhanced representation of a part of a technical documentation.

A clean-operator assigns a pixel the value 0 or 1 depending on the distribution of 0's and 1's in a (3*3)-environment: if more than 4 pixels are 0's the pixel is assigned the value 0 and if more than 4 are 1's the pixel is assigned the value 1. It is easy to see that these local operations may be generalized to images with more than two gray values.

3.2 Fourier Transform and Digital Filters

As mentioned above the local operations may be interpreted as convolutions or, more precisely, digital filtering in the spatial (x,y)-domain. In this section we will give an overview of two dimensional fourier transforms and digital filtering.

Let s(x,y) be a continuous function of two independent variables x and y; its fourier transform is (under certain circumstances / 43/) defined by

$$S(f_x,f_y) = \int_{-\infty}^{+\infty}\int_{-\infty}^{+\infty} s(x,y)\ \exp(-2\pi i(xf_x+yf_y))\ dx\ dy.$$

The inverse fourier transform is defined by

$$s(x,y) = \int_{-\infty}^{+\infty}\int_{-\infty}^{+\infty} S(f_x,f_y)\exp(2\pi i(xf_x+yf_y))\,df_x\,df_y$$

and s(x,y) and $S(f_x,f_y)$ are a fourier transform pair. In general $S(f_x,f_y)$ is a complex function of f_x and f_y which can be interpreted as the signal frequencies in x- and in y-direction of the original signal s(x,y). The fourier transform is a linear shift-invariant operation. Filtering in the frequency (or fourier) domain is to multiply the fourier transform $S(f_x,f_y)$ with a transfer function $H(f_x,f_y)$:

$$S'(f_x,f_y) = H(f_x,f_y)\,S(f_x,f_y).$$

The inverse fourier transform applied to $S'(f_x,f_y)$ will give us the filtered image s'(x,y) in the spatial domain.

Let us consider now the filtering in the spatial domain. A two dimensional linear filter in the spatial domain is described by the output of the filter applied to the dirac delta function $\delta(x,y)$ ($\delta(x,y)=1$ if $(x,y)=(0,0)$ and $\delta(x,y)=0$ else):

$$h(x,y) = A(\delta(x,y)).$$

If we write s(x,y) as

$$s(x,y) = \int_{-\infty}^{+\infty}\int_{-\infty}^{+\infty} s(u,v)\delta(u-x,y-v)\,du\,dv \text{ we have:}$$

$$s'(x,y) = A\big(s(x,y)\big) = A\Big(\int_{-\infty}^{+\infty}\int_{-\infty}^{+\infty} s(u,v)\delta(u-x,v-y)\,du\,dv\Big)$$

$$= \int_{-\infty}^{+\infty}\int_{-\infty}^{+\infty} s(x,y)\,h(u-x,v-y)\,du\,dv.$$

The last formula is the convolution of the two continuous functions s and h which is often abbreviated as s*h.

The relations between the filtering in the spatial and the frequency domain are represented in figure 3.4. We can see that h(x,y) and $H(f_x,f_y)$ are a fourier pair. Filtering in the frequency domain is achieved by the multiplication of the transform of s(x,y) with the fourier transform of h(x,y). Filtering in the spatial domain is achieved by the convolution of the function s(x,y) with the function h(x,y).

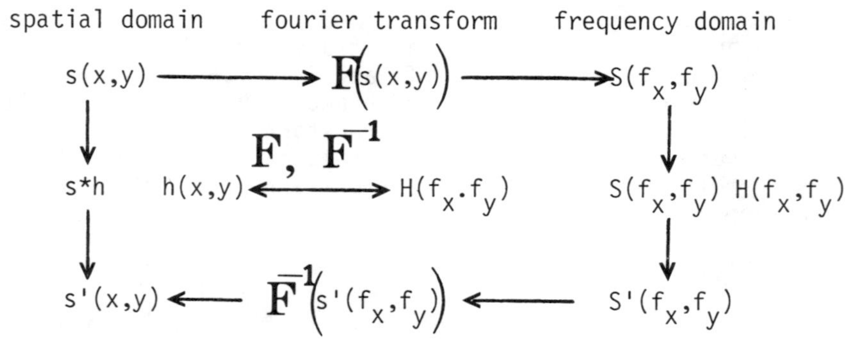

Figure 3.4. Relations between the spatial and the frequency (fourier) domain

If s(x,y) is a discrete image function as defined in section 1 the fourier transform can be described as a discrete image transformation. Essentially we replace the integrals by sums which have the range of the image field.

Digital filters are often described by means of their transfer function $H(f_x,f_y)$. Filters which reduce high frequencies are called low pass filters. The effect of a low pass filter we have discussed with the example of the moving average. Filters which reduce low frequencies are called high pass filters. Their effect we have described with the Laplacian or the vector gradient. A band pass filter reduces all frequencies except those within a certain interval. Figure 3.5. shows an original image (part of a finger print) and a pictorial representation of its fourier transform.

We finish the discussion of digital filters with an other type of filters. In a line scanner image it sometimes occurs that the image contains pixels or even whole lines which had been recorded incorrectly. The isolated pixels could e.g. have the gray value 255 and the incorrect lines could have an arbitrary random pattern. To detect and correct these lines we compute the correlation between two consecutive image lines. If the correlation is greater than a given threshold we accept that the last line we have processed is correct. If the correlation is lower than the threshold we presume to have an incorrect line which we may replace by the preceeding line. In the case of isolated incorrect pixels we compute the average gray value of the surrounding 8 neighbors and compare this value with the gray value of the pixel.

If the difference is greater than a given threshold we replace the incorrect pixel by the average value of the neighbors.

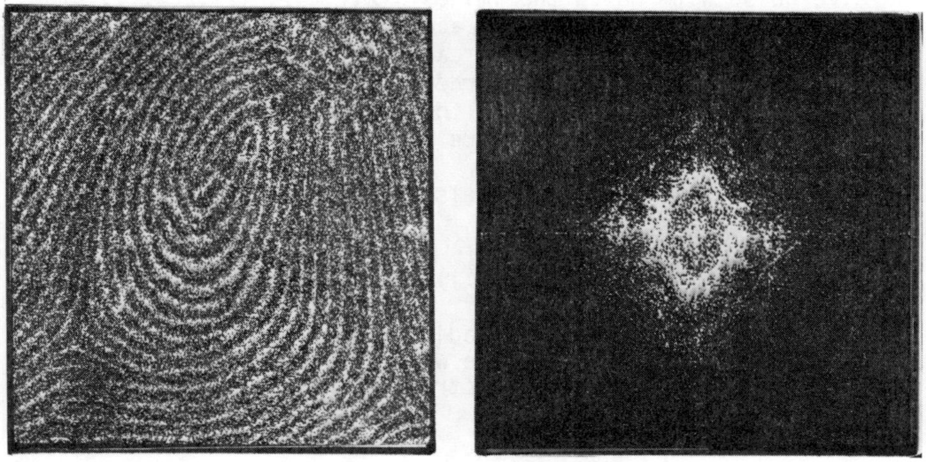

Figure 3.5. Part of a finger print; pictorial representation of the fourier transform.

3.3. Operations with Multichannel Images

Let s(x,y,n) be an image with N channels. The mean value of the gray values of a pixel averaged over the different channels is

$$s'(x,y) = 1/N \sum_{i=1}^{N} s(x,y,n_i).$$

If the image S is a multispectral image each pixel of S' is its intensity obtained by averaging over all spectral channels, i.e. its panchromatic intensity. If the first channel of S is an arbitary gray tone image and the second a line art image, e.g. a regular grid of white lines (gray value 255) on a black background (gray value 0), the sum is the gray tone image with a "transparent" grid intergrated into it.

The difference of two channels

$$s'(x,y) = s(x,y,n_1) - s(x,y,n_2) + c, \text{ where}$$

c is a constant which prevents to produce negative gray values, is also used in image processing. The case of spectral bands we compute the differences between two bands. If the two channels are registrated images from the same object taken at times t_1 and t_2 the difference computation is a simple approach to change detection. Generally, if S is a time sequence of a single channel image S=s(x,y,t) the accumulation may be used e.g. to reduce the additive noise with which the images $s(x,y,t_i)$, i=1,2,3,... are distorted. With a combination of accumulation and difference the relative movement of image parts can be discovered. In medicine similar methods are applied in order to obtain enhanced images, e.g. images of inner organs with their boundaries emphasized, c.f. the angiocardiogram of figure 3.6.

Another combination of the channels of an image s(x,y,n) is the ratio:

$$s'(x,y)=(s(x,y,n_1)-s(x,y,n_2))/(s(x,y,n_1)+s(x,y,n_2))*c$$

where c is scaling factor. The radio is invariant against the multiplication of the input image with a constant. The results of differences or ratio between spectral channels yields similar output images.

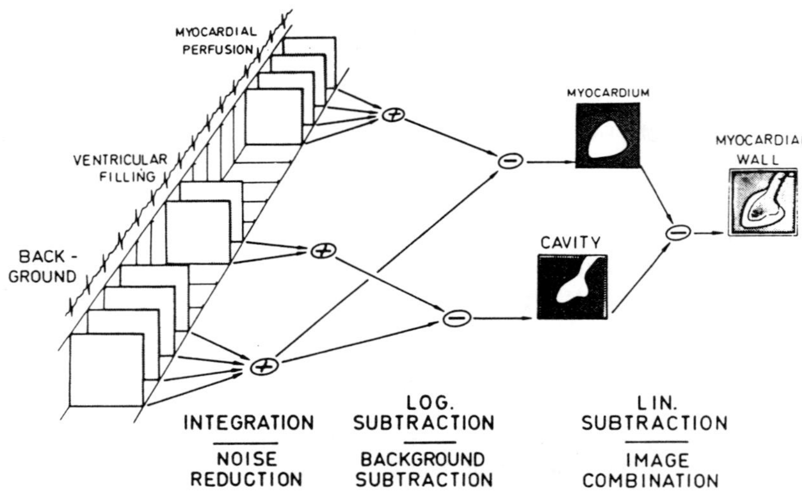

Figure 3.6. Schematic drawing of the basic image processing steps for the visualization of the myocardial wall. (From: Digital Cardiovascular Radiology, Medical motivations and technical developments. P.H.Heintzen, R. Brennecke, J.H.Bürsch, Digital Image Processing in Medicine, Springer, Hamburg, 1981)

3.4. Orthonormal Expansions

Let s(x,y) be a real function over a region S of the (x,y)-plane, which is square integrable. Suppose that $\phi_{mn}(x,y)$, m=0,1,2,...; n=0,1,2,... is an orthonormal basis (orthogonal, normalized and complete) of functions. Then s(x,y) can be expanded as

$$s(x,y) = \sum_{m=0}^{\infty} \sum_{n=0}^{\infty} a_{mn}\phi_{mn}(x,y), \text{ where}$$

$$a_{mn} = \iint_S s(x,y)\phi^*_{mn}(x,y)\, dx\, dy.$$

The coefficients a_{mn} are determined so that they minimize the mean square error

$$e^2_{MN} = \iint_S \left(s(x,y) - \sum_{m=0}^{M-1} \sum_{n=0}^{N-1} a_{mn}\phi_{mn}(x,y) \right)^2 dx\, dy.$$

Since $\phi_{mn}(x,y)$ is an orthonormal basis it holds

$$\lim_{\substack{M \to \infty \\ N \to \infty}} e^2_{MN} = 0.$$

One example of a problem independent expansions of s(x,y) is the fourier expansion. In this case the orthonormal basis are the functions

$$\phi_{mn}(x,y) = 1/\sqrt{AB}\ \exp(2\pi i(mx/A+ny/B)),$$

where the region S is a rectangle in the (x,y)-plane with sides A and B (e.g. the region of the image s(x,y)). Other examples of orthonormal problem independent expansions are expansions with Walsh-functions or with Legendre-polynomials.

The expansion may be generalized to cases where s(x,y) is a continuous or discrete random field (a digitized image s(x,y) may be interpreted as the outcome of the experiment ω: "digitization", where ${}^{\omega}s(x,y)$ is a two dimensional random field).

We get a problem dependent expansion if we assign more constraints to the orthonormal basis. In the Karhunen-Loewe expansion we take the coefficients a_{mn} of the expansion as new (image) samples and require that the orthonormal basis yields samples which are uncorrelated. An example: let s(x,y,n) be a multichannel image with N channels and the covariance matrix C which describes the covariance of pairs of channels. We denote a pixel of $s(x,y,n)$ as a N-dimensional vector $\underline{g}(x,y)=(g_1(x,y),\ g_2(x,y),\ldots,\ g_N(x,y))^T$. Then the above requirement yields as orthonormal basis the N eigenvectors $\{\underline{\phi}_i\}$, $i=1,2,\ldots,N$ of the covariance matrix C which are ordered with respect to the ordered eigenvalues $\lambda_1 \geqq \lambda_2 \geqq \ldots \geqq \lambda_N$. The uncorrelated samples are

$$a_i(x,y) = \underline{g}^T(x,y)\underline{\phi}_i,\ i=1,2,\ldots,N.$$

$\underline{g}^T(x,y)\underline{\phi}_i$, $i=1,2,\ldots,N$ is often called the i-th principal component of the image $s(x,y,n)$ which has as variance the value of the i-th eigenvalue λ_i. If we define a threshold for the variances we may compute only the principal components with a variance greater than the threshold. Doing so we have achieved a data reduction by eliminating correlated samples. Figure 3.7. shows the first three principal components of a M²S image.

Other important applications of the Karhunen-Loewe transformation are to be found in image transmission where one determines uncorrelated samples in the spatial (x,y)-domaine.

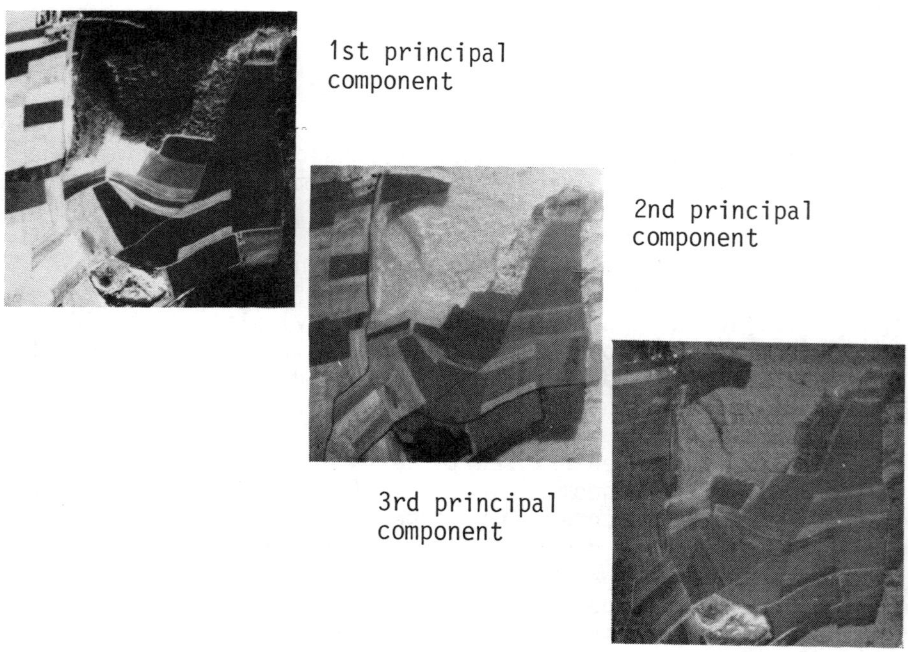

Figure 3.7. First three principal components of a M²S image.

3.5. Geometrical Transformations

A simple example of a geometrical transformation is the enlargement of an image by a factor $f>1$. The coordinates (x,y) are transformed by f to $(x',y')=(fx,fy)$. The gray value $s(x,y)$ of the original image occurs at position (x',y'): $s(x,y)=s'(x',y')$. With the direct method we compute for every pixel $s(x,y)$ the position (x',y') in the enlarged image. Obviously in the grid of the enlarged image there will be many positions to which no gray value can be assigned. We avoid this with the indirect method: for every position (x',y') of the enlarged image we compute a position $(\alpha,\beta)=(x'/f,y'/f)$ in the original. In general α and β will be no integers. Thus the position (α,β) will not coincide with any position of the original and we have to make a desicion which gray value we choose (resampling). With the nearest neighborhood method the gray value of the closest position to (α,β) is chosen: $s(x',y')=s(x,y)=s\left(\mathrm{int}(\alpha+0.5),\mathrm{int}(\beta+0.5)\right)$. Other methodes for resampling are bilinear interpolation or cubic convolution.

If we allow the scaling factor f to be in the range $0<f\leqq 1$ we have a reduction in size of the original image. An arbitrary geometrical distortion is defined by equations that specify the transformation from the undistorted reference coordinate system (x,y) to the distorted system (x',y'). These equations are of the general form $x'=f_1(x,y)$ and $y'=f_2(x,y)$. In the case of perspective distortions the transformation is linear, e.g. of the form $x'=A_0+A_1x+A_2y$, $y'=B_0+B_1x+B_2y$. If the mapping functions f_1 and f_2 are known (e.g. they may be derived from the system which records the data) we call the geometrical transformation a system dependent rectification. If the correct mapping functions are unknown we may use approximating polynomials of a certain order, e.g.

$$x'=A_0+A_1x+A_2y+A_3x^2+A_4xy+A_5y^2\,,$$

$$y'=B_0+B_1x+B_2y+B_3x^2+B_4xy+B_5y^2\,.$$

The parameters A_i and B_i are computed by means of control points which are corresponding points in an undistorted reference image (e.g. a map) and the distorted image. The least squares method is applied if more than the necessary number of control points have been determined in order to improve the quality of the rectification. Figure 3.8. is an example of the rectification of a Landsat image with respect to the Gauss-Krueger-grid of the German topographical maps.

The computation of the three dimensional coordinates of an object from the pixel coordinates of its image is another geometrical transformation. It is the fundament of photogrammetry where the three dimensional topography of a terrain is recontructed from

Figure 3.8. Rectification of a Landsat image by means of the control point method with respect to the Gauss-Krueger-grid of the German topographical maps.

stereo image pairs. Recently methods have been described which allow even the three dimensional evaluation of line scanner images. The outcome of such a three dimensional evaluation of remote sensing data may be e.g. a digital terrain model. In pattern recognition the three dimensional evaluation finds applications in tracking and in industry robots. Here the three dimensional information is derived from real time stereo image pairs or from a temporal sequence of images, i.e. the movement of an object's image in the image plane.

References for section 3 (basic text books and tutorial papers are underlined):

2, <u>5</u>, 6, 11, 14, 24, <u>27</u>, 29, 31, 32, <u>33</u>, 34, <u>35</u>, 39, 40, <u>41</u>, 42, <u>43</u>, <u>45</u>, 46, <u>48</u>, 49, <u>51</u>, 52, <u>53</u>, 54, <u>55</u>, 58, <u>59</u>.

4. SCENE ANALYSIS AND SEGMENTATION

4.1. Preprocessing, Segmentation, Image Understanding

Scene Analysis is concerned with the extraction of information from an image field by automatic or semiautomatic methods. In Pattern Recognition a pattern as a whole is usually assigned to a certain class of pattern (e.g. a representation of the letter "A" to the class "letter A"). In Scene Analysis the image field is devided into regions of pixels which have similar properties on the basis of a given uniformity predicate (e.g. all regions of a color image which are red). The next step is the interpretation of the regions or of combinations of regions and the assignment of a higher meaning to them. The image as a whole is not cataloged to a certain class of images.

Scene Analysis may be devided into preprocessing (feature extraction) which includes some techniques described in sections 1 and 3, segmentation (sometimes called classification) and image understanding (interpretation). In sections 4.2. until 4.4. we will first discuss some techniques of segmentation. The image features that we will use in these sections will be the simple gray value or multispectral information. But the described techniques are not limited to features of these types. Thus in sections 4.5. and 4.6. we will discussed several other feature extraction methods. Finally we close with some examples in the field of image understanding.

4.2. Thresholding

If an image $s(x,y)$ contains light objects on a dark background these objects may be extracted by simple thresholding:

$$s'(x,y) = \begin{cases} 1, & s(x,y)>t, \\ 0, & s(x,y)\leq t. \end{cases}$$

All pixels which belong to the object are coded with 1's, the background is coded with 0's. Sometimes it is appropriate to smooth the images before thresholding them. If we define more than one threshold we get the technique of gray level slicing which may be applied to images with several objects of different gray values on a dark background. In semithresholding we take for all pixels with gray values greater than the threshold the original gray value and for all pixels less or equal the threshold the gray value 0:

$$s'(x,y) = \begin{cases} s(x,y), & s(x,y)>t, \\ 0, & s(x,y)\leq t. \end{cases}$$

In this case the object gray values are preserved and may be used in further processing steps.

The selection of the threshold may be done interactively with a raster image display by modifying the internal look-up-table depending on the movement of a trackball. The threshold may also be derived from the histogram. If the image contains a light object on a dark background the histogram should be bimodal and we may place the threshold at the minimum between the two maxima. If we know the approximate size of the object we can define a threshold which produces p% of object pixels where p is the ratio of the object size and the image size.

In cases where the background gray values are not very homogenuous over the whole image and even the objects have oscillating gray values a technique with dynamic threshold selection must be applied. Here the threshold for a pixel is derived from a certain environment of it. An example is shown in figure 4.1. where for a digitized finger print (a) a binary image of good quality (b) had to be produced. In this case the dynamic thresholding had been applied. Figure 4.1. (c) is a pictorial representation of the oscillation of the threshold.

(a) (b) (c)

Figure 4.1. Dynamic thresholding of a digitized finger print. (a) original image, (b) binary image, (c) pictorial representation of the oscillating threshold.

4.3. Multivariate Classification

Let S=s(x,y,n) be an image with N channels which contains three objects o_1, o_2 and o_3 (we write object as abbreviation for "pixels which are recorded parts from the original object", e.g. an image contains no water but pixels which have been generated by the remitted radiance of water). For simplification we take N=2 for the moment. Suppose that $p_S(i_1,i_2)$ is the two dimensional histogram of S as defined in section 1.3. If the two channels contain significant information of our objects and if the channels are uncorrelated the pixels of the image will tend to form clusters in certain regions of the two dimensional feature space which is formed by channel 1 and 2 (figure 4.2.).

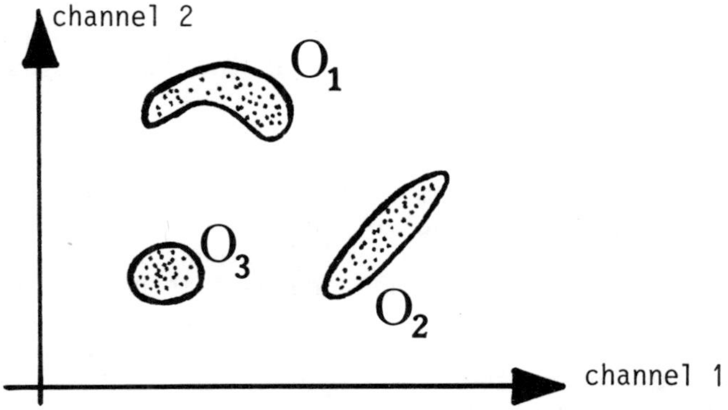

Figure 4.2. Graphical representation of the two dimensional histogram of an image with three objects. The pixels tend to form clusters.

Now the following questions arise:

- how can we determine the clusters in the feature space,
- how can we describe the clusters with simple mathematical parameters and
- how can we make a decision to assign an unknown pixel to one of the three classes?

Sometimes it is possible to establish a random test for every object class from which we get the appropriate clusters. E.g. in medical image processing if we want to classify three different cell types we can establish a random set of cells for each type. In landuse classification of remotely sensed data training areas for the different classes are used to approximate the clusters. In these cases where a random test is used to approximate the object clusters, classification is called supervised.

In unsupervised classification the clusters of the different objects have to be found automatically by the classification system. We give an example of a simple clustering algorithm (Minimum-Distance clustering). Let the two dimensional vectors $\underline{g}_i=(g_{i1},g_{i2})^T$,

$i=1,2,\ldots,LR$ be the pixels of the image $s(x,y,n)$. Define an arbitrary pixel, e.g. $\underline{g}_1$, as the center c_1 of the first cluster in the feature space. Classify $\underline{g}_i$, $i=2,3,\ldots$ to the cluster c_1 if the Euclidian distance in the feature space

$$d_i^{(1)} = \sqrt{(g_{i1}-g_{11})^2+(g_{i2}-g_{12})^2}$$

is less or equal than a given threshold t. If for a certain i_2 this condition does not hold define $\underline{g}_{i_2}$ as the second cluster center c_2. Classify $\underline{g}_i$, $i=i_2+1,\ldots$ to c_1 if $d_i^{(1)}<t$ and to c_2 if $d_i^{(2)}<t$. If for a certain i_3 no one of these two conditions holds define $\underline{g}_{i_3}$ as the third cluster center c_3, etc. At the end of this algorithm every pixel is classified into one cluster. Now clusters may be merged together or devided depending on special problemdependent circumstances. It is obvious that the threshold t has great influence on this clustering algorithm. The algorithm has some disadvantages, e.g. it is possible that a pixel $\underline{g}_i$ is classified to a cluster c_j and after a new cluster c_k has been generated it would be classified to c_k. Thus many other unsupervised clustering algorithms are described in the literature which avoid these disadvantages. Figure 4.3. is the result of the above algorithm applied to a Landsat image. It shows the two dimensional feature space of the first Landsat channel (green) and the third channel (near infrared) and its separation into eleven clusters. The cluster centers are marked.

Now the second and third questions, how to approximate the clusters and how to classify unknown pixels will be discussed. In a multithreshold technique we define for every cluster and for every channel a pair of thresholds which may be derived from the histogram $p_S(i_1,i_2)$. With this technique the clusters are approximated by axis parallel rectangles and in the N-dimensional case with hyperboxes. An unknown pixel $\underline{g}_i$ is classified to class o_j if it lies in the box of o_j.

Pixels which lie in no box are classified to the rejection class. Some problems may arise if the boxes overlap. In this case the "Hyperbox-classifier" may be combined with other methods.

In "Minimum-distance classification" we determine the vector of the mean values for every class. An unknown pixel $\underline{g}_i$ is classified to class o_j if the Euclidian distance in feature space to the class o_j is less than those to all other classes. In addition it is required that the distance is less than a certain threshold. With this method the clusters are approximated by (N-dimensional) spheres.

A statistical approach to classification is the "Maximum-likelihood classifier" which is a special case of a Bayes'ian classifier. Here the conditional average loss for assigning an unknown pixel $\underline{g}_i$ to a class o_k is determined and $\underline{g}_i$ is assigned to class o_j with the minimum conditional average loss. If we assume that the membership of the pixels to the different classes may be described by a Gaussian density function we get the maximum-likelihood classifier. The distance between the pixel $\underline{g}_i$ and the center of class o_j is called the Mahalanobis distance and the clusters are approximated by (N-dimensional) ellipsoids.

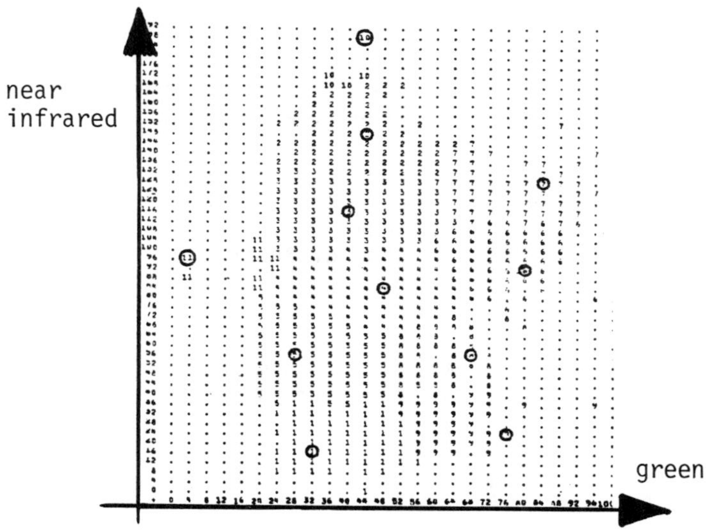

Figure 4.3. Unsupervised Minimum-Distance clustering applied to a landsat image. The graphic shows a two dimensional feature space and its separation into eleven clusters.

4.4. Contour and Line Following, Region Growing

With the methods described above a typical segmentation algorithm would be: establish an approximation of the different classes in feature space, define a classification rule and classify all pixels of the image. In many applications, especially with large images, it is not necessary to evaluate all pixels of the image but only those which are candidates on the basis of a special algorithm. We distinguish between edge oriented and region oriented approaches.

In contour following we start with a given pixel which is on the gray value contour of an object. Now we investigate the pixels of a certain neighborhood and try to find a new candidate of the contour. To make the decision we could use the magnitude of the gray value gradient which we threshold.

A line, e.g. a dark line on a light background, has two edges, one on each side. Following a line from a given point we may compute the mean gray value along straight lines in several directions starting at the given point. We take the next line point in the direction which has the lowest mean value. When we have found all points of the contour or the line we could apply some algorithms of contour smooting. Finnaly we accept the region to the left (or the right) of the contour (or line) to be the searched object (e.g. a cell in a medical preparation).

In region growing we go the opposite way. Starting with a given point within a certain region we evaluate the 8-neighbors whether they belong to the object or not. In the next step we evaluate all surrounding points of the already found area, etc. (figure 4.4.).

Another region oriented approach is to overlay the image with a initial, arbitrary segmentation, e.g. with a square grid of n*n pixels. We evaluate every one of these initial regions and merge those adjacent regions which have similar properties (the question what features we could use will be discussed in the next sections). We proceed iteratively until we find no adjacent regions which we could merge. For segmentation we may also apply a technique similar to the quadtrees as described in section 1.4. In this case we end up with a tree structure whose leaves provide a segmentation of the image.

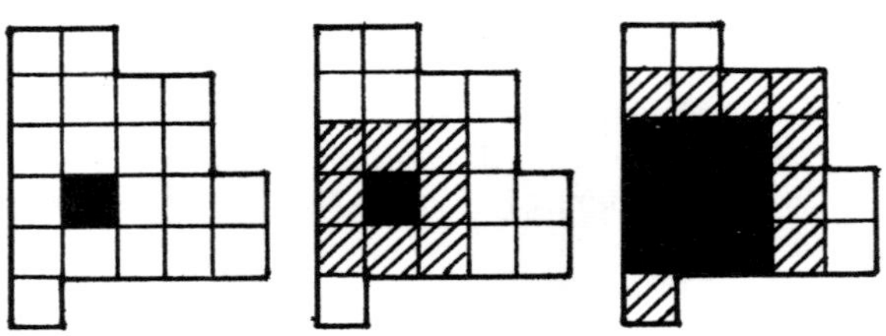

Figure 4.4. Region growing: starting with a given pixel of an object the area is expanded step by step.

4.5. Features for Segmentation: Area Oriented

In the segmentation algorithms of the previous sections we used as features the gray value of a single image or the multispectral information of a multichannel image. If the original image is a digitized color image with a blue, green and red channel the feature space is three dimensional and the feature which we use for segmentation is the color of the objects. If the image has been recorded with a multispectral scanner we have generally more than three channels available. The channels may include spectral information from ultra violet to the near and thermal infrared. The feature which we use is the spectral signature of the objects. Examples of multispectral images are Landsat scenes (four channels: green, red and two infrared) or scenes recorded with the M^2S airborn scanner (Modular Digital Multispectral Data Acquisition and Data Processing System) which has eleven spectral bands from ultra violet to thermal infrared. It is obvious that digital evaluation is not restricted to certain spectral bands if appropriate sensors establish digital image information.

With the example of Landsat scenes we can explain the use of multitemporal image data: from a certain area under investigation we could collect Landsat data from e.g. two different seasons of the year. If we merge them we obtain an eight channel image, four channels from the first date and four from the second. The features which we use in this eight dimensional feature space are the multispectral signatures of the two different seasons which have often good discriminatory capability.

Generally, if s(x,y,t) is a time sequence of images we may extract features which are time dependent, such as in the multitemporal case. Another time dependent feature is the movement of an object in the sequence of images. The feature "movement" could be applied in an environment where we want to track a vehicle which moves in the landscape or in the example of section 2 where we described a system which tracks bottles on a conveyor-belt. A simple method to establish a measure for the feature "movement" is to compute the differences of two images at time t_1 and t_2:

$$s'(x,y) = s(x,y,t_1)-s(x,y,t_2).$$

A large magnitude of a pixel s'(x,y) is a hint that there was a change.

Instead of differences of multichannel images we may also use other functional combinations (e.g. the ratio, section 3)in order to extract discriminatory features for the objects which we want to classify.

In section 3 the Fourier transform and orthonormal expansions, especially the Karhunen-Loewe expansion of an image have been described. The transform coefficients which generally span a feature space of lower dimension than the original yield useful features for segmentation. The Fourier domain intensity pattern $S(f_x,f_y)$ is normally examined in specific regions (e.g. horizontal or vertical slit, ring, sector) to isolate image features. The first m principal components of a multichannel image which contain uncorrelated samples provide usually good discriminatory capabilities.

Many portions of natural images are devoid of significant detail over large areas. These areas can often be characterized as exhibiting a repetitive structure called texture. There are many attempts to find measurements for texture. The Fourier spectrum should express the texture coarseness: regions of coarse texture should have their spectral energy concentrated at low spatial frequencies while fine texture should exhibit high spatial frequencies. The spatial autocorrelation is another function which has been suggested as the basis of a texture measure:

$$A(\varepsilon,\eta,x',y') = \frac{\sum_m \sum_n s(m,n)\ s(m-\varepsilon,n-\eta)}{\sum_m \sum_n (s(m,n))^2}$$

A square window of the original image is correlated at each image point (x,y) with offset values $\varepsilon,\eta=0, \pm1,\pm2,\ldots,\pm T$. A region of coarse texture will exhibit a higher correlation for a fixed shift (ε,η) than a region of fine texture.

A number of textural measures are based on the gray tone spatial dependence matrix $p(a,b,\Delta x,\Delta y)$ which describes the cooccurrence of the gray values a and b in the direction of the vector $(\Delta x,\Delta y)$ for all pixels (figure 4.5.).

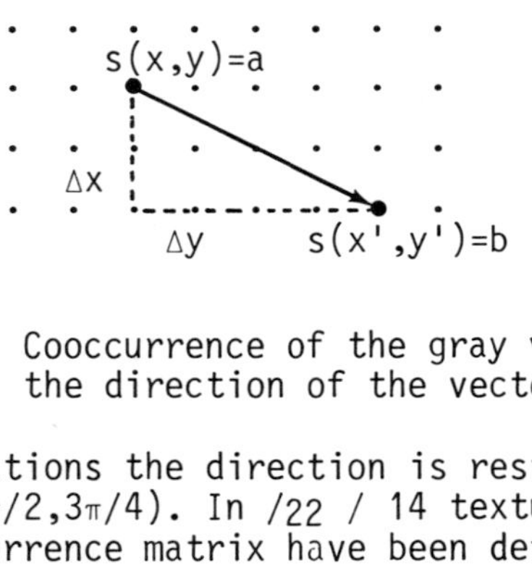

Figure 4.5. Cooccurrence of the gray values a and b in the direction of the vector $(\Delta x,\Delta y)$.

In practical applications the direction is restricted to a few values, e.g. $0,\pi/4,\pi/2,3\pi/4$). In /22 / 14 texture measures on the basis of this cooccurrence matrix have been defined.

Run length coding may also be applied to texture analysis. Coarse texture tends to create long runs of consecutive pixels of the same gray value lying in a particular direction while fine texture results in short runs.

Finally we should mention that textural features are often applied together with the technique of region growing as described in section 4.4, since texture is not a feature of an isolated point but of the surrounding region.

As the last example of area oriented features for segmentation we will discuss matching of patterns against an image. The pattern to be matched may be an edge, a template which represents a known object which is searched in the image or a piece of the image itself as discussed above in texture analysis. There are many possible ways of measuring the degree of match or missmatch between an image s(x,y) and a template g(x,y). The use of the similarity measure

$$\sum_x \sum_y \big(s(x+u,y+v)-g(x,y)\big)^2$$

leads to computing the expression

$$\frac{\sum_x \sum_y s(x+u,y+v)g(x,y)}{\sum_x \sum_y \big(s(x+u,y+v)\big)^2}$$

which is the well-known normalized cross correlation. After we have evaluated the normalized cross correlation for each pixel we could threshold the result thus obtaining the positions with high correlation.

4.6. Features for Segmentation: Edge Oriented

The area oriented features of section 4.5 may be applied to extracting objects that exhibit some type of uniformity. Another approach is based on the detection of discontinuity, i.e. of places which indicate the end of a region and the beginning of another. Therefore we are going to give a short review of edge detection techniques.

In section 3.1.we explained two local operations, especially the Laplacian operator and the vector gradient. Both operators are applied to edge detection. Generally, derivative operators can be used as gray value edge detectors. If we compute for each pixel its "edge value" (e.g. the magnitude of the gray value gradient) we can perform segmentation as described above in the two dimensional feature space subtended by the edge value and by the gray value.

With edge fitting methods we match a known course of and edge, i.e. a model of the edge, against the original image. The edge values obtained with these algorithms are basically values of the correlation coefficient.

Texture edges can be discovered with methodes of texture analysis and color or signature edges can be evaluated with weighted sums of the different (spectral) channels.

4.7. Image Understanding, Image Interpretation

The interpretation of a segmented image is the final processing step in Scene Analysis that we will discuss within this paper. Suppose that we have the result of a supervised multispectral classification of remote sensing data: the image is segmented into regions of similar properties which have been defined with the training areas for the different object classes. The interpretation of the regions has been made implicitly when we defined the training areas for a certain class, e.g. saying "these are the properties of the object water". Thus from the point of view of Scene Analysis a supervised classified image can be regarded as interpreted. Figure 4.6. is an example of a supervised Landsat data classification.

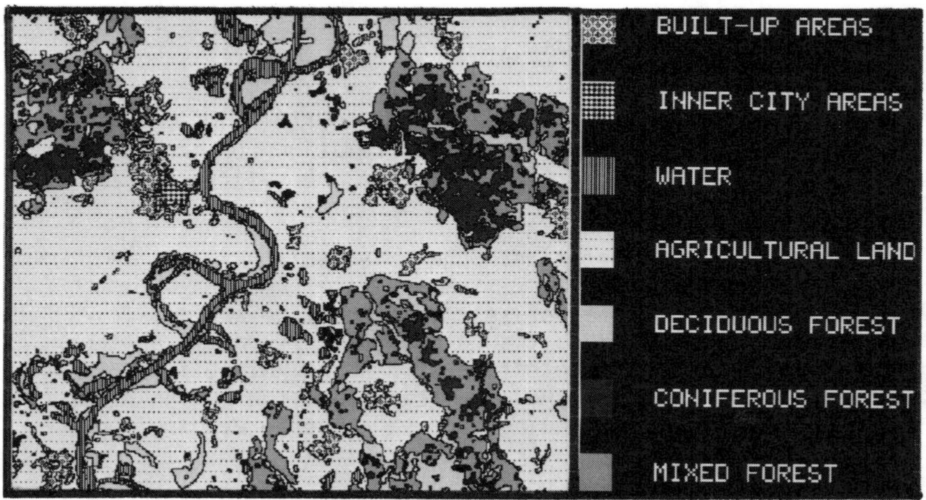

Figure 4.6. Supervised classification of Landsat data covering the area of the German topographical map L6716, Speyer, original scale 1:50000; seven classes plus rejection class.

In contrast to this approach it is often aspired to achieve an automatic interpretation. A first step is to investigate the spatial arrangement of the different areas, e.g. with an adjacency graph (figure 4.7.). Here the nodes of the graph are the regions and two nodes are connected if they lay adjacent. If we establish a data base (a model of the "world") with the graphs of all objects that may occur we can compare the new graph with those of the data base and try to find one which fits best.

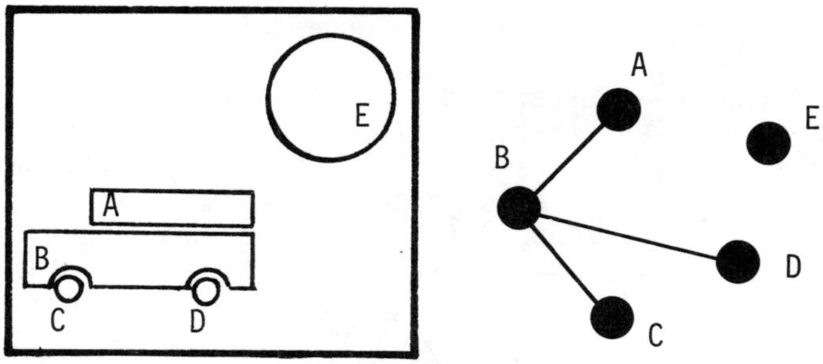

Figure 4.7. Example of an adjancy graph

Another approach is to establish a parsing tree (figure 4.8.) which is evaluated with syntactic methods which are applied in the field of formal languages: we define a grammar with which all "sentences" (all parsing tree structures) which may occur can be derived. With the actual tree we make a bottom-up or top-down analysis in order to determine whether the tree is a sentence of the grammar or not.

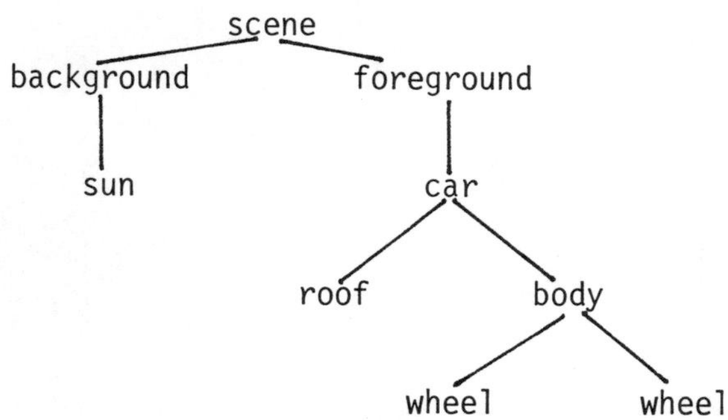

Figure 4.8. Example of a parsing tree

Finally, with relational methods we provide more sense to the connecting lines of the nodes, if we label them with connectivity and spatial relationships (figure 4.9.). In this case we compute a directional graph which we might compare with models of objects or even whole scenes stored in a data base.

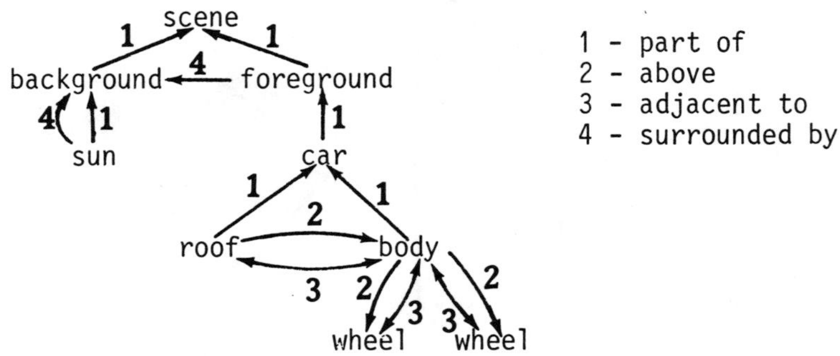

Figure 4.9. Example of a directional graph

References for section 4 (basic text books and tutorial papers are underlined):

2, 3, <u>5</u>, 7, 8, <u>10</u>, 11, 15, 16, 17, 19, 20, 21, 22, 24, 26, 30, 34, <u>35</u>, 39, <u>41</u>, <u>43</u>, 46, <u>48</u>, <u>53</u>, 54, <u>55</u>, 56, 57, 60.

5. REFERENCES

/ 1/ Albrecht, R. and Schaller, S., Tololo-Vienna Interactive Image Processing System, Institute for Astronomy of the University Vienna, 1979

/ 2/ Andrews, H.C., Computer Techniques in Image Processing, Academic Press, New York, 1970

/ 3/ Browning, J.D. and Tanimoto, S.T., Segmentation of Pictures into Regions with a Tile-by-Tile Method, Pattern Recognition, Vol. 15, No. 1, pp 1-10, 1982

/ 4/ Cederberg, R.L.T., Chain-Link Coding and Segmentation for Raster Scan Devices, Computer Graphics and Image Processing, Vol. 10, pp 224-234, 1979

/ 5/ Duda, R.O. and Hart, P.E., Pattern Classification and Scene Analysis, John Wiley & Sons, New York

/ 6/ Dyer, C.R. and Rosenfeld, A., Fourier Texture Features: Suppression of Aperture Effects, IEEE Transactions on Systems, Man, and Cybernetics, 1976

/ 7/ Fiedler, R., Beitrag zur automatischen Untersuchung und Erkennung von Texturen in digitalisierten Bildern aus der Fernerkundung, DFVLR-Forschungsbericht FB 79-08, Porz-Wahn, 1979

/ 8/ Foith, J.P., ed., Angewandte Szenenanalyse, DAGM Symposium, Karlsruhe, 10.-12. Oct. 1979, Springer, Berlin, 1979

/ 9/ Freeman, H., Computer Processing of Line-Drawing Images, Computing Surveys, Vol. 6, No. 1, pp 57-97, 1974

/10/ Fu, K.S. and Mui, J.K., A Survey on Image Segmentation, Pattern Recognition Vol. 13, pp 3-16, 1981

/11/ Gardner, W.E., ed., Machine-aided Image Analysis, The Institute of Physics, Conference Series Number 44, Bristol, 1978

/12/ Goodenough, D.G., The Canada Centre for Remote Sensing's Image Analysis System (CIAS), 4th Canadian Symposium on Remote Sensing, Québec City, 1977

/13/ Goodenough, D.G., Programming hardware for remote sensing image analysis, AFIPS-Conference Proceedings Vol. 47, National Computer Conference, pp 119-129, 1978

/14/ Goßmann H. and Haberäcker, P., Image Processing of HCMM-satellite thermal images for superposition with other satellite imagery and topographic and thematic maps, 14. Kongress der Int. Society for Photogrammetry, Vol XXIII, Part B3, pp 256-271, Hamburg, 1980

/15/ Gupta, J.N. and Wintz, P.A., A Boundary Finding Algorithm and its Applications, IEEE Transactions on Circuits and Systems, Vol. Cas 22, No. 4, pp 351-362, 1975

/16/ Haberäcker, P., Multispektrale Klassifizierung in DIBIAS, Digitale Bildverarbeitung, GI/NTG Fachtagung, Informatik Fachberichte 8, pp 71-79, 1977

/17/ Haberäcker, P., Untersuchung zur Klassifizierung multispektraler Bilddaten aus der Fernerkundung, DLR-Forschungsbericht 77-72, DFVLR, Porz-Wahn, 1977

/18/ Haberäcker, P., DIBIAS - The Digital Image Processing System at DFVLR, System Design and Applications, International Conference "Earth Observation from Space and Management of Planetary Resources" Proceedings, Toulouse, 1978

/19/ Haberäcker, P., On the Classification of Multitemporal Image Data, Machine-aided Image Analysis, Conference Series No. 44, Institut of Physics, Oxford, 1978

/20/ Haberäcker, P., et al., Auswertung von Satellitenaufnahmen zur Gewinnung von Flächennutzungsdaten, Schriftreihe "Raumordnung" des Bundesministers für Raumordnung, Bauwesen und Städtebau 06.039, Bad Godesberg, 1979

/21/ Haken, H., ed., Pattern Formation by Dynamic Systems and Pattern Recognition, Springer Series in Synergetics, Springer, Berlin, 1979

/22/ Haralik, R.M., Shanmugam, K. and Dinstein, I., Textural Features for Image Classification, IEEE Transactions on Systems, Man and Cybernetics, Vol. SMC-3, No. 6, 1973

/23/ Herbst, N.M. and Will, P.M., An Experimental Laboratory for Pattern Recognition and Signal Processing, Communications of the ACM, Vol. 15, No. 4, pp 231-244, 1972

/24/ Höhne, K.H., ed., Digital Image Processing in Medicine, Lecture Notes in Medical Informatics Vol. 15, Springer, Berlin, 1981

/25/ Hofmann, O., Digitale Aufnahmetechnik, Bildmessung und Luftbildwesen 50, Heft 1, pp 16-32, 1982

/26/ Horowitz, S.L. and Pavlidis, T., Picture Segmentation by a Tree Transversal Algorithm, J. Assoc. Comput. Mach. 23, No. 2, pp 368-388, 1976

/27/ Huang, T.S., ed., Picture Processing and Digital Filtering, Topics in Applied Physics Vol. 6, Springer, Berlin, 1979

/28/ Hunter, G.M. and Steiglitz, K., Linear Transformation of Pictures Represented by Quad Trees, Computer Graphics and Image Processing Vol. 10, pp 289-296, 1979

/29/ Itten, K.I. and Fasler, F., Thematic Adaptive Spatial Filtering of Landsat Landuse Classification Results, Proceedings 13th Internation Symposium on Remote Sensing of Environment Ann Arbor, Michigan, 1979

/30/ Kraus, W., Zur Theorie der Klassifizierung multispektraler Bilder, Bildmessung und Luftbildwesen, 4/79, pp 119-128, 1979

/31/ Lev, A., Zucker, S.W. and Rosenfeld, A.; Iterative Enhancement of Noisy Images, IEEE Transactions on Systems, Man, and Cybernetics, Vol. Smc-7, No. 6, pp 435-442, 1977

/32/ Lichtenegger, J., Seidl, K. and Kübler, O., Methoden zur Überlagerung von Landsat-Bildern für multispektrale Landnutzungskartierung, Bildmessung und Luftbildwesen No. 2, pp 53-61, 1978

/33/ Liu, B., Digital Filters and the Fast Fourier Transform, Halsted Press, 1975

/34/ Marr, D. and Hildreth, E., Theory of Edge Detection, Proc. R. Soc. Lond. B 207, pp 187-217, 1980

/35/ Meisel, W.S., Computer-Oriented Approaches to Pattern Recognition, Academic Press, New York, 1972

/36/ Milgram, D.L., Constructing Trees for Region Description, Computer Graphics and Image Processing Vol. 11, pp. 88-99, 1979

/37/ Moik, J.G., User's Guide for Batch Operation of the SMIPS/VICAR Image Processing System, Goddard Space Flight Center, Greenblet, 1976

/38/ Morrin, T.H., Chain-Link Compression of Arbitrary Black-White Images, Computer Graphics and Image Processing, Vol. 5, pp 172-189, 1976

/39/ Nagel, H.-H.,ed., Digitale Bildverarbeitung, Digital Image Processing, Informatik-Fachberichte Nr. 8, Springer, Berlin, 1977

/40/ Ney, H., Dynamic Programming as a Technique for Pattern Recognition, 6th International Conference on Pattern Recognition, Munich, 1982

/41/ Niemann, H., Methoden der Mustererkennung, Akademische Verlagsgesellschaft, Frankfurt am Main, 1974

/42/ Nowak, P., Application of Two-dimensional Digital Filerts to Multispectral Scanner Imagery, Proceedings of 5th Canadian Symposium on Remote Sensing, Victoria, 1978

/43/ Pratt, W.K., Digital Image Processing, John Wiley & Sons, New York, 1978

/44/ Proffitt, D., and Rosen, D., Metrication Errors and Coding Efficiency of Chain-Encoding Schemes for the Representation of Lines and Edges, Computer Graphics and Image Processing, Vol. 10, pp 318-332, 1979

/45/ Rabiner, L.R., and Gold, B., Theory and Applications of Digital Signal Processing, Inc.Englewood Cliffs, Prentice-Hall, 1975

/46/ Radig, B.,ed., Modelle und Strukturen, DAGM Symposium, Hamburg, Oct. 1981, Springer, Berlin, 1981

/47/ Ranade, S., and Rosenfeld, A., Shape Approximation using Quad Trees, Pattern Recognition Vol. 15, No.1, pp 31-40, 1982

/48/ Rosenfeld, A., and Kak, A.G., Digital Picture Processing, Academic Pree, New York, 1976

/49/ Schuhr, W., Digitale Entzerrung multispektraler Bilder, Bildmessung und Luftbildwesen No.5, pp 202-208, 1976

/50/ Shapiro, L.G., Data Structures for Picture Processing: A Survey, Computer Graphics and Image Processing Vol.11, pp. 162-184, 1979

/51/ Simon, H., Automatische Bildverarbeitung in der Medizin, Verlag Theodor Steinkopf, Dresden, 1975

/52/ Steiner, D., and Kibby, M.E., Geometrical Referencing of Landsat Imagery by Afine Transformation and Overlaying of Map Data, Photogrammetria, No.33, pp 41-75, 1977

/53/ Tou, J.T., and Gonzalez, R.C., Pattern Recognition Principles, Addison-Wesley Publishing Company, London, 1974

/54/ Triendl, E.,ed., Bildverarbeitung und Mustererkennung, Informatik-Fachberichte Nr. 17, Springer, 1978

/55/ Ullmann, J.R., Pattern Recognition Techniques, Butterworth, London, 1973

/56/ Wong, R.Y., and Hall, E.L., Sequential Hierarchical Scene Matching, IEEE Transactions on Computers, Vol. C-27, No. 4, pp 359-366, 1978

/57/ Zucker, S.W., Toward a Model of Texture, Computer Graphics and Image Processing, Vol. 5, 1976

/58/ Zucker, S.W., and Hummel, R.A., An Application of Relaxation to Line and Curve Enhancement, IEEE Transactions on Computers, Vol. C-26, No. 4, pp 394-403, 1977

/59/ Zucker, S.W., Krishnamurthy, E.V., and Haar, R.L., Relaxation Processes for Scene Labeling: Convergence, Speed, and Stability, IEEE Transactions on Systems, Man, and Cybernetics, Vol. SMC-8, No. 1, pp 41-48, 1978

/60/ Zucker, S.W., and Terzopoulos, D., Finding Structure in Cooccurrence Matrixes, Dept. of Electrical Engineering McGill University, Technical Report 79-13R, Montreal, 1979

CHAPTER 7

SELECTED TOPICS OF COASTAL ZONE COLOR SCANNER (CZCS) DATA EVALUATION

B. Sturm

Joint Research Centre (EEC), 21020 Ispra (Va), Italy

ABSTRACT

Problems related to the quantitative evaluation of nature and quantity of suspended matter in the oceans and their coastal regions from digital imagery taken by the Coastal Zone Color Scanner (CZCS) on Nimbus-7 are treated. Special consideration is given to the correction of the sensitivity loss of the instrument which became evident recently. The methods of evaluating suited atmospheric parameters for the correction of aerosol scattering light and procedures for locating picture elements geographically are presented. Some examples of CZCS data evaluation of scenes from various parts of worlds oceans are presented briefly.

1. INTRODUCTION

The Coastal Zone Color Scanner (CZCS) on Nimbus-7 was launched in October 1978 and has since then aquired a huge amount of imagery over coastal zones and open ocean. A summary of Nimbus-7 CZCS history, characteristics, data collection and processing has been given by W.A. Hovis (1980) and first results have been discussed at the COSPAR/SCOR/ IUCRM Symposium "Oceanography from Space", held in May 1980, in Venice, Italy, and published in ref. (1). In spite of the fact that in the framework of the Eurasep project (ref. (2)) JRC-Ispra has distributed a considerable amount of computer compatible tapes with CZCS imagery received at NASA Madrid

A. P. Cracknell (ed.), Remote Sensing Applications in Marine Science and Technology, 137–167.

station and processed to calibrated radiance tapes (CRT) at NASA/GSFC and in spite of the fact that CZCS data over Europe are being received in two places (ESA Nimbus-7 receiving station at Lannon, France and Satellite receiving station of Dundee University, Dundee, Scotland), the use of CZCS data for current environmental and marine biological research in European coastal waters remained scarce.
Obviously this is mainly due to the lack of satellite data processing facilities capable to treat CZCS data for atmospheric effects in order to make full use of its information content on the nature and quantity of suspended matter in the sea.
The problem of atmospheric correction of CZCS data has been discussed already in the very early history of CZCS (ref. (3)) and in Europe JRC has organized within the Eurasep project two international Workshops on that topic (refs. (4), (5)). Recent publications on the subject can be found in ref. (6) and (7). There are however other problems besides that of atmosphere correction that make CZCS data evaluation a difficult task:

- problems with the inaccuracy of sensor calibration;
- sensitivity loss;
- determination of "right" atmospheric parameters when no ground measurements are available;
- algorithms to be used for evaluating the desired parameters;
- geolocation of CZCS picture elements.

In the following we try to give to these problems answers which can directly be translated in software. Therefore the following text assumes sometimes the character of a "cooking book" or "recipe collection". Nevertheless the author is convinced that such type of presentation is especially suited to give the necessary information for practical use.

2. CZCS CALIBRATION PROBLEMS

2.1 CZCS inflight calibration

A schematical set up of the CZCS internal calibration for the visible channels 1 to 5 is shown in Fig. 1. During each scan, a voltage calibration staircase is switched into the amplifier during the non-active part of the scan and the resulting count rates are recorded. From the values C_i and V_i the constants a_V and b_V in a linear relation

$$C_i = a_V V_i + b_V \qquad (2.1)$$

are determined by a least square fit. Once every 16 scans the calibration lamp with known radiance is viewed by the scan

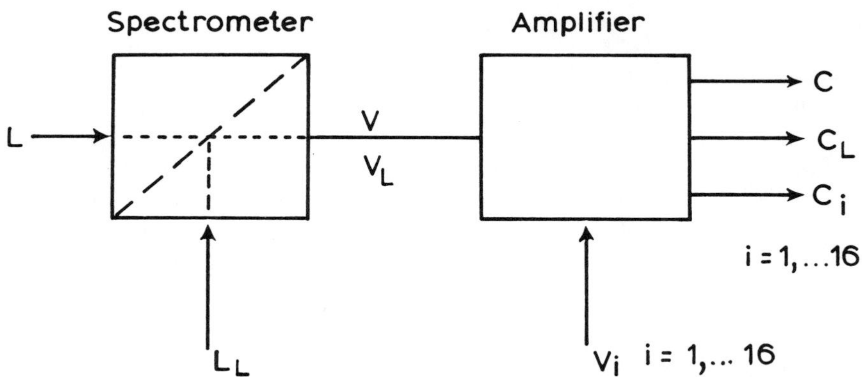

Fig. 1 - Schema of CZCS internal calibration procedure.

L = radiance from earth view (mW/cm^{2}sr μ)
L_L = radiance from calibration lamp (mW(cm^{2}sr μ)
V = voltage from earth view
V_L = voltage from calibration lamp
V_i = voltage of voltage calibration step i
C = digital count for earth view
C_L = digital count for calibration lamp
C_i = digital count for voltage calibration step i

mirror and the corresponding digital count C_L is detected. Since the resulting voltage V_L is related to C_L by (2.1)

$$V_L = \frac{C_L - b_v}{a_v} = a_L \cdot L_L \tag{2.2}$$

where the characteristic of the lamp $a_L = V_L/L_L$ is a known constant for each visible channel, we have

$$C_L = a_v a_L \cdot L_L + b_v \tag{2.3}$$

The value of $a_v a_L$ and b_v has been found rather constant for all 970 scan lines of a CZCS scene and therefore the radiance of the earth view is evaluated from

$$L = \frac{1}{a_v a_L} \cdot C - \frac{b_v}{a_v a_L} \tag{2.1}$$

$$L = \text{SLOPE} \cdot C + \text{INTERCEPT}$$

Table 2.1 - Preflight values of $a_v a_L$ and b_v for gain 1 (from Ball Brothers final report, ref. (8))

Chan.	$a_v a_L$	b_v	slope	intercept
1	22.46	- 0.89	0.04452	0.03963
2	32.23	- 2.05	0.03103	0.06361
3	40.54	- 3.24	0.02467	0.07992
4	88.00	- 1.00	0.01136	0.01136
5	10.62	+0.30	0.09416	- 0.02825

Preflight calibration of the CZCS gave the slopes and intercepts given in Table 2.1. During flight recorded values listed on the calibrated radiance tapes (CRT) distributed by NASA or elaborated from CZCS data received at Dundee (ref. (9)) and Lannion (ref. (10)) show that the slopes and intercepts did not vary in time and have not varied more than $\pm$ 6% (with the exception of channel 3 which changed by 8 to 10%).
Table 2.2 gives some examples of slope and intercepts taken from NASA CRT's and from Lannion and Dundee elaborations.

Table 2.2 - Slopes and intercepts from inflight calibration for gain 1 and ch. 1 to 4

	NASA ON 2684		NASA ON 6595		ESA ON 16533		Dundee ON 3583	
Ch.	Slope	Intercept	Slope	Intercept	Slope	Intercept	Slope	Intercept
1	.04485	.02104	.04491	.01737	.04473	.01800	.04665	.02244
2	.03134	.04371	.03137	.03945	.03102	.04086	.03293	.04359
3	.02634	.02569	.02637	.02335	.02637	.01763	.02732	.01756
4	.01146	.00984	.01148	.00728	.01148	.00632	.01164	.01297

2.2 CZCS sensitivity loss

From the monitoring in time (Orbit No.) of pixels with low chlorophyll content and low aerosol path radiance, it was found (ref. (13)) that CZCS sensitivity decreases with time (Orbit No.)

$$C^{\lambda}(N) = f(N)\ (a_v a_{L'}\ L + b_v)$$

N = orbit number.
Proposed correction functions f:

$f(N) = 1 - S \cdot N$ (Gordon (ref. 11), Viollier(ref. 12))

$f(N) = e^{-S \cdot N}$ (Austin (ref. 13), JRC)

S-values reported are listed in Table 2. 3.

Table 2. 3 - Proposed CZCS sensitivity loss correction factors

Ch.	Austin	Gordon	Viollier	JRC
1	2.12	2.80	2.10	1.66
2	1.22	1.18	1.42	1.28
3	0.78	0.83	1.22	0.44
4	0.0	0.01	0.60	0.0
5	0.0	0.0	0.0	0.0

2. 3 Conclusion

Regarding the calibration constants of CZCS there is:

- disagreement about correction factors for slope and intercept due to use of different methods for atmospheric corrections and probably also different parameters (E_o, τ_R, β_A, etc.);
- agreement about sensitivity change with orbit number.

Recommendation

- Use "Your" atmospheric correction method with one of the sets of S-values;
- Use preflight calibration slopes and intercepts and adjust intercept so as to give best results over clear water and spatially smooth values of $Lss(\lambda)$ in turbid water. However, adjustments should not exceed $\pm$ 11%.

3. DETERMINATION OF $\epsilon(\lambda, 670)$

Atmospheric correction equations can be written as (ref. (3), (5), (6), (7)):

$$L(\lambda) = L_R(\lambda) + L_A(\lambda) + L_w(\lambda) \qquad (3.1)$$

where L is the radiance at the aperture of the scanner and L_R the Rayleigh pathradiance.

$$L_w(\lambda) = Lss(\lambda) \cdot \frac{1-\rho(\lambda)}{n^2(\lambda)} \cdot T_{RAOZ}(\lambda, \mu) \qquad (3.2)$$

Lss being the upwelled subsurface radiance, ρ the Fresnel reflectance and n the refractive index of sea water.

$$T_{RAOZ}(\lambda, \mu) = e^{-(1/2\tau_R(\lambda) + \beta_A \cdot \tau_A(\lambda) + \tau_{OZ}(\lambda))/\mu}$$

is the irradiance transmittance through the atmosphere τ_R, τ_A and τ_{OZ} being the Rayleigh, aerosol and ozone optical thicknesses, and μ the cosine of the satellite zenith angle.

$$L_A(\lambda) = L_A(670)\, \epsilon(\lambda, 670) \frac{E_o(\lambda) T_{OZ}(\lambda, \mu, \mu_o)}{E_o(670) \cdot T_{OZ}(670, \mu, \mu_o)} \qquad (3.3)$$

is the aerosol path radiance at wavelength λ, where $\epsilon(\lambda, 670)$ is defined as ratio of aerosol optical thickness τ_A at wavelength λ and 670 nm (Ch. 4 of CZCS).

$$\epsilon(\lambda, 670) = \tau_A(\lambda)/\tau_A(670)$$

E_o is the extraterrestrial sun irradiance and μ_o the cosine of the sun zenith angle.

$$T_{OZ}(\lambda, \mu, \mu_o) = e^{-\tau_{OZ}(\lambda)(1/\mu + 1/\mu_o)}$$

If $L_w(\lambda)$ is known, then

$$\epsilon(\lambda, 670) = \frac{(L(\lambda) - L_R(\lambda) - L_w(\lambda))}{(L(670) - L_R(670) - L_w(670)} \cdot \frac{E_o(670) \cdot T_{OZ}(670, \mu, \mu_o)}{E_o(\lambda) T_{OZ}(\lambda, \mu, \mu_o)} \qquad (3.4)$$

In clear water zones ($C < 0.3$ μg/l) (ref. (14)):

$$\text{Lss}(520) = 0.90 \cdot \mu_o T_{RAOZ}(520, \mu_o)$$
$$\text{Lss}(550) = 0.52 \cdot \mu_o T_{RAOZ}(550, \mu_o) \qquad (3.5)$$

From this equation we can now find a strategy for the determination of $\epsilon(\lambda, 670)$:

1) Assume $\tau_A(\lambda) = \text{const} \approx 0.1$ ($\epsilon(\lambda, 670) = 1$!)

2) Process your CZCS scene using eq. (3.1) to (3.3) assuming Lss(670) = 0 for each pixel (Gordons method)

3) Identify clear water zones in the resulting Lss(λ)-image (Lss(520)/Lss(550) $\approx$ 1.8). They must be not too near to east side of bright objects (clouds, land) and possibly in the centre part of the entire CZCS scene ($\mu \geq 0.82$).

4) Calculate for groups of at least 3 x 3 clear water pixels from eq. (3.4) $\epsilon(520, 670)$ and $\epsilon(550, 670)$ with

$$L_w(520) = 0.90 \cdot \mu_o T_{RAOZ}(520, \mu_o) \frac{1-\rho(\mu)}{n^2(520)} T_{RAOZ}(520, \mu)$$

$$L_w(550) = 0.52 \mu_o T_{RAOZ}(550, \mu_o) \frac{1-\rho(\mu)}{n^2(550)} T_{RAOZ}(550, \mu)$$

and then v(520) and v(550) from:

$$\epsilon(520, 670) = \left(\frac{670}{520}\right)^{\nu(520)}$$

$$\epsilon(550, 670) = \left(\frac{670}{550}\right)^{\nu(550)}$$

and

$$\epsilon(443, 670) = \left(\frac{670}{443}\right)^{\frac{v(520)+\nu(550)}{2}}$$

if:

$$\boxed{\begin{array}{l} \nu(443) \approx \nu(520) \approx \nu(550) \quad \text{and} \\ \approx 0.3 \leq \nu(443),\ \nu(520),\ \nu(550) \leq \approx 3.0 \end{array}} \qquad (3.6)$$

5) then: use the average $L_A(670)$ from the clear water pixels to calculate

$$\tau_A(670) = \frac{4\pi\,\mu L_A(670)}{E_o(670)T_{OZ}(670,\mu_o)\cdot p_A(\psi)}$$

$$\approx 0.26\;\frac{\mu L_A(670)}{T_{OZ}(670,\mu_o)}$$

and the average of the $\epsilon(\lambda, 670)$ values to calculate

$$\tau_A(\lambda) \;=\; \tau_A(670)\cdot\epsilon(\lambda,670)$$

With the so-obtained $\tau_A(\lambda)$ and $\epsilon(\lambda, 670)$ process your scene for final result.

6) else: check if your clear water zone is not too near to the east of clouds or land (distance must be at least 50 to 100 pxls to avoid blending effect of bright objects), in case choose other clear water pixels and go to 4).

7) If 6) does not work, check whether your clear water zones have not too low sun elevation ($\mu_o > 0.64$). If not try to change the correction factors for slope and intercept of channel 2 or/and 3 within the range of no more than $\pm 10\%$ in order to satisfy condition (3.6). Then go to 4).

4. ATMOSPHERIC CORRECTION OVER TURBID WATER

4.1 The iteration procedure

The atmospheric correction equations give

$$Lss(\lambda) = \frac{n^2(\lambda)}{1-\rho(\mu)}\cdot\frac{1}{T_{RAOZ}(\lambda,\mu)}(L(\lambda)-L_R(\lambda)-L_A(\lambda)) \quad (4.1)$$

where

$$L_A(\lambda) = \epsilon(\lambda,670)\cdot\frac{E_o(\lambda)T_{OZ}(\lambda,\mu,\mu_o)}{E_o(670)T_{OZ}(670,\mu,\mu_o)}\cdot$$

$$(L(670) - L_R(670) - L_w(670)) \quad (4.2)$$

and

$$L_w(670) = Lss(670)\,\frac{1-\rho(\mu)}{n^2(670)} \cdot T_{RAOZ}(670,\mu) \qquad (4.3)$$

For turbid water where Lss(670) ≠ 0, Smith and Wilson (ref. 15) have proposed the following iteration procedure:

1) $Lss^{(1)}(670) = 0 \rightarrow L_w^{(1)}(670) = 0$
   Calculate $L_A^{(1)}(\lambda)$ from (4.2) and $Lss^{(1)}(\lambda)$ from (4.1), if $Lss^{(1)}(443) \leqslant 0$, $Lss^{(1)}(443) = 0.001$.

2) Calculate $Lss^{(2)}(670)$ from

   $Lss^{(2)}(670 = f(Lss^{(1)}(443),\ Lss^{(1)}(520),\ Lss^{(1)}(550))$

   and with this from (4.3), (4.2) and (4.1)
   $Lss^{(2)}(\lambda)$

3) If $|Lss^{(2)}(670) - Lss^{(1)}(670)| <$ THR result for thin pixel is $Lss^{(2)}(\lambda)$ and go to next pixel. THR should be < 0.02.

4) Else $Lss^{(2)}(\lambda) \rightarrow Lss^{(1)}$ and go to 2).

Generally most pixels converge with less than 49 iterations. A typical histogram of iterations is shown in Table 4.1.

Table 4.1 - Iteration histogram

No. Iter.	No. of pxls	No. Iter.	No. of pxls	No. Iter	No. of pxls	No. Iter	No. of pxls
1	7988	11	122	21	23	31	12
2	390256	12	84	22	16	32	4
3	102831	13	95	23	19	33	7
4	12898	14	47	24	23	34	4
5	3366	15	47	25	16	35	8
6	885	16	35	26	13	36	10
7	448	17	34	27	16	37	3
8	300	18	39	28	31	38	7
9	248	19	25	29	21	39	8
10	183	20	22	30	14	40	1

Number of pixels with more than 40 and less than 50 iterations was 45 and more than 49: 1461. Total number of pixels processed: 521,670, scene size: 905 lines, 1968 elements = 1,781,040 pixels. Computing time: ≈ 68 min. on PDP11/70 System RSX11/M.

4.2 The function f(Lss(443), Lss(520), Lss(550)) for calculating Lss(670)

Austin, Petzold (ref. (26)) have given relations between ratios Lss(670)/Lss(550), Lss(520)/Lss(550) and Lss(443)/Lss(550) for oceanic waters from which one can derive the following relation (ref. (16)):

$$Lss(670) = Lss(550) \cdot 10^{-(1.048+0.764\log\xi+0.07(\log\xi)^2)} \tag{4.4}$$

where

$$\xi = 0.642\frac{Lss(443)}{Lss(550)} + 0.892\frac{Lss(520)}{Lss(550)} - 0.533$$

for turbid near coastal waters a somewhat different relation has been proposed by Sturm (ref. (16)):

$$Lss(670) = Lss(550) \cdot 10^{-(1.177+1.168\log\xi-0.292(\log\xi)^2)} \tag{4.5}$$

For turbid near coastal waters in the northern Adriatic we have deduced from experiments (ref. (16)):

$$Lss(670) = Lss(550) \cdot 10^{-(0.881+0.816\log\xi)} \tag{4.6}$$

The right formula to be used can be found by comparing the results of $L_A(670)$ and Lss(670): if there is strong correlation between them (visual inspection), then there is probably a bad discoupling between $L_A(670)$ and $L_w(670)$ (respectively Lss(670) in equation (4.2)).

Recommendation

Use (4.4) in open ocean and (4.5) in estuaries and other areas where suspended matter is high (> 4-5 mg/l).

5. ALGORITHMS FOR CHLOROPHYLL-a, TOTAL SUSPENDED MATTER AND DIFFUSE ATTENUATION COEFFICIENT

5.1 Relations between subsurface upwelling radiance Lss and subsurface diffuse reflectance R

The subsurface diffuse reflectance R is defined as ratio of upwelling and downwelling subsurface irradiance $E_u^-(o)$ and $E_d^-(o)$:

$$R = \frac{E_u^-(o)}{E_d^-(o)} \tag{5.1}$$

The downwelling irradiance abcve the water surface $E_d^+(o)$ is given by

$$E_d^+(o) = E_o \mu_o e^{-(1/2\tau_R + \beta_A \tau_A + \tau_{OZ})/\mu_o} \tag{5.2}$$

where τ_R, τ_A and τ_{OZ} are the optical thicknesses of molecular and aerosol scattering and absorbtion by gases (ref. (7)). β_A is the backscattering coefficient of aerosols and is approximately $\beta_A \approx 0.1$ (ref. (17)), μ_o the sun zenith angle and E_o the extraterrestrial incoming sun irradiance. E_d^+ is related to the subsurface up- and downwelling irradiances by

$$E_d^-(o) = (1-\bar{\rho})E_d^+(o)+\bar{\rho}E_u^-(o) = \frac{1-\bar{\rho}}{1-\bar{\rho}R} E_d^+(o) \tag{5.3}$$

where ρ is the Albedo of the water surface which is slightly dependent on wavelength λ and cosine of the sun zenith angle μ_o (ref. (18)).

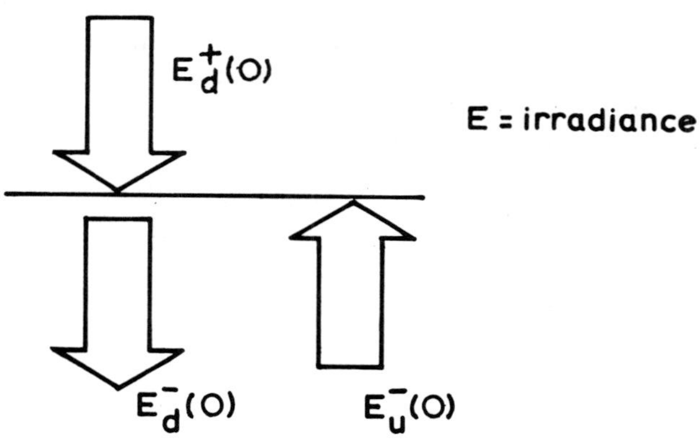

Fig. 5.1 - Irradiances at the water surface

Since $\rho \approx 0.06$, $R \leqslant 0.15$, we can approximately write $1 - \rho R \approx 1$.
Upwelling subsurface irradiance is found proportional to the upwelling subsurface radiance

$$E_u^-(o) = Q \cdot Lss \qquad (5.4)$$

$Q \approx \pi$ for isotropic upwelling radiance. From experimental evidence (ref. (19)) it is found that Q is somewhat higher than π, Gordon and Clark (ref. (14)) propose Q = 4.55.
From (5.4), (5.3) and (5.1)

$$Lss = (1 - \bar{\rho}) \frac{E_d^+(o)}{Q} \cdot R \qquad (5.5)$$

Since ρ, $E_d^+(o)$ are wavelength dependent and Q (probably) dependent from R:

$$\frac{Lss(\lambda_1)}{Lss(\lambda_2)} \neq \frac{R(\lambda_1)}{R(\lambda_2)} \qquad (5.6)$$

5.2 Algorithms for chlorophyll-a

Table 5.1 shows algorithms using CZCS channel ratios for Lss and R for deriving chlorophyll-a concentration (chlorophyll-a + phaeophytin-a).
All except that of ref. (22) are derived from underwater radiance or reflectance measurements.
Smith and Bakers algorithm was derived from atmospherically corrected CZCS radiances and ship chlorophyll measurements.

Algorithm used by NASA for evaluating pigment (chlorophyll-a + phaeophytin-a) concentration in derived photographic products (ref. (25)):

$$C_1 = 1.13 \left(\frac{Lss(443)}{Lss(550)}\right)^{-1.71}$$

$$C_2 = 3.326 \left(\frac{Lss(520)}{Lss(550)}\right)^{-2.439}$$

$$C(mg/m^3) = \begin{cases} C_1 \text{ for } C_1 \leqslant 1.5 \vee C_2 \leqslant 1.5 \\ C_2 \text{ for } C_1 > 1.5 \wedge C_2 > 1.5 \end{cases}$$

Table 5.1 - CZCS chlorophyll-algorithms $C = A\left(\frac{X(i)}{X(k)}\right)^B$

Type i/K X	A	B	Ref.
1/3 Lss	.505	- 1.269	(20)
1/3 Lss	.766	- 1.329	(21)
1/3 Lss	.783	- 2.120	(15)
1/3 Lss	1.262	- 2.589	(22)
1/3 R	1.92	- 1.8	(23)
1/3 R	.565	- 2.725	(24)
1/3 R	.30	- 1.86	(16)
2/3 Lss	.843	- 3.975	(20)
2/3 Lss	1.694	- 4.449	(21)
2/3 Lss	2.009	- 5.93	(15)
2/3 R	1.135	- 3.469	(24)
2/3 R	.729	- 3.37	(16)

This avoids the use of Lss(443) in high chlorophyll (turbid) water.

5.3 Algorithms for total suspended matter S

Table 5.2 gives algorithms for S. They are derived in very different types of water: class 1 water in Golf of Mexico (ref. (21)), and class 2 water near the Lagoone of Venice in the Adria (ref. (16)).

Table 5.2 - CZCS total suspended matter algorithms

$$S = A\left(\frac{X(i)}{X(k)}\right)^B$$

Type i/K X	A	B	Ref.
1/3 Lss	0.40	- 0.88	(21)
1/2 Lss	0.33	- 1.09	(21)
2/3 Lss	0.76	- 4.38	(21)
1/3 R	2.61	- 1.30	(16)
2/3 R	4.48	- 2.53	(16)

5.4 Algorithms for diffuse attenuation coefficient $K(\lambda)$

The diffuse attenuation coefficient $K(\lambda)$ is defined by:

$$K(\lambda) = -\frac{1}{E(\lambda, Z)} \cdot \frac{dE(\lambda, Z)}{dZ}$$

where
Z = depth, E = irradiance
K can be measured for up and downwelling irradiance.
If K = const between Z_1 and Z_2, then

$$K = \frac{1}{Z_2 - Z_1} \ln \frac{E(\lambda, Z_2)}{E(\lambda, Z_1)}$$

following algorithms have been reported for deriving K from CZCS upwelling radiances:

$$K_u(490) = 0.0883\left(\frac{Lss(443)}{Lss(550)}\right)^{-1.491} + 0.022 \qquad \text{(ref. (26))}$$

$$K_u(520) = 0.0663\left(\frac{Lss(443)}{Lss(550)}\right)^{-1.398} + 0.044 \qquad \text{(ref. (26))}$$

$$K_d(550) = 0.053\left(\frac{R(443)}{R(550)}\right)^{-1.740} + 0.063 \qquad \text{(ref. (16))}$$

$$K_d(550) = 0.085\left(\frac{R(520)}{R(550)}\right)^{-4.683} + 0.063 \qquad \text{(ref. (16))}$$

The latter two were obtained from 13 profiles measured near the Lagoone of Venice in the northern Adriatic.

5.5 Conclusions

The variety of algorithms reflects (probably) the variety of waters and phytoplankton types measured. In coastal water the coefficient A in Table 5.1 is found lower than in class 1 waters, probably due to the presence of yellow substance. High values of A and B are found from CZCS radiances; this may be explained with an "under-correction" for atmospheric effects (too high upwelling subsurface radiances).

Recommendation

For open ocean water use NASA's algorithm together with Lss(670) = 0 for each pixel (no iteration on Lss(670) = Gordon's method for atmospheric correction). For "exotic" phytoplankton species (dinoflagelates, cocolithophores, red tide, etc.), neither of the listed algorithms in Table 5.1 will work, due to special spectral signatures of such algae-blooms.

Remark

The chlorophyll-a concentration which can be detected by CZCS is the optically weighted chlorophyll (ref. (21)):

$$\overline{C}_f = \frac{\int_0^{Z_{90}} C(z) e^{2K \cdot z} dz}{\int_0^{Z_{90}} e^{-2K \cdot z} dt}$$

$Z_{90} = 1/K$; K = downwelled irradiance attenuation coefficient.
In the most cases $C_f \approx C_o$ = surface value. In cases where an average value of chlorophyll is calculated over a certain depth

$$\overline{C} = \frac{\int_0^D C(Z)\, dz}{D}$$

This value might be greater than $\overline{C}_f$ due to stratification of phytoplankton with higher values in deep layers.

6. CZCS GEOLOCATION ALGORITHMS

Angles (respectively their cosines) needed for atmospheric corrections (ref. (7)) are:

θ = view angle or zenith angle of satellite at pixel location
ϕ = satellite azimut in pixel location
θ_0 = sun zenith angle at pixel location
φ_0 = sun azimut at pixel location
ψ_- = scattering angle sun to satellite
ψ_+ = scattering angle reflected sun to satellite

To calculate θ_0 and φ_0, Long., Lat., time at pixel and day of year are needed. Long. and Lat. may be interpolated from the long. and lat. of anchor points of which 77 are given on the NASA-CRT for each line. Also time in GMT is given for each line. Then

$$\theta_0 = \text{arc cos } (\sin \text{lat.} \cdot \sin\delta + \cos \text{lat} \cos\delta \cos\varphi_t) \quad (6.1)$$

$$\varphi_0 = \text{arc sin } (\cos \delta, \sin\varphi_t / \sin \theta_0) \quad (6.2)$$

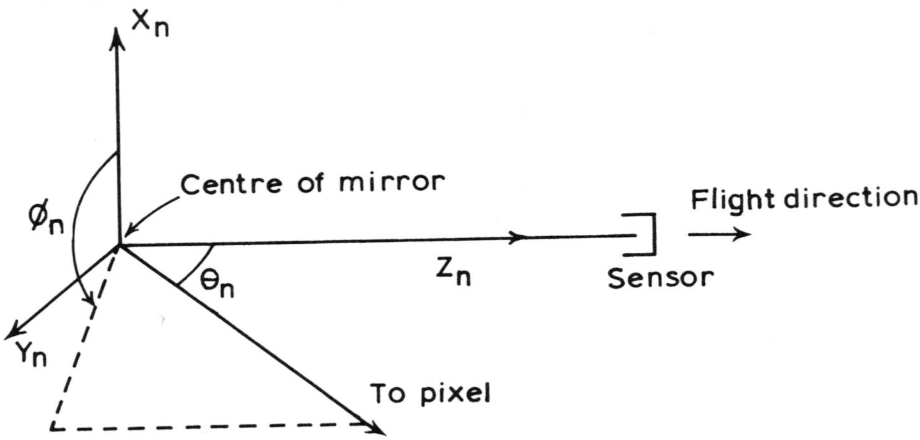

Fig. 6.1 - "Normal" coordinate system of satellite

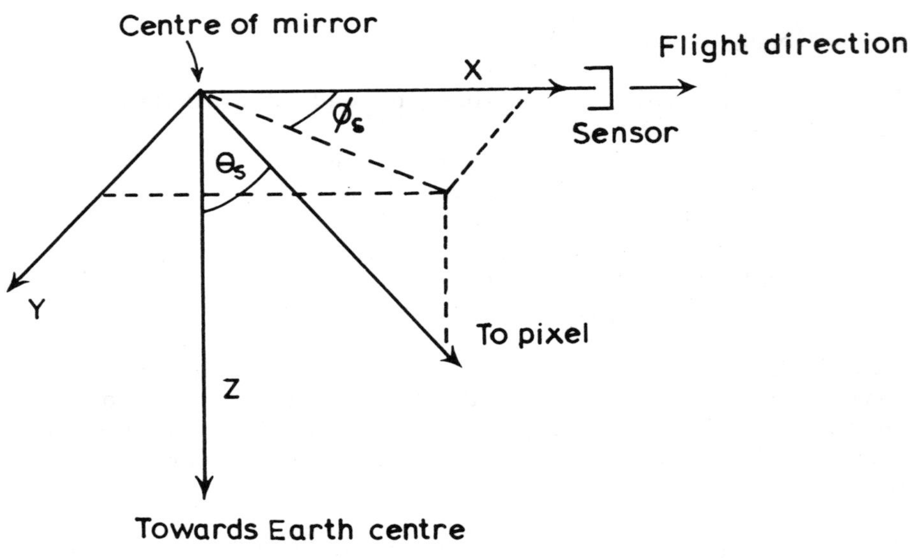

Fig. 6.2 - Satellite coordinate system
$X = Z_n, \quad Y = Y_n, \quad Z = -X_n$

$$\delta = \text{declination} = 23.49 \sin (0.968 \cdot D - 78) \tag{6.3}$$

φ_t = Hour angle = $(12 - t_s) \cdot 15$

t_s = solar time (hours) = $\frac{\text{Long}}{15} + \frac{\text{Eq}}{3600} + T_{GMT}$

E_q = time equation (sec) = $-461 \sin(D - 6.28) - 605.5 \sin(1.94 \cdot D + 23.4)$; D = day of year (6.4)

T_{GMT} = time at which pixel is seen (hours) = time of line in CRT

θ and φ can be obtained when the scan-angle (mirror shaft rotation angle) and the tilt are known at the moment the pixel is seen (ref. 27, 28).

η = rotation angle = $-39.04 + 0.04 \cdot E$

E = Element number of pixel in line

$$\theta_n = 2 \cdot \text{arc cos}(1/\sqrt{2}\ (\cos t/2 + \sin t/2 \cos \eta)) \tag{6.5}$$

$$\phi_n = 180 + \text{arc tg}(\sin\eta / \sin t/2 - \cos t/2 \cos \eta)) \tag{6.6}$$

θ_n and ϕ_n are the coordinates of the ray satellite-pixel in the "normal" coordinate system of the satellite (Fig. 6.1). Transforming θ_n and ϕ_n to the satellite coordinate system (Fig. 6.2) one obtains the nadir angle to pixel θ_s and the satellite azimut with respect to the flight direction ϕ_s:

$$\theta_s = \text{arc cos } (-\sin \theta_n \cos \phi_n) \tag{6.7}$$

$$\phi_s = \text{arc tg } (\tan \theta_n \sin \phi_n) \tag{6.8}$$

In order to obtain from this the satellite zenith angle and the satellite azimut at pixel location, we have to correct these angles for parallaxe due to earth curvature and for inclination of subsatellite track direction at pixel location. The parallaxe correction is easily made (Fig. 6.3):

$$\theta = \text{arc sin } (\frac{R+H}{R} \sin \theta_s) \tag{6.9}$$

R = earth radius (km)
H = satellite height (km)

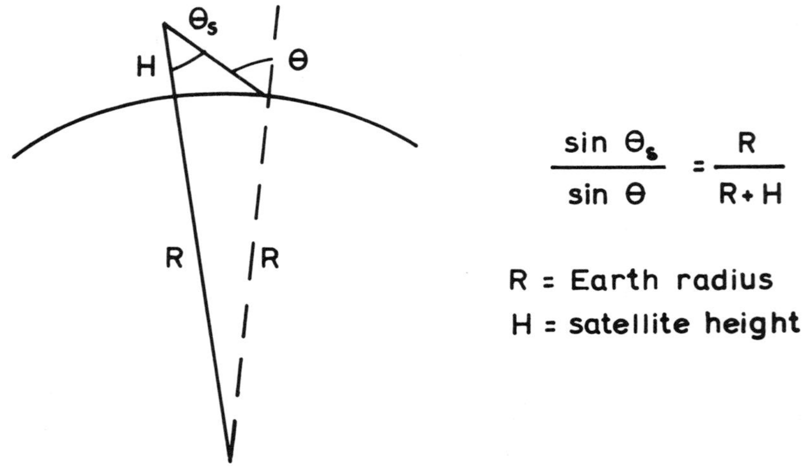

Fig. 6.3 - Parallax correction

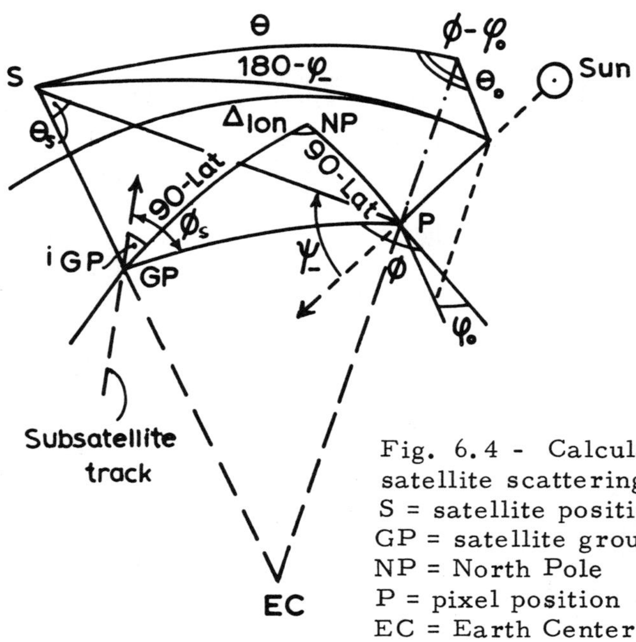

Fig. 6.4 - Calculation of sun-to-satellite scattering angle
S = satellite position
GP = satellite ground point
NP = North Pole
P = pixel position
EC = Earth Center

For correcting for satellite track direction at pixel location we need the inclination of satellite track at pixel location lat, this is given by

$$i = \arc\sin(\sin i_o / \cos lat) \tag{6.10}$$

where i_o = inclination at equator = 9.28° west of north. Subtracting i from ϕ_s we find the satellite azimut (west from south) ϕ

$$\phi = \phi_s - \arc\sin(\sin 9.28/\cos lat) \tag{6.11}$$

Now we can calculate the sun to satellite scattering angle ψ_- (see Fig. 6.4)

$$\cos(180 - \psi_-) = \cos\theta\cos\theta_0 + \sin\theta\sin\theta_0\cos(\phi - \varphi_0)$$

$$\cos\psi_- = -\cos\theta\cos\theta_0 - \sin\theta\sin\theta_0\cos(\phi - \varphi_0) \tag{6.12}$$

and (see ref.7)

$$\cos\psi_+ = \cos\psi_- + 2\cos\theta\cos\theta_0 \tag{6.13}$$

If the pixel coordinates(Lon, Lat) are not known but the coordinates of the groundpoint GP Lon_o and Lat_o then the pixel coordinates can be calculated from the spherical triangle GPNPP in Fig. 6.4:

$$\beta = \phi_s - \arc\sin(\sin 9.28/\cos lat_o) \tag{6.14}$$

$$lat = \arc\sin(\sin lat_o \cos\theta + \cos lat_o \sin\theta \cos\beta) \tag{6.15}$$

$$\Delta lon = \arc\sin(\sin\beta \sin\theta/\cos lat) \tag{6.16}$$

and

$$lon = lon_o + \Delta lon \tag{6.17}$$

In these calculations the roll pitch and yaw of the satellite is neglected which is possible when position of pixels are needed to calculate the angles needed in the atmospheric correction routines. For calculations of exact pixel positions in order to identify seatruth positions in a CZCS scene, roll pitch and yaw and exact height of the satellite are needed and the exact geographic position of a reference pixel in the scene. For this a programme package written by Wilson, Smith and Nolten (ref. (27)) is available in Fortran and HPL for a Hewlett Packard 9825.

A translation of these programmes into PL1 is also available which allows to calculate the position of pixels with high precision when the geographical coordinates of one near-by pixel are known (ref. (29)).

7. EXAMPLES OF CZCS DATA EVALUATION

7.1 Northern Adriatic, Orbit No. 3541

Date: July 7, 1979.
Processed with:
correction factors for slope and intercept (5 channels): 1.029, 0.988, 0.989, 1.007, 1.008
sensitivity loss constant S: 0 for all channels (ref. (30))
$\tau_A(\lambda)$ = 0.854, 0.641, 0.533, 0.398 (from measurement)
$\epsilon(\lambda, 670)$ = 2.146, 1.611, 1.339, 1.0
chlorophyll algorithm:

$$C(mg/m^3) = 0.30 \left(\frac{R(443)}{R(550)}\right)^{-1.86}$$

sediment algorithm:

$$S(mg/l) = 2.61 \left(\frac{R(443)}{R(550)}\right)^{-1.30}$$

Lss(λ) was transformed into diffuse subsurface reflectance by:

$$R(\lambda) = \frac{\pi \cdot Lss(\lambda)}{E_d^+(o)}$$

Figs. 7.1 to 7.7 show the results of the evaluation.

7.2 Western British Channel

Calciferous Algae Bloom, Orbit No. 17997. Date: May 18, 1982

This processing was done in order to locate an algal bloom observed by scientists from CNEXO (France) and Marine Biological Assoc. (UK) on Quick-looks from Dundee CZCS receiving station. A calibrated radiance tape (CRT) was produced by Earthnet from data received at ESA receiving station at Lannion and brought directly toJRC-Ispra. We have corrected the subscene shown in Fig. 7.8 for atmosphere using:
Sensitivity loss constant: $2.3 \cdot 10^{-5}$, $1.45 \cdot 10^{-5}$, $0.87 \cdot 10^{-5}$, 0.0.
Correction factors for slope and intercept: (5 channels) 0.992, 0.968, 0.983, 1.006, 1.016.

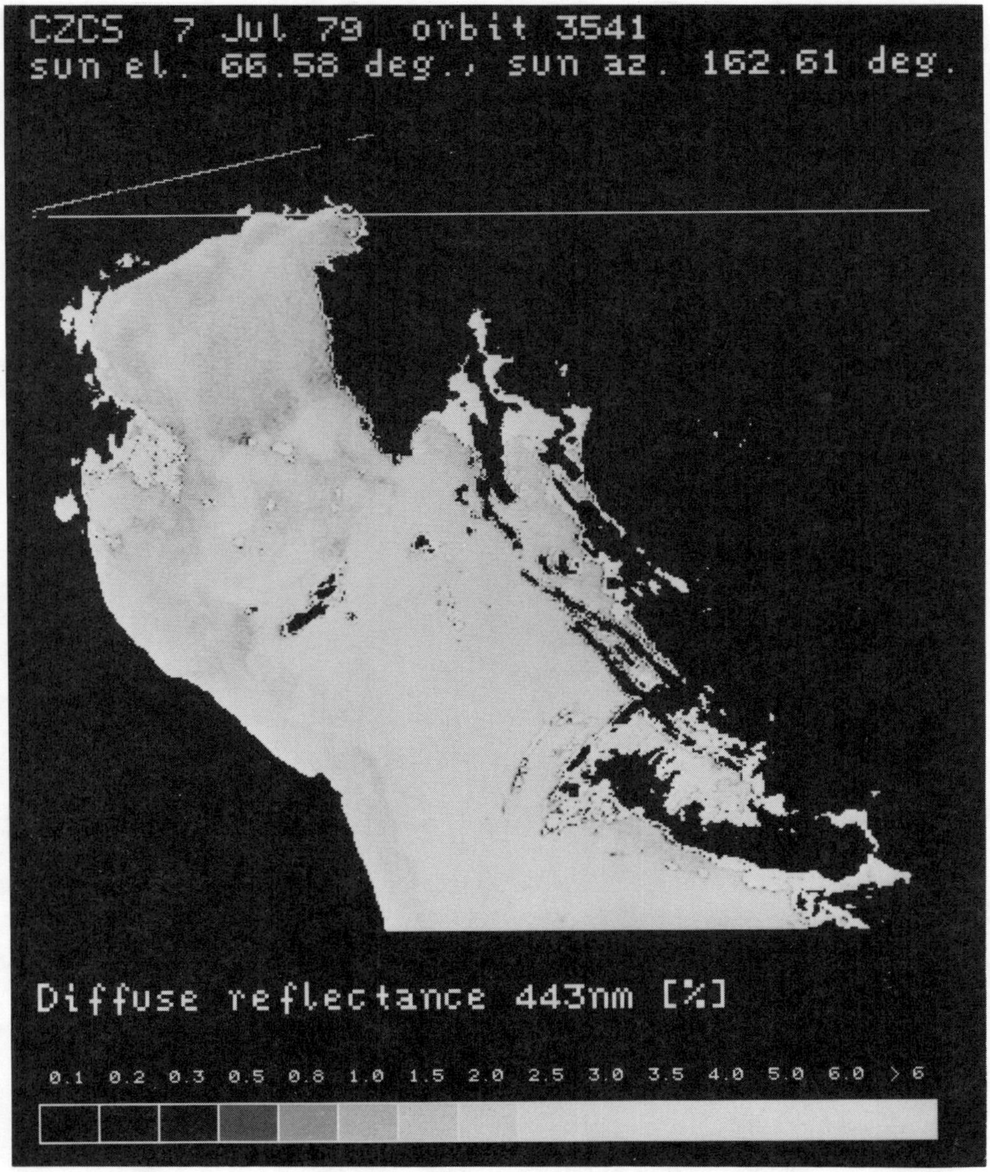

Fig. 7.1 - Diffuse subsurface reflectance in channel 1 (λ = 443 nm).
The image has been geometrically corrected.
Note low reflectance along the coast of Italy and high reflectance in the open Adriatic.

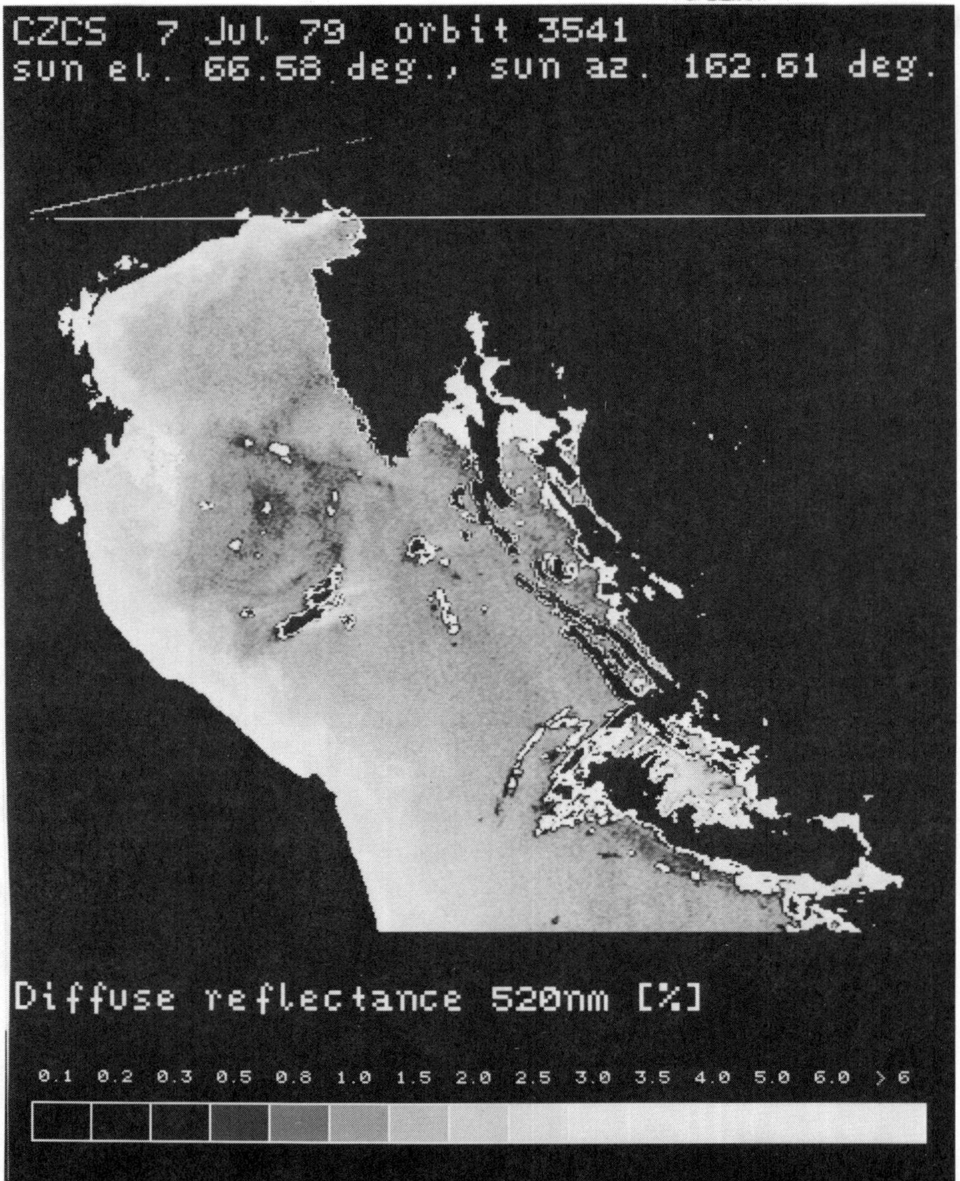

Fig. 7.2 - Diffuse subsurface reflectance in channel 2 (λ = 520 nm).
Note high reflectance along the coast of Italy and lower reflectance in the open Adriatic.

Fig. 7.3 - Diffuse subsurface reflectance in channel 4 (670 nm).
Note the high values (> 1%) near to the coast of Italy due to high content of total suspended matter.

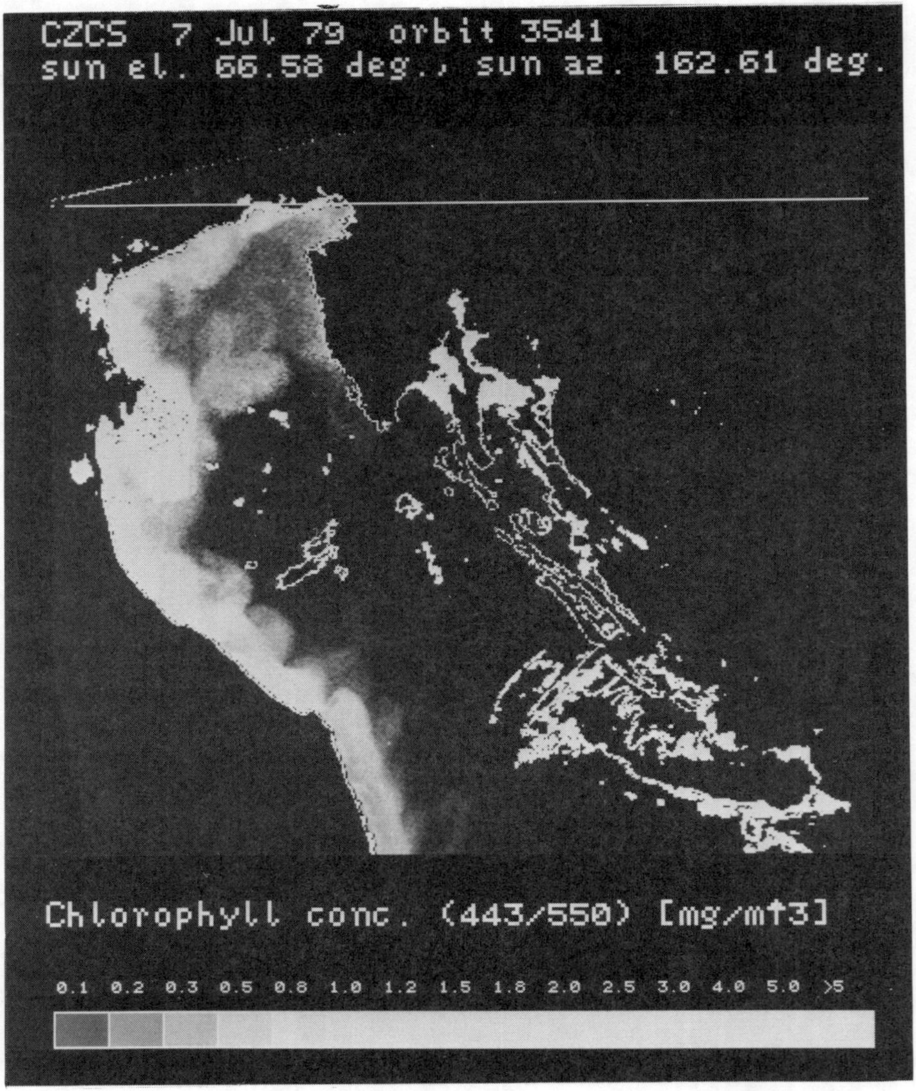

Fig. 7.4 - Chlorophyll-a concentration calculated from

$$C = 0.30 \left(\frac{R(443)}{R(550)}\right)^{-1.86} \qquad \text{(ref. 16)}$$

Note the plume of chlorophyll-rich water half way between Po-estuary and Istria.

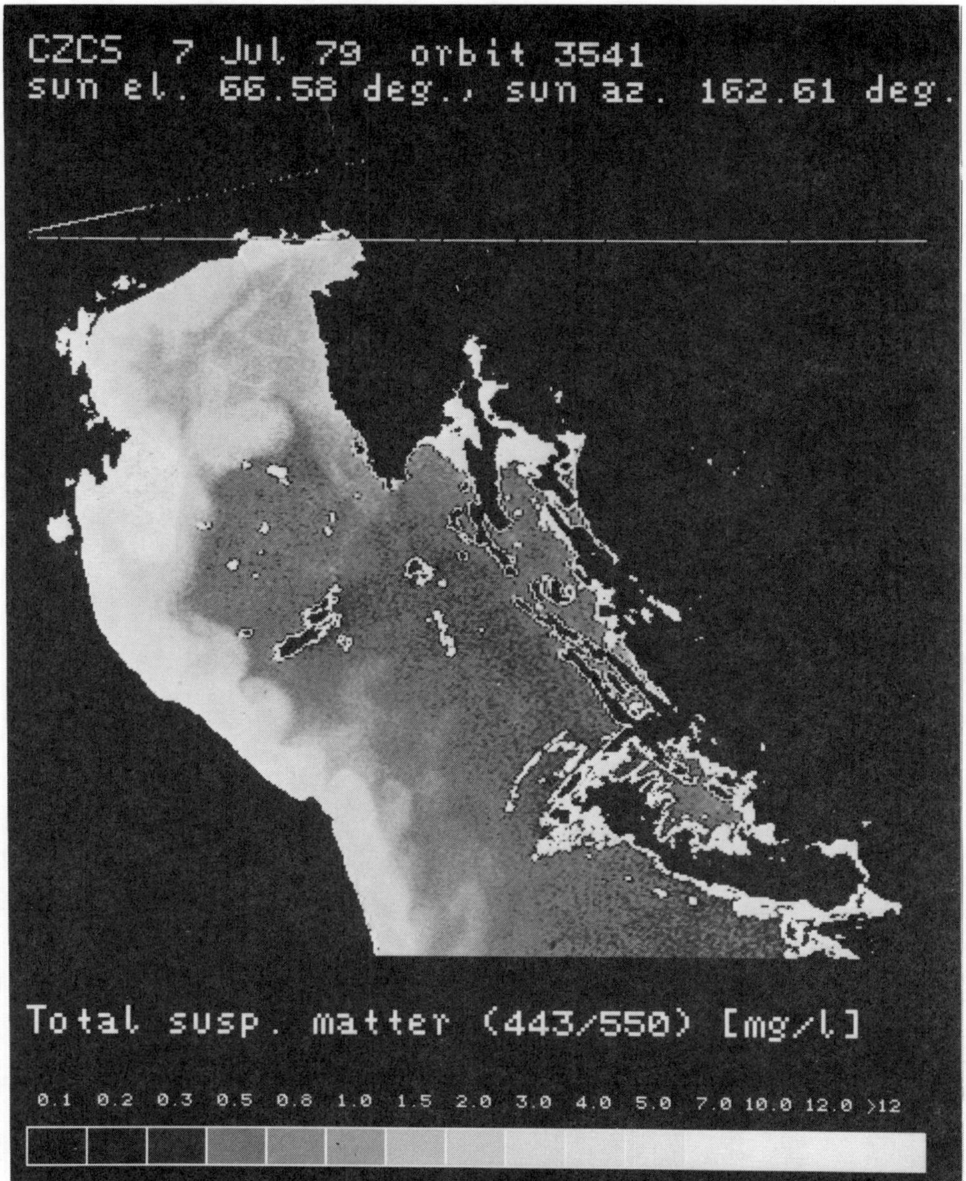

Fig. 7.5 - Total suspended matter concentration (mg/l).
Note high values (> 7 mg/l) along the coast of Italy and low values in the open Adriatic.
Algorithm used:

$$S = 2.61 \left(\frac{R(443)}{R(550)}\right)^{-1.3} \qquad \text{(ref. 16)}$$

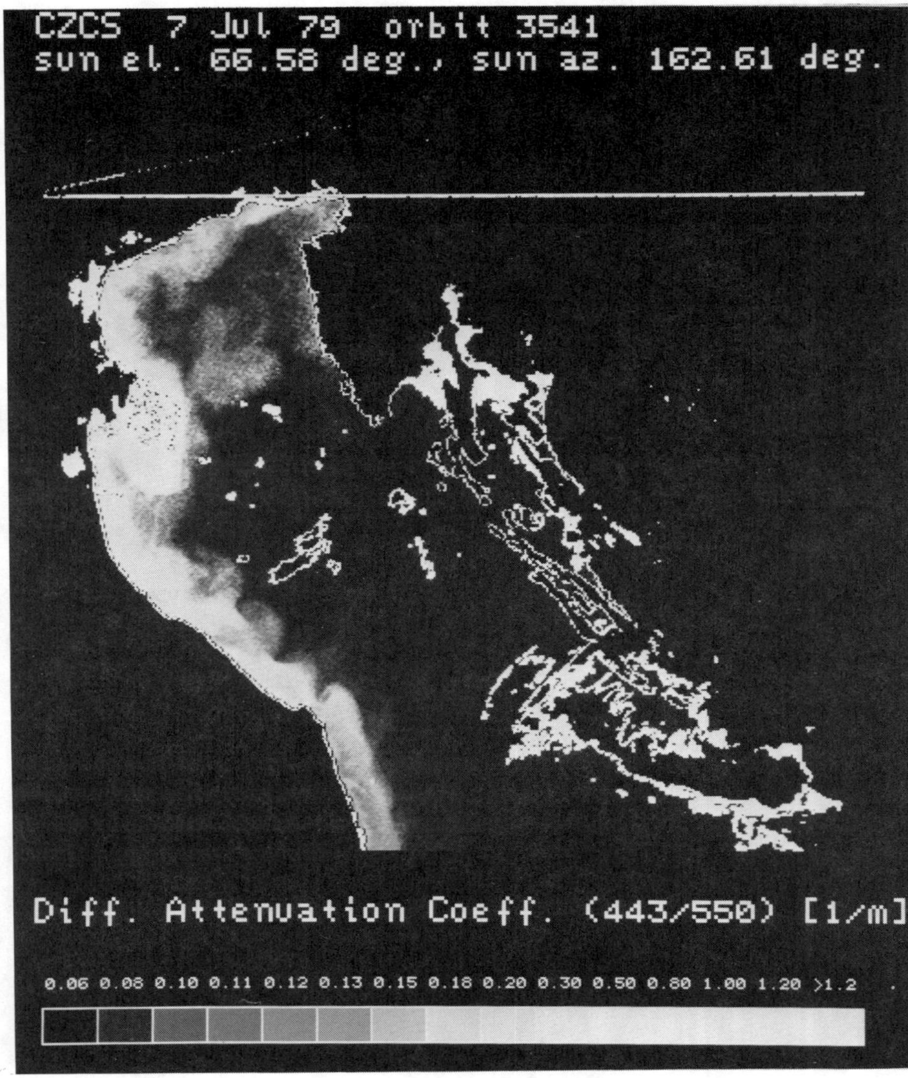

Fig. 7.6 - Diffuse attenuation coefficient at λ = 550 nm derived from algorithm (ref. 16)

$$K_d(550) = 0.053 \left(\frac{R(443)}{R(550)}\right)^{-1.740} + 0.063$$

High values along the Italian coast and the plume half way between Po-estuary and Istria correlate with chlorophyll and sediment maps.

Fig. 7.7 - Seasurface black body temperature as calculated from channel 6 of CZCS.
Note the plume of warmer water half way between Po-estuary and Istria which correlates with high turbidity water seen in the preceeding figures and small warm water plumes along the coast of Rimini south of the Po-delta.

$\tau_A(\lambda) = 0.181,\ 0.157,\ 0.124,\ 0.1,\ 0.089$ (obtained from ϵ-search over clear water).
$\epsilon(\lambda, 670) = 1.81,\ 1.57,\ 1.24,\ 1.0.$

After that a geographical grid was overlayed to it in order to find the position of the two algae bloom patches seen on Fig. 7.8.

From inspection of another image taken over the same area on May 15, 1982, it was found that the same two algae blooms were present there nearly at the identical position. The results were then communicated to the local investigators for planning of seatruth campaigns in that area. All the procedure: inspection of quick-looks, production of CRT, shipping to JRC and processing there and informing the local investigators took all together 13 days of which only one day was needed for processing the final result.

7.3 Upwelling off shore Senegal (Dakar)

This data elaboration example was made in the framework of the preparation of a project to study in collaboration with the Sénegal Government Secretariat for the Scientific and Technical Research (SGSSTR), the French Office de la Recherche Scientifique et Technique Outre Mer (ORSTOM), its research centre in Senegal CRODT and ESA, the use of CZCS and Meteosat data for fishery along the Senegal coast. If approved by NASA/NOAA, CZCS data will be received at ESA Maspalomas station beginning November 1982, and processed together with seatruth and Meteostat data to chlorophyll-a and temperature maps. An attempt will be made to process the data within 21 days in order to direct periodic seatruth campaigns to areas of greatest interest. Figs. 7.9 and 7.10 show examples of processing data taken at Maspalomas and processed at Earthnet/Frascati to CRT's. These were then brought to JRC-Ispra and corrected for atmospheric effects. Fig. 7.9 shows the scene taken on orbit No. 11875 on March 1, 1981 as it appears after atmospheric corrections. The parameters used were:
sensitivity loss constants: as in 7.2, correction factors for slope and intercept: 1.036, 0.985, 0.968, 0.999, 1.016.
$\tau_A(\lambda) = 0.1,\ 0.1,\ 0.1,\ 0.1,\ 0.1$
$\epsilon(\lambda, 670) = 1.0,\ 1.0,\ 1.0,\ 1.0,\ 1.0$
Gordons method (no iteration on Lss(670)) was applied.
Fig. 7.10 shows the results for orbit No. 16533 of Feb. 1, 1982. This scene was processed using $\epsilon(\lambda, 670)$ determined in clear water zones. It was found that ϵ was varying over the scene: in the south west part of the scene ϵ-values were found which

characterize typically marine aerosol with ν of the order of 1.1 to 1.2, while near the Sahara dust streak ν-values between 2.2 and 2.5 were obtained. The result shown in Fig. 7.10 was obtained using the following parameters:
Sensitivity loss constant: $S \cdot 10^{-5}$ = 2.35, 1.45, 1.0, 0.18.
Slope and intercept correction factors: 1.027, 0.978, 0.990, 1.011, 1.014.
$\tau_A(\lambda)$ = 0.364, 0.317, 0.277, 0.230, 0.207
$\epsilon(\lambda, 670)$ = 1.583, 1.378, 1.204, 1.0, .900.

The atmospheric correction method used the iteration procedure on Lss(670) explained in chapter 4, eq. (4.5) because the other methods (Gordons method (Lss(670) = 0), Austin/Petzold method (eq. (4.4)) failed to give a satisfactory discoupling between subsurface radiance and path radiance in the very turbid waters along the coast of Guinea.

Figures 7.8–7.10 will be found in the Colour Section, pp. 450, 451, 452.

REFERENCES

1. Hovis W.A. (1980), The Nimbus-7 Coastal Zone Color Scanner (CZCS) Program, Oceanography from Space (J.F.R. Gower, editor), Plenum Press, 1981, pp. 213-225.
2. Sturm B., Sorensen B. M. (1978), Satellite Surveillance of Coastal Pollution, EEC Programme, Proc. Sem. on Coastal Poll. Assessment, April 20-21, 1978, Cork, Ireland.
3. Gordon H. R. (1978), Removal of atmospheric effects from satellite imagery of the oceans, Appl. Optics, 17, No. 10, pp. 1631-1636.
4. Sorensen B. M. (Editor) (1978), Recommendations of the International Workshop on Atmospheric Correction of Satellite Observation of Sea Water Color, 20-22 March, 1978, Ispra, Italy.
5. Sorensen B. M. (Editor (1981), Recommendations of the 2nd International Workshop on Atmospheric Correction of Satellite Observation of Sea Water Color, 30 March - 1 April, 1981, Ispra, Italy.
6. Gordon H. R. (1980), A preliminary assessment of the Nimbus-7 CZCS atmospheric correction algorithm in a horizontally inhomogeneous atmosphere, Oceanography from Space (J. F. R. Gower, editor), Plenum Press, 1981, pp. 257-265.
7. Sturm B. (1980), The atmospheric correction of remotely sensed data and the quantitative determination of suspended matter in marine water surface layers, Remote Sensing in Meteorology, Oceanography and Hydrology (ed. A. P. Cracknell), Ellis Horwood Ltd., 1981.
8. Ball Brothers, Aerospace System Division (1979), Development of the coastal zone color scanner for Nimbus-7, Vol. 2, Test and performance data, Final report, F78-11, Rev. A.
9. Cracknell A. P., Singh S. M. (1981), Coastal zone research work at Dundee university using remote sensing techniques, Proc. of an EARSEL/ESA Symp., Voss Norway, May 19-20, 1981 (ESA SP-167, July 1981).
10. Fusco L. (1981), private communication.
11. Gordon H. R. (1982), 19. NIMBUS experimental team (NET) meeting, May, 1982.
12. Viollier M. (1982), 18. NET, Jan. 1982.
13. Austin R. W. (1982), 18. NET, Jan. 1982.
14. Gordon H. R., Clark D. (1981), Clear water radiances for atmospheric correction of coastal zone color scanner imagery, 17. NET, Sep. 1981.

15. Smith R.C., Wilson W.H. (1980), Ship and satellite bio-optical research in the California Bight, Oceanography from Space (J.F.R. Gower, editor), Plenum Press, 1981 pp. 281-294.
16. Sturm B. (1982), Developments in CZCS data evaluation at JRC Ispra during 1981, 18. NET, Jan. 1982.
17. Tanre D., Herman M., Deschamps P.Y., De Leffe A., (1979), Atmospheric modelling for space measurements of ground reflectances including bidirectional properties, Appl. Optics 18, 21 pp. 3587-3594.
18. Plass G.N., Kattawar G.W., Guinn J.A. jr., Appl. Optics 14, pp. 1924-1936.
19. Austin R.W. (1980), Coastal zone color scanner radiometry, Ocean Optics VI (S.Q. Duntley, editor), 208, SPIE, pp. 170-177.
20. Gordon A.R., Clark D.K., Mueller J.L., Hovis W.A. (1980), Phytoplankton pigments derived from the Nimbus-7 CZCS: initial comparisons with surface measurements, Science, 210, pp. 63-66.
21. Clark D.K. (1980), Phytoplankton pigment algorithms for the Nimbus-7 CZCS, Oceanography from Space (J.F.R. Gower, editor), Plenum Press, 1981, pp. 227-237.
22. Smith R.C., Baker K.S. (1982), Oceanic chlorophyll concentrations as determined by satellite (Nimbus-7 Coastal Zone Color Scanner), Marine Biology, 66, pp. 269-279.
23. Morel A. (1980), In-water and remote measurements of ocean color,Boundary-layer Meteorology, 18, pp. 177-201.
24. Holligan P.M., Viollier M., Dupony C., Aiken J.,(1982) Satellite studies on the distribution of chlorophyll and dinoflagellate blooms in the Western English Channel (to be published).
25. SASC (1981), CZCS derived products validation scene production, adopted pigment algorithm, 17. NET, Sep. 1981.
26. Austin R.W., Petzold T.J. (1980), The determination of the diffuse attenuation coefficient of sea water using the coastal zone color scanner, Oceanography from Space, (J.F.R. Gower, editor), Plenum Press, 1981, pp. 239-256.
27. Wilson W.H., Smith R.C., Nolten J.W. (1981), The CZCS geolocation algorithms, SIO Ref. 81-32 Vis. Laboratory, La Jolla, Cal. 92093.
28. Viollier M. (1982), private communication.
29. Sturm B. (1982), CZCSA, a PL1 program for geolocation of CZCS imagery (unpublished).
30. Sturm B., Ferrari G., Maracci G., Schlittenhardt P., Alberotanza L., Chlorophyll-a and total suspended matter concentration in the North Adriatic Sea determined from Nimbus-7 CSCS. Paper presented at the 69th Statutory Meeting of the ICES at Woods Hole, Mass., 5-14 Oct.,1981.

CHAPTER 8

SURFACE WIND VECTOR MEASUREMENTS FROM SATELLITES

D. Offiler

Meteorological Office, Bracknell

1. INTRODUCTION

This paper discusses the measurement of surface wind speed and direction over the ocean, using the technique of "scatterometry", with particular reference to the SEASAT-A Satellite Scatterometer (SASS). The theoretical and experimental background to the technique is outlined, and how this translates into a practical instrument. Some results from an assessment of the measurement accuracy of the SASS and some synoptic examples are also given.

Although the SEASAT mission was experimental, similar techniques will be used for semi-operational satellites during the late 1980's. Some of the potential uses and problems with such data are briefly discussed.

2. THEORETICAL AND EXPERIMENTAL BACKGROUND

As soon as microwave radar became widely used in the 1940's, it was found that at low elevation angles, surrounding objects and terrain caused large echos, obliterating signals from targets of interest. There is a similar effect at sea caused by reflections from waves, and this unwanted signal became known as sea (or ground) "clutter". Ever since, designers and users of radar equipment have sought to reduce this noise [1]. However researchers investigating the effect found that the backscattered echo became larger with increasing wind speed, opening the possibility of remotely measuring wind speed and so encouraging further investigation [2].

A. P. Cracknell (ed.), Remote Sensing Applications in Marine Science and Technology, 169–182.

The backscattering is due to in-phase reflections from a rough surface; for incidence angles of more than about 20 deg., this occurs when the Bragg condition is met:

$$\Lambda \sin \theta_i = n\lambda / 2 \qquad (1)$$

where Λ is the surface roughness wavelength, λ is the radar wavelength, θ_i the incidence angle (measured from the vertical) and n=1,2,3.... First order Bragg scattering (n=1) at microwave wavelengths ($\lambda \sim 2$cm) arises from small capillary waves superimposed on the larger gravity waves. These capillary waves are generated by the instantaneous surface wind stress.

The level of backscatter from an object is usually expressed as the Radar Cross-section (RCS or σ) which is defined as "the area intercepting that amount of power which, when scattered isotropically, produces an echo equal to that of the object" [3]. For extended targets, such as the sea surface, the backscatter is expressed as the Normalised Radar Cross-section (NRCS or σ^o), which is the RCS per unit area. In terms of other known or measurable radar parameters,

$$\sigma^o = \frac{64\,\pi^3\,R^4}{\lambda^2\,L_s\,G_o^2\,(G/G_o)^2\,A} \cdot \frac{P_R}{P_T} \qquad (2)$$

where R is the slant range to the target of area A, λ is the radar wavelength, L_s includes atmospheric attenuation and other system losses, G_o is the peak antenna gain with G/G_o the relative gain in the target direction. P_T is the transmitted power and P_R the received power. This is often referred to as the Radar Equation. (σ^o is usually measured in decibels - ie $\sigma^o\,(\text{dB}) = 10\log_{10}\sigma^o$)

Experimental evidence from wind-wave tanks and scatterometers on static platforms, aircraft and satellites operating over the ocean show that σ^o increases with surface wind speed, decreases with incidence angle and is also dependent on the beam azimuth angle relative to the wind direction [4]. It is generally lower for horizontal polarisation than for vertical and the dependence of σ^o on radar frequency appears to be small over the range 5-20GHz.

Figure 1 is a smoothed plot of combined Skylab and aircraft scatterometer σ^o (vertical polarisation), against relative wind direction for various wind speeds, at 30 deg. incidence angle. Direction 0 deg. corresponds to looking upwind, 90 deg. crosswind and 180 deg. downwind. As yet, no theoretical basis exists which explains the σ^o signature with wind speed and direction, incidence angle, polarisation and radar frequency. Empirical σ^o model functions can be fitted to these curves of the form:

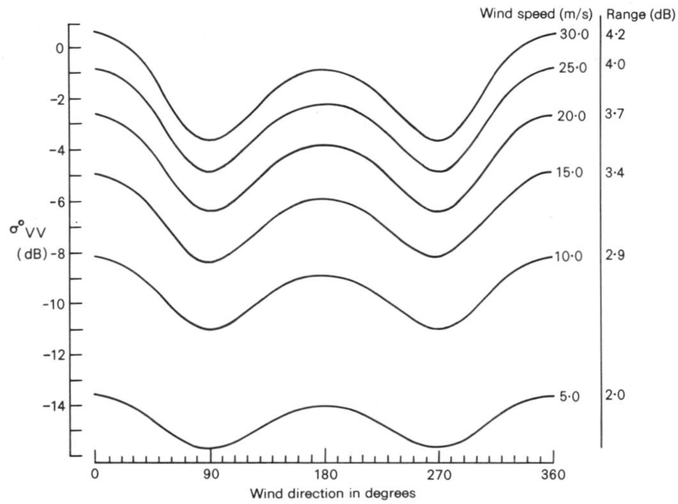

Figure 1. Backscatter cross-section σ° against relative wind direction for various wind speeds. Vertical polarisation, 30 deg. incidence angle.

$$\sigma^\circ = a_0(U,\theta_i,P) + a_1(U,P)\cos\phi + a_2(U,P)\cos 2\phi \quad \text{dB} \qquad (3)$$

or:

$$\sigma^\circ = G(\phi,\theta_i,P) + H(\phi,\theta_i,P)\log_{10} U \quad \text{dB} \qquad (4)$$

where U is the wind speed, ϕ the relative wind direction , θ_i the incidence angle and P is vertical or horizontal polarisation. The coefficients a_0, a_1 and a_2 are proportional to $\log_{10}$ U, and they (or G and H) are derived by regression from measured backscatters in areas where the wind vector is known. Originally, the values of these coefficients were based on aircraft measurements, with the conventionally measured wind speeds referenced to a common height of 19.5m (though σ° is more closely related physically to the surface friction velocity, u_*). New data from SEASAT have enabled these coefficients to be optimised [4].

Since σ° shows anistropy with wind direction, in principle measuring σ° at two or more different azimuth angles allows both speed and direction to be retrieved. This is the method used by the SASS and other designs of wind measuring scatterometer.

3. THE SASS INSTRUMENT

The SEASAT scatterometer used two fore and two aft radar beams, directed 45 deg. either side of the sub-satellite track to form a cross-shaped pattern of illumination on the ground, as shown in Figure 2. The beam widths (half-power) were 0.5 deg. in the horizontal and about 25 deg. in the vertical plane, forming a swath 500km wide on each side. (There were actually two extensions to this 'primary' swath: a central one of 0-8 deg. incidence angle, not direction sensitive, and an outer one of 55-65 deg., expected to be sensitive only to high winds.)

Using the motion of the spacecraft, Doppler filtering of the return signal was employed to divide the total swath into 15 cells. For each beam measurement period (1.89s), 61 RF pulses at 14.6GHz were integrated for about 300ms, with a further 500ms noise integration measurement. The value P_R in the Radar Equation (2) is then:

$$P_R = P(\text{signal+noise}) - P(\text{noise only}) \tag{5}$$

Several modes of antenna/polarisation switching were possible, for example mode 1 sequence was antenna 4 V-pol, 1V, 3V, 2V. In all cases, a backscatter measurement made in a particular cell by a forward beam would be followed a few minutes later by one from a

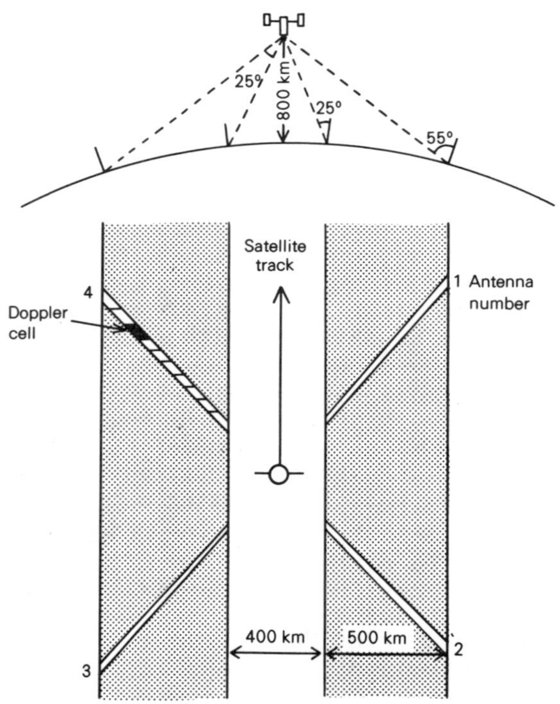

Figure 2. The SASS viewing geometry: section in the plane of beams 1 and 3 (top diagram); beam illumination pattern and ground swath (bottom diagram).

rearward beam on the same side at (almost) the same geographical position. The registration is not exact due to the Earth's rotation in the time between the two orthogonal measurements [5].

4. RETRIEVAL OF WIND VECTOR

The derivation of wind vectors from the raw instrument telemetry involves a series of complex algorithms. In general terms, these perform the tasks of:

* Earth locating each cell from the orbital data and Doppler filter frequencies and determining the cell area and slant range.
* calculating the ratio of received signal power to transmitted power.
* determining the values of system losses and antenna gain in the cell direction from internal and pre-flight calibration.
* calculating σ° from the Radar Equation and correcting this for atmospheric attenuation derived from the passive microwave radiometer (SMMR) also on SEASAT [6,7] and for other instrumental biases.

Further algorithms match up orthogonal σ°'s from the fore and aft beams and, using a σ° model function (usually in the form of G & H look-up tables), perform an iterative fitting technique to derive a wind vector. However, due to the harmonic nature of σ° with wind direction, there can be up to four 'solutions' of wind speed and direction, each with similar speeds but widely different in direction.

5. SASS ACCURACY ASSESSMENTS

The SASS was designed to measure the surface wind vector over the ocean to an accuracy of 2m/s (or 10%, whichever is greater) in speed and 20 deg. in direction over the range 4-25m/s. In order to validate the actual accuracies achieved by each instrument, a special exercise, the Gulf of Alaska SEASAT Experiment (GOASEX) was undertaken, involving several oceanographic ships and buoys and an aircraft carrying a similar scatterometer [8]. Comparisons with this surface data showed that the SASS generally measured the winds to the required accuracy but some large biases remained. These results were used to update the processing algorithms [4].

Coincidentally, the Joint Air-Sea Interaction (JASIN) project took place in the North Atlantic (between Scotland and Iceland) [9], within the lifetime of SEASAT and when all of its instruments were operating correctly. This set of surface observations had a high spatial and temporal resolution, was of sufficiently high quality and included a reasonable range of meteorological conditions to properly assess the SASS performance. Results from a small number of orbits passing over the JASIN area (using the updated GOASEX coefficients) showed the SASS accuracy to be around 1.6m/s and 18 deg. with much smaller biases than with GOASEX [10]. After the SASS-JASIN Workshop which produced these results, the wind derivation algorithms were again refined. The Jet Propulsion Laboratory (JPL), who were responsible for SEASAT data, have processed the complete set of scatterometer winds using this latest "SASS-1" algorithm, including atmospheric corrections from the SMMR where the swaths overlap.

The Satellite Meteorology Branch of the Meteorological Office has made a thorough investigation of the SASS winds [11], using all the SEASAT orbits which passed over the JASIN area during the two months of that experiment; a brief description of that study is given below.

The Institute of Oceanographic Sciences (IOS) have collated much of the JASIN data, and quality-controlled it as far as possible, but the winds still refer to the height of each anemometer - from 2.5m asl to 23m with the platforms used. In order to be compatible with the SASS data, all wind speeds were

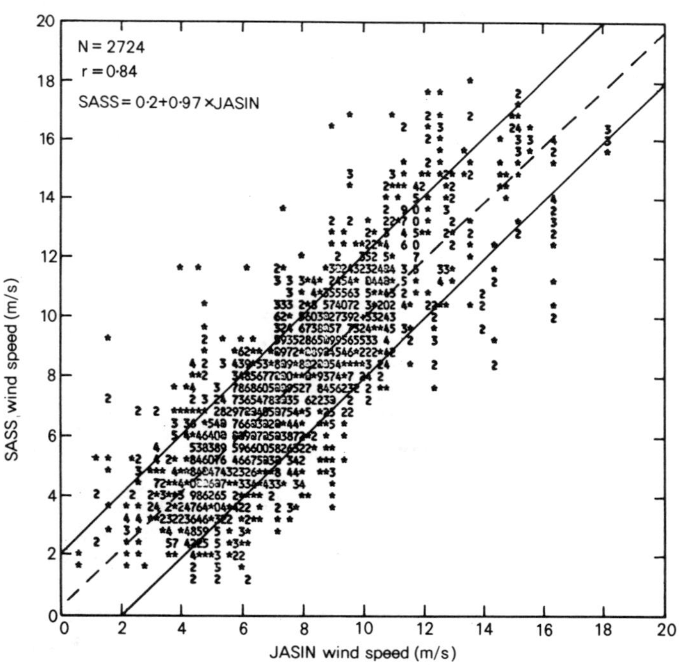

Figure 3a. Scatter diagram of wind speed measurements made by SASS against colocated JASIN observations. The design rms limits of 2m/s are indicated by the solid parallel lines and the least-squares fit by the dashed line.

first adjusted, assuming neutral stability, to a common height of 19.5m. Each SASS wind was then paired, if possible, with a JASIN observation within 60km and 30 minutes; there were a total of 2724 such pairs. As previously mentioned, each SASS measurement may produce up to four solutions for wind direction; for this study, the one closest in direction to the JASIN observation was taken.

From each 'colocated' pair of wind estimates, the standard deviation of the SASS-JASIN differences ("SDD") in speed and direction were calculated. Figure 3 shows scatter plots of SASS vs. JASIN wind speed (a) and direction (b), for the whole colocation set. Overall, the SASS agrees with the surface observations to 1.7m/s and 17 deg., well within the design goal of 2m/s and 20 deg. The mean differences are small, being less than 0.1m/s and 3 deg.

Further analyses were performed, dividing the data according to the values of various parameters such as swath position or wind speed. Figure 4 shows the variation of SDD with (a) incidence angle (inner to outer edges of the swath), (b) wind quadrant and (c) wind speed. Except in the speed range 12-16m/s (Figure 4c - due particularly to one anemometer having a short averaging period and also overspeeding) and in one quadrant (Figure 4b), the 2m/s and

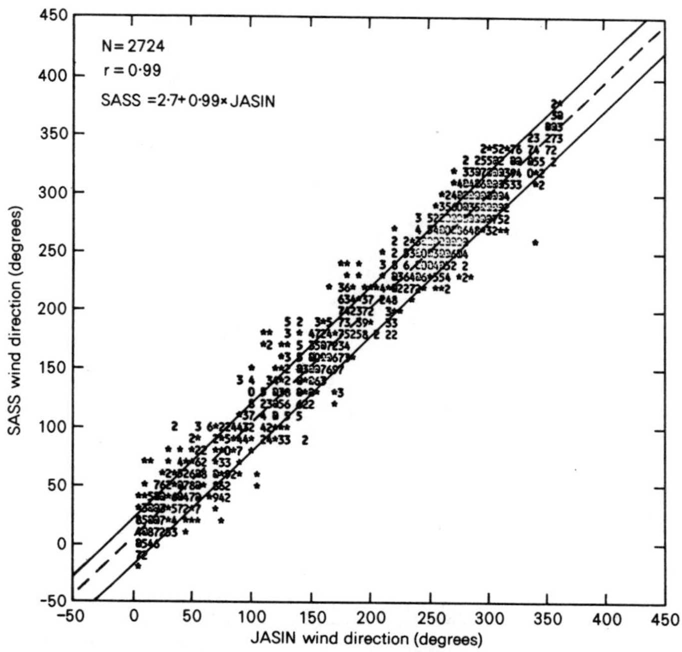

Figure 3b. As for 3a but wind direction. The design rms limits of 20 deg. are shown.

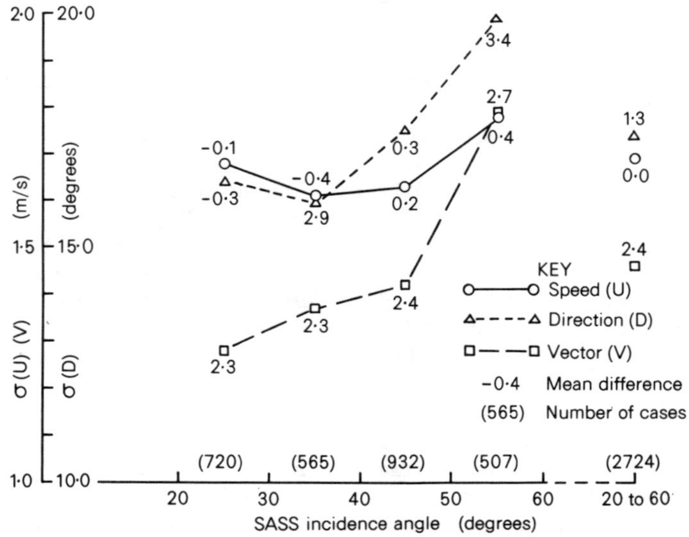

Figure 4a. Standard deviations of SASS-JASIN wind speeds, directions and vector differences against bands of SASS incidence angles.

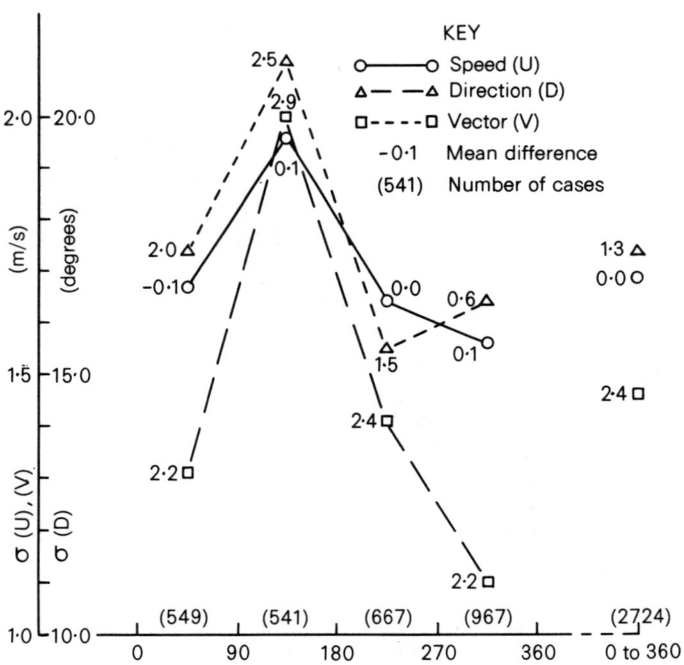

Figure 4b. As for 4a, against JASIN wind quadrant.

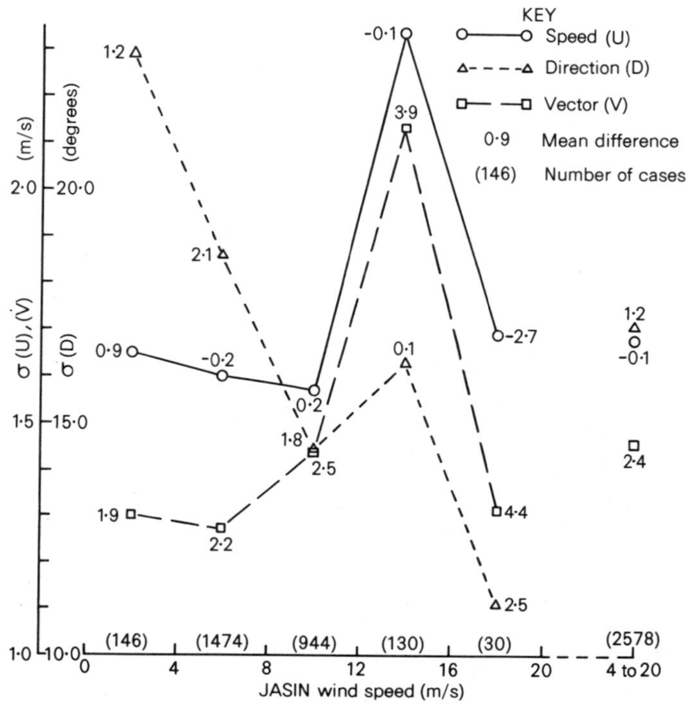

Figure 4c. As for 4a, against bands of JASIN wind speed.

20 deg. design criteria were met in all these sub-categories. A similar analysis with respect to the radar polarisation used for each wind measurement (both beams horizontal polarisation, vertical or mixed) showed very little difference between these modes of operation.

6. SYNOPTIC EXAMPLES

6.1 4 August: thunderstorm case

There was only one occasion during JASIN when the wind speeds measured by SASS were clearly in error by more than 10m/s. This occurred during orbit 557 at 2315 GMT on 4 August 1978, near the southern vertex of the JASIN triangle. This swath for horizontal polarisation is shown in Figure 5, together with the nearby JASIN observations for 2300 GMT, other surface measurements taken at midnight, and the analysis (based on the Bracknell Central Forecasting Office midnight subjective analysis). SASS values ranged from 6 to 22m/s (most vertical polarisation measurements were less than 10m/s), while JASIN wind speeds were about 4m/s,

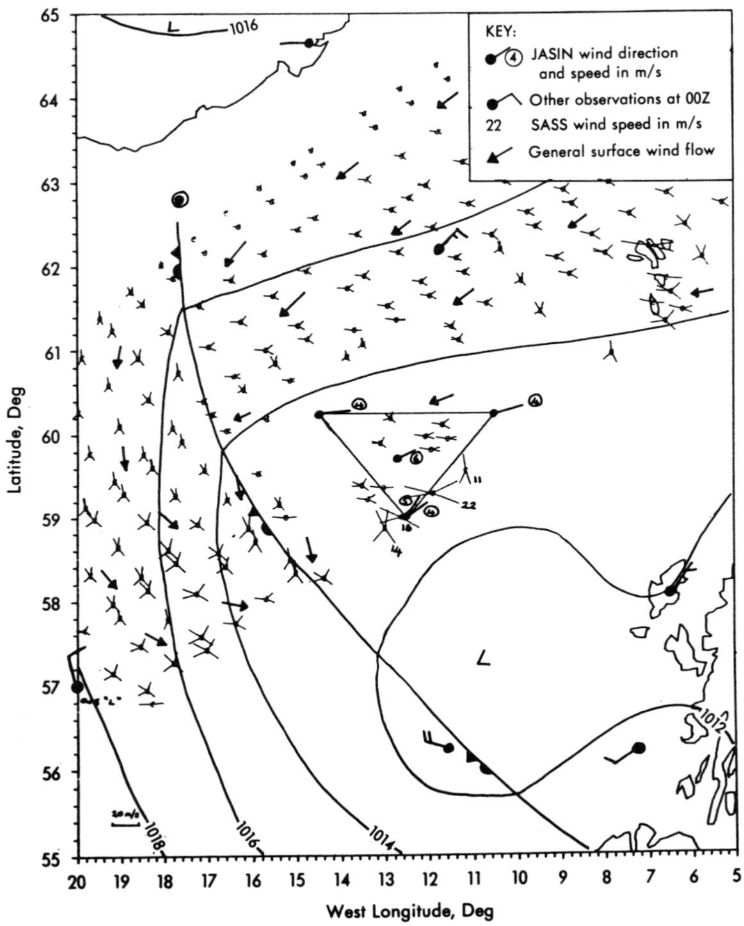

Figure 5. SASS anomaly, 2315 GMT, orbit 557 (horizontal polarisation).

with little variation before and after the satellite pass. At 2300 GMT, ship Meteor reported light drizzle, with thunder during the previous hour.

An investigation by IOS and JPL [12] shows that other nearby JASIN ships Discovery and Shackleton also reported thunderstorm activity, with Discovery logging heavy rain just after the pass. Their analysis showed that during the previous pass 100 minutes earlier, there were high backscatter values in this area, coincident with deep convective clouds in the SEASAT VIRR imagery. Radiosonde observations indicated potentially unstable air behind the surface occlusion. It seems likely, therefore, that the SASS measurements were affected by a thunderstorm associated with deep convection on the occlusion, and may have several causes:

* SASS was operating correctly and measuring locally high winds, such as from a gust front or downdraughts from the convecting cell(s).
* Large raindrops or hail stones caused increased sea-surface roughness, invalidating the backscatter/wind speed relationship.
* Raindrops, usually thought to attenuate microwaves, were large enough to cause additional backscatter, with possibly some contribution from a 'bright band'.

Since none of the many ships and buoys in the area (some sampling every minute) recorded any high winds, the first point is improbable. Considering the likely rainfall rates and cell dimensions, either of the other two causes - or even both - could have been responsible for these anomalous SASS winds.

6.2 31 August: cold front case

This case demonstrates the usefulness of satellite-derived winds because of their density of observations over a swath. A low pressure centre over Iceland had been moving north-eastwards, bringing its associated fronts over the JASIN area by midnight on 31 August 1978. SEASAT orbit 930 passed just south of Iceland at 0050 GMT, enabling the port swath of the scatterometer to measure winds across JASIN, as shown in Figure 6. Also marked are the observations (at 0100 GMT) and analysis. Meteor's winds had veered by 30 deg. during the previous hour and Tydeman's by 50 deg. in the following hour, showing the cold front to be between them. In this case, the surface frontal zone shows clearly in the SASS winds as a line of cells with only two solutions, marking the transition between the wind directions either side of the front.

With experience, synoptic features such as fronts, and especially low pressure centers, can be accurately positioned, even with the present level of ambiguity. Subjective analysis of the wind field has been successfully demonstrated by meteorologists working experimentally on SASS data in the United States [13].

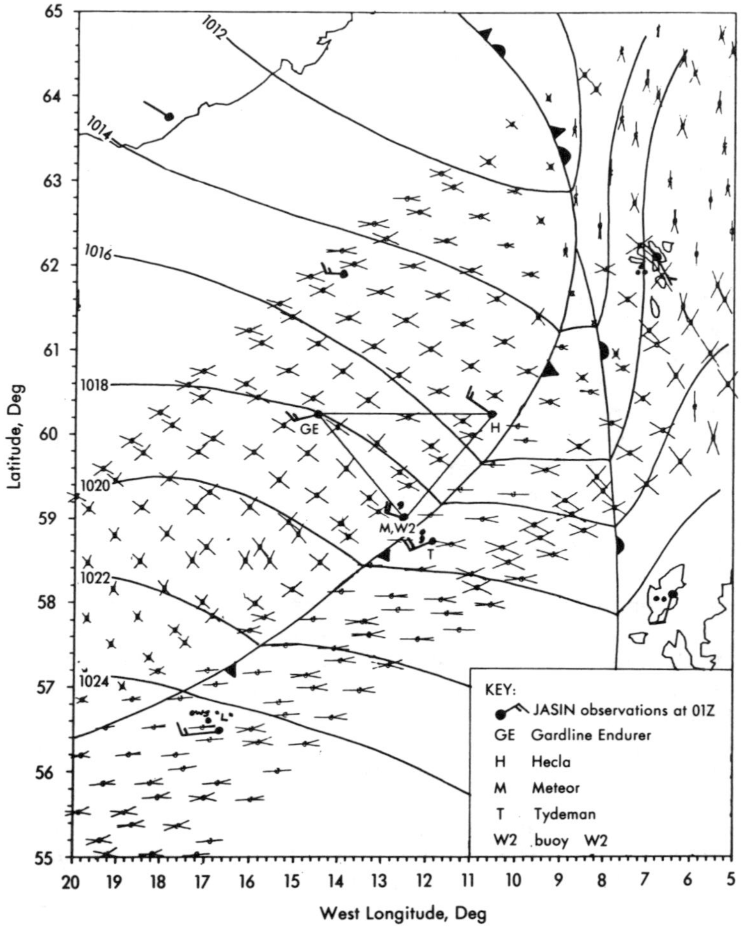

Figure 6. Cold front example, 0050 GMT, orbit 930 (vertical polarisation).

7. CONCLUSIONS AND DISCUSSION

As other studies using GOASEX and small amounts of JASIN data have shown, the SASS instrument on SEASAT met its design goal of measuring near-surface wind vectors to 2m/s and 20 deg. rms. Using data from the whole JASIN period, we have confirmed that overall (and in nearly all sub-categories) SASS was well within these limits, with the latest retrieval algorithms producing very small biases.

We have also made a comparison between the JASIN conventionally observed data which indicates variability of a similar size to that between the two systems: the SASS measurements are therefore at least as 'accurate', on the average, as this special surface data. It should be remembered that the JASIN data set provided a high density of surface measurements in both time and space, and that a special effort was devoted to maintaining the quality of those observations. The normal day-to-day coverage is very sparse in some areas, particularly in the Pacific and southern oceans; these are the areas where satellite data are potentially most useful.

However, comparison of the direction measurements, involved a prior estimate of the true wind direction in order to select one of the four possible solutions. Whilst experienced meteorologists have demonstrated that the wind field can be deduced from SASS data alone, for operational use, satellite derived winds should be available with minimum delay and preferably be non-ambiguous. An ambiguous wind field could be resolved using analysed or forecast fields from numerical models, but small errors in these fields might lead to the incorrect choice of solution. However, if mobile areas such as low pressure centres and frontal zones were avoided, parts of the swath could be made unambiguous by referring to the model field, and the rest of the area processed with existing techniques based on pattern recognition.

Even if such techniques were operationally feasible, it would obviously be desirable for future scatterometers to provide information from which wind speed and direction could be obtained without ambiguity. Reducing the level to only a two-fold(180 deg.) ambiguity would be a great improvement to a forecaster. To this end, much effort is being expended on new wind scatterometers, such as the one proposed for the European satellite ERS-1, due for launch in 1987. It is hoped that by using dual polarisation, or more beams at different azimuths, the extra information in each cell will enable many of the wind determinations to be unique.

REFERENCES

1. Harrold, T W: 1974. Ground clutter observed in the Dee weather radar project. Met. Mag., 103, p.140.

2. Jones, W L and Schroeder, L C: 1978. Radar backscatter from the ocean: dependence on surface friction velocity. Boundary Layer Met., 13, p.133.

3. Long, M W: 1975. Radar reflectivity of land and sea. Lexington Books, Lexington, Mass., USA.

4. Schroeder, L C, Boggs, D H, Dome, G, Halberstam, I M, Jones, W L, Pierson, W J and Wentz, F J: 1982. The relationship between wind vector and normalised radar cross section used to derive SEASAT-A satellite scatterometer winds. J. Geophys. Res., 87 (C5), p.3318.

5. Johnson, J W, Williams, L A, Bracalente, E M, Beck, F B, Grantham, W L: 1980. SEASAT-A Satellite Scatterometer instrument evaluation. J. Ocean. Eng., OE-5, p.138.

6. Bracalente, E M, Boggs, D H, Grantham, W L and Sweet J L: 1980. The SASS scattering coefficient $\sigma°$ algorithm. J. Ocean. Eng., OE-5, p.145.

7. Moore, R K, Birrer, I J, Bracalente, E M, Dome, G J and Wentz, F J: 1982. Evaluation of atmospheric attenuation from SMMR brightness temperatures for the SEASAT satellite scatterometer. J. Geophys. Res., 87 (C5), p.3337.

8. Jones, W L, Black, P G, Boggs, D H, Bracalente, E M, Brown, R A, Dome, G, Ernst, J A, Halberstam, I M, Overland, J E, Peteherych, S, Pierson, W J, Wentz, F J, Woiceshyn, P M and Wurtele, M G: 1979. SEASAT scatterometer: results of the Gulf of Alaska Workshop. Science, 204, p.1413.

9. Royal Society: 1979. Air-Sea Interaction project: summary of the 1978 field experiment. London.

10. Jones, W L, Boggs, D H, Bracalente, E M, Brown, R A, Guymer, T H, Shelton, D and Schroeder, L C: 1981. Evaluation of the SEASAT wind scatterometer. Nature, 294, p.704.

11. Offiler, D: 1982. Surface wind measurements from satellites - a comparison of SEASAT scatterometer data with JASIN surface winds. Met 0.19 Branch Memorandum No. 64, Meteorological Office (unpublished manuscript - copy available in the National Meteorological Library, Bracknell).

12. Guymer, T H, Businger, J A, Jones, W L and Stewart, R H: 1981. Anomalous wind estimates from the SEASAT scatterometer. Nature, 294, p.735.

13. Wurtele, M G, Woiceshyn, P M, Peteherych, S, Borowski, M and Appleby, W S: 1982. Wind direction alias removal studies of SEASAT scatterometer-derived wind fields. J. Geophys. Res., 87 (C5), p.3365.

CHAPTER 9

THE MEASUREMENT OF OCEANOGRAPHIC PARAMETERS USING DEKAMETRIC RADAR.

Lucy Wyatt

Dept. of Electronic & Electrical Engineering,
Birmingham University

1. INTRODUCTION

At the 1980 summer school in Dundee, Shearman (1) presented an overview of the HF (high frequency) radar technique for observing the sea surface. A description of the scattering mechanisms and some general comments on the nature of the received signal and its use in wind direction and current measurement are presented there and will not be repeated here. Some of the features of the spectrum described by Shearman are indicated in figure 1.1

Three topics will be covered here: firstly the mathematical formulation of the spectrum which I hope will give an insight into methods for extracting sea-state information; secondly a new technique for measuring spectra and its advantages over existing HF techniques and finally some approximate solutions to the mathematical problems and their application to experimental results.

2. THEORY

The theory and much of the data extraction work has been developed by Barrick (2,3,4,5,6,7) I do not intend to dwell for long on the mathematical problem that he solved, i.e. the interaction of electromagnetic waves with a random, moving rough surface. I shall spend more time looking at the structure of the solution that describes the relationship between the radar backscatter cross-section and the ocean wave spectrum.

A. P. Cracknell (ed.), Remote Sensing Applications in Marine Science and Technology, 183–205.

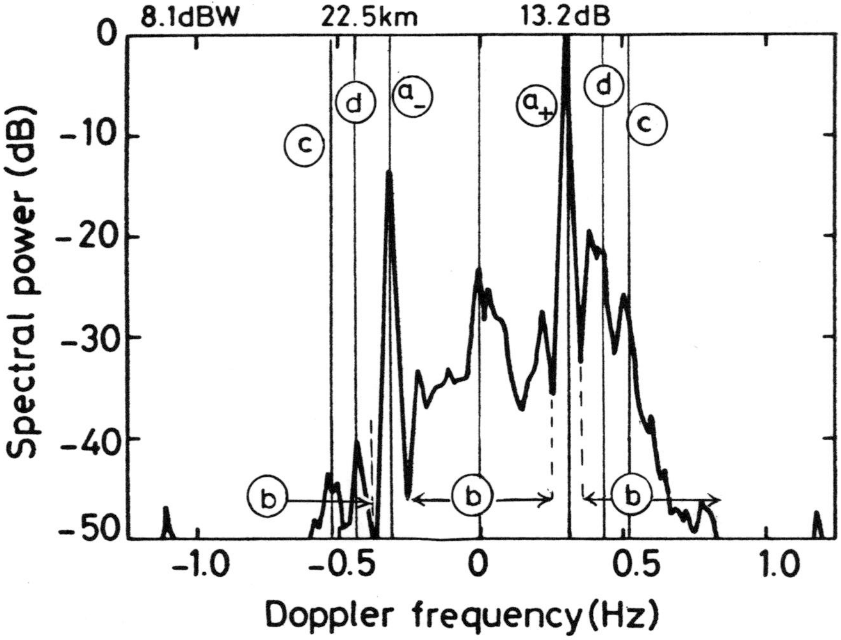

Figure 1.1 A typical ground-wave Doppler spectrum.

(a+) - approaching first order Bragg line at $\omega_B = \sqrt{2gk_o}$ where $\underline{k}_o$ is the radar wavenumber.

(a−) - receding first order Bragg line at $-\omega_B$.

(b) - second order continuum.

(c) - corner reflector at $\pm 2^{3/4} \times \omega_B$.

(d) - 2nd harmonic of ocean waves of wavenumber $\underline{k}_o$ at $\pm \sqrt{2} \times \omega_B$.

2.1 Method

The method used by Barrick involves a Fourier series expansion with random coefficients of the sea surface height. A similar expansion but with unknown coefficients is applied to the electromagnetic field and the coefficients are determined by enforcing boundary conditions at the surface. The fields at the boundary and hence the coefficients in the electromagnetic field expansion are expanded in perturbation series with the following constraints:

1 - The height of the surface must be small in terms of the radio wavelength.
2 - Surface slopes must be small compared with unity.
3 - The impedance of the surface must be small compared with the free space wave impedance.

All these conditions are satisfied at HF except possibly in very extreme seas. To first order in these perturbation parameters the solution describes exactly the Bragg resonant interaction described above. To second order, electromagnetic interactions of the corner reflector type are described. The random Fourier series representation of the sea surface facilitates the derivation of equations describing second order ocean wave interactions in terms of the first order (linear) ocean wavefield. A perturbation expansion is applied to the Fourier coefficients with the same constraints as before. Clearly (although the algebraic complication is all too obvious!) the two expansions can be carried out simultaneously and the second order electromagnetic field due to the second order ocean wavefield determined.

2.2 Solution

The expression Barrick found is presented in various ways in the literature. I will use the following formulation:

$$\sigma(\omega) = \sigma^{(1)}(\omega) + \sigma^{(2)}(\omega)$$

σ = backscatter cross-section. The superscript denotes the order of the contribution.

$$\sigma^{(1)}(\omega) = 2^6 \pi k_o^4 \sum_{m=\pm 1} S(-2m\underline{k}_o)\,\delta(\omega - m\omega_B) \quad (1)$$

$\underline{k}_o$ = radar wavenumber
ω = Doppler frequency
$\omega_B = \sqrt{2gk_o}$ = Bragg resonant frequency
$S(\underline{k})$ = sea-wave directional spectrum

$$\sigma^{(2)}(\omega) = 2^6 \pi k_o^4 \sum_{m,m'=\pm 1} \int_0^\infty \int_{-\pi}^{\pi} |\Gamma|^2 S(m\underline{k})\, S(m'\underline{k}') \times \delta(\omega - m\sqrt{gk} - m'\sqrt{gk'})\, k\, dk\, d\theta \quad (2)$$

$\underline{k}, \underline{k}'$ = wavenumbers of two interacting waves where $\underline{k} + \underline{k}' = -2\underline{k}_o$
k, θ = polar coordinates of $\underline{k}$
Γ = coupling coefficient = $\Gamma_H + \Gamma_{EM}$

Γ_H = hydrodynamic coupling coefficient

$$= -\frac{i}{2}\,\frac{k + k' - (kk' - \underline{k}.\underline{k}')}{mm'\sqrt{kk'}}\,\frac{(\omega^2 + \omega_B{}^2)}{\omega^2 - \omega_B{}^2} \qquad (3)$$

Γ_{EM} = electromagnetic coupling coefficient

$$= \frac{1}{2}\,\frac{(\underline{k}.\underline{k}_o)(\underline{k}'.\underline{k}_o)/k_o^2 - 2\,\underline{k}\,.\,\underline{k}'}{\sqrt{k.k'} + k_o\Delta} \qquad (4)$$

Δ = normalised electrical impedance of the sea surface. To inspire confidence that equations (1) and (2) are indeed solutions to the problem, figure 2.1 shows (b) a measured Doppler spectrum compared with (a) a spectrum obtained by evaluating equations (1) and (2) using ocean wave spectra that roughly model wave conditions at the time of observation. There is an indication of a small current component away from the radar in the measured spectrum. This spectrum was obtained with a radar with a broad beamwidth and 7.5km range resolution and so even if the current was constant in magnitude and direction in the radar cell the radar is integrating first order returns from different directions over the beamwidth leading to a broadening of the peak. Any current variability over the measuring cell and any third order ocean wave interactions will also contribute to broadening. The calculation does not take into account any beamwidth or current effects and is restricted to second order interactions. Thus the amplitude of the first order peak is determined only by equation (1) and the frequency resolution and is larger relative to the second order amplitude then in the measured spectrum. Differences in detail in the second order spectrum are due to differences between the model and the actual wave spectrum and will be discussed later.

In evaluating equation (2) it is possible to look in more detail at the way the integral is built up - to see for example which interactions are more important cases and whether any interactions are important in all cases. One reason for doing this will become clearer when I discuss using equation (2) to determine the ocean wave spectrum. For the moment we will concentrate on understanding the manner in which various interactions contribute to the integral.

Because of the delta function in equation (2), for a given value of ω the integral is restricted to an integral round a curve in the k,θ plane given by $\omega = m\sqrt{gk} + m'\sqrt{gk'}$. These integrations contours for $\omega > \omega_B$ are shown in figure 2.2. The frequencies are normalised by ω_B. Also shown on the figure are the relative amplitudes of various wave combination contributions to the integral. The corner reflector is indicated as are strong hydrodynamic interactions between waves with the same wavenumber but different direction. This type of interaction has a maximum for colinear waves

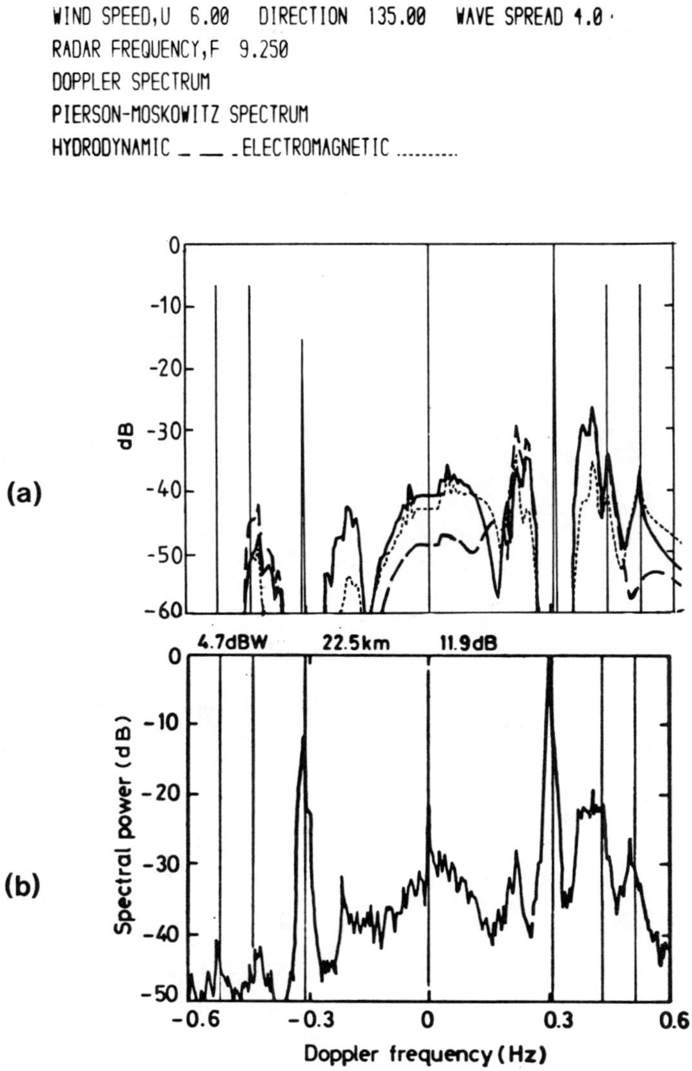

Figure 2.1 Observed Doppler spectrum (b) compared with simulated Doppler spectrum (a) obtained by using the sea wave spectrum inferred from (b) as input to the simulation calculation. (a) is plotted to -60dB to account for the difference in peak amplitude as discussed in the text.
Note that the vertical and horizontal scales are slightly different.

WAVENUMBER SPACE, DOPPLER FREQUENCY CONTOURS
CONTRIBUTIONS TO THE SECOND ORDER INTEGRAL * MAX o 50% x 75% ◆ 90%
PIERSON-MOSKOWITZ SPECTRUM

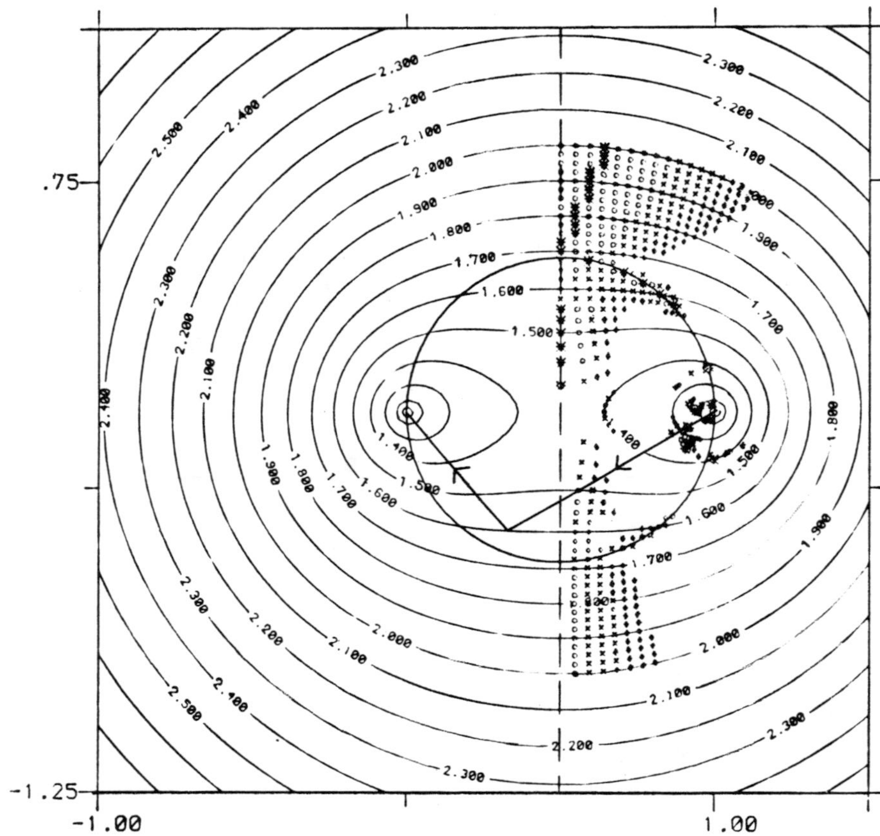

Figure 2.2 Normalised Integration Contours for $\omega > \omega_B$
An example of an interaction between two waves producing a contribution at $1.6 \times \omega_B$ is shown. Vertical and horizontal axes are the wavenumber components which have been normalised with respect to the first order Bragg wavenumber.
The solid circle through both wavenumber origins indicates wavenumber combinations producing corner reflection. The vertical dashed line indicates hydrodynamic interactions between waves of the same wavenumber.

with the same wavenumber at $\sqrt{2} \times \omega_B$. The radar is 'seeing' the second harmonic of waves of the radar wavenumber $\underline{k}_o$. This local maximum is often masked in measured Doppler spectra by interactions involving the longer more energetic waves at smaller Doppler frequencies. (See figure 2.1(b)) Equation (2) has been evaluated for a large variety of wind and wave conditions and in all cases similar patterns are found for $\omega > \sqrt{2} \times \omega_B$. Variations due to the different conditions only occur for $\omega < 1.4\ \omega_B$. It should be noted that as the radar frequency increases for a given wind speed (or vice versa) more interactions are observed with Doppler shifts close to ω_B. The contributions to the other regions of the spectrum: $\omega < -\omega_B$ and $-\omega_B < \omega < \omega_B$ are more dependent on conditions.

Finally, figure 2.1(a) shows the resulting spectrum for the integration illustrated in figure 2.2. The way in which the hydrodynamic and electromagnetic interactions contribute to the amplitude of the spectrum is indicated.

The problem we shall be returning to later is how to invert integral equation (2) to evaluate $S(\underline{k})$.

3. RADAR INSTALLATIONS AND TECHNIQUES

There are skywave and groundwave programs in Europe, North America and Australia, but I shall be concentrating here on the British program. Two groups have been collaborating in this field for several years. One of these, at Rutherford and Appleton Laboratories is now concentrating mainly on skywave work and the other, at Birmingham, is concentrating on groundwave work.

3.1 Skywave radar

This is located near Cricklade in Wiltshire. It operates monostatically with 100kW., 500-microsecond pulses giving 75km range resolution and uses an electronically steerable 49-element antenna with a 6 degree beamwidth. A map indicating the area of coverage is shown in figure 3.1. On-line computer processing has been developed which allows real time processing of wind direction. A year long observation program is currently underway to look at both the accuracy of wind direction estimates and at the proportion of time when good data can be obtained. This is necessary because skywave radar operates by reflecting from the ionosphere which is itself unstable and can introduce its own Doppler shifts or introduce multiple propagation paths making it difficult to interpret the Doppler spectrum. Signal processing techniques have been developed by the US team (8) to minimise ionospheric distortion and these are being implemented.

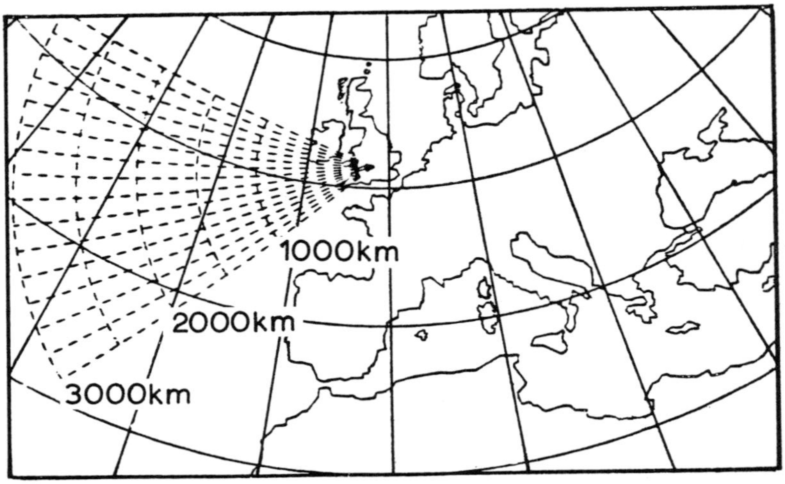

Figure 3.1 Skywave radar coverage

3.2 Ground-wave radar

Our ground-wave work is based at East Blockhouse, Pembrokeshire. The map in figure 3.2 indicates the location and area of coverage. Until recently we were only operating with an omnidirectional 1.95MHz pulse radar which was only able to give qualitative information about sea-state and in fact at this frequency was mostly only measuring swell to first order with very little second order return (see figure 3.3).

We have been experimenting with a new technique for monostatic ground-wave work, frequency modulated continuous wave (FMCW) radar (9,10) and have recently obtained results that have proved the technology. The FMCW radar operates by sweeping repetitively over a frequency band of tens of kilohertz, the swept bandwidth determining the range resolution. In general, ground-wave pulse radars, for sea-state measurement, are restricted to less than 50km range whereas we were able to get good signal to noise out to over 80km in our first experiments with very low transmitter powers (see figure 3.4). We envisage obtaining results out to 150km or so within the next few months.

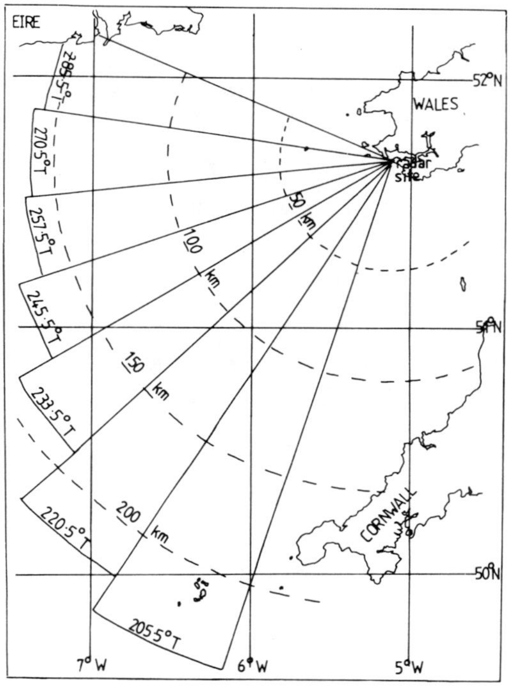

Figure 3.2 Groundwave coverage indicating beamwidths.

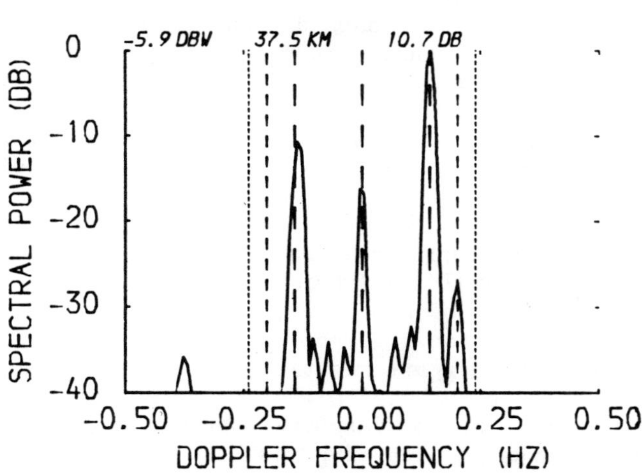

Figure 3.3 A 1.95MHz Doppler spectrum with very little second order structure.

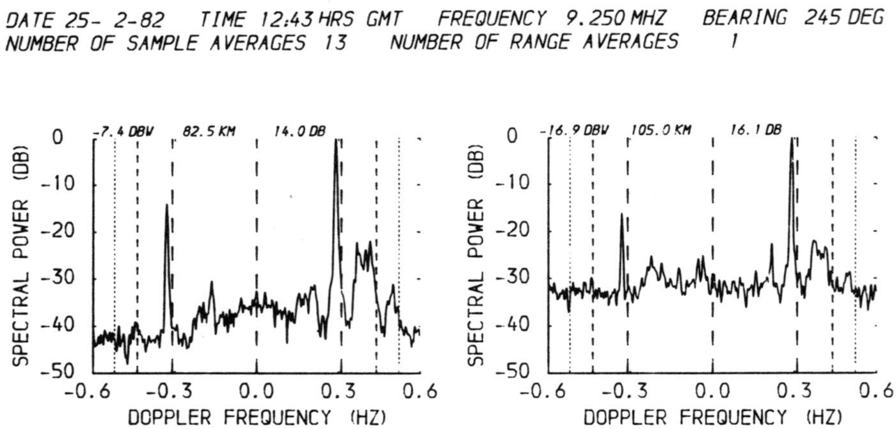

Figure 3.4 9.25MHz spectra measured at long ranges

We will be operating over a range of frequencies from 7 - 15MHz with very easy switching from one to another so that hopefully we can avoid interference as far as possible. However, FMCW radars can operate in interference conditions which would be unworkable with pulse radar. A further advantage with FMCW is the ease with which range resolution can be changed. In pulse radar this depends on the pulse width and in general only one or two choices are available for this. The swept bandwidth can be changed very easily with very fine resolution. We operate generally with 7.5km resolution corresponding to the pulse width of the 1.95MHz pulse radar but we have done experiments with as little as 1.5km resolution as can be seen in figure 3.5.

Most of the oceanographic results I will be discussing later have been obtained using this FMCW radar.

4. DATA PROCESSING

The received signal from the ground-wave radar is recorded on magnetic tape for processing on the computer facilities at Birmingham University. Because of the nature of the signal from an FMCW radar two techniques of analysis are possible: either one long or two shorter fast Fourier transforms. In the first case the signal for a number of consecutive sweeps is Fourier analysed producing a composite spectrum containing Doppler spectra for the individual ranges

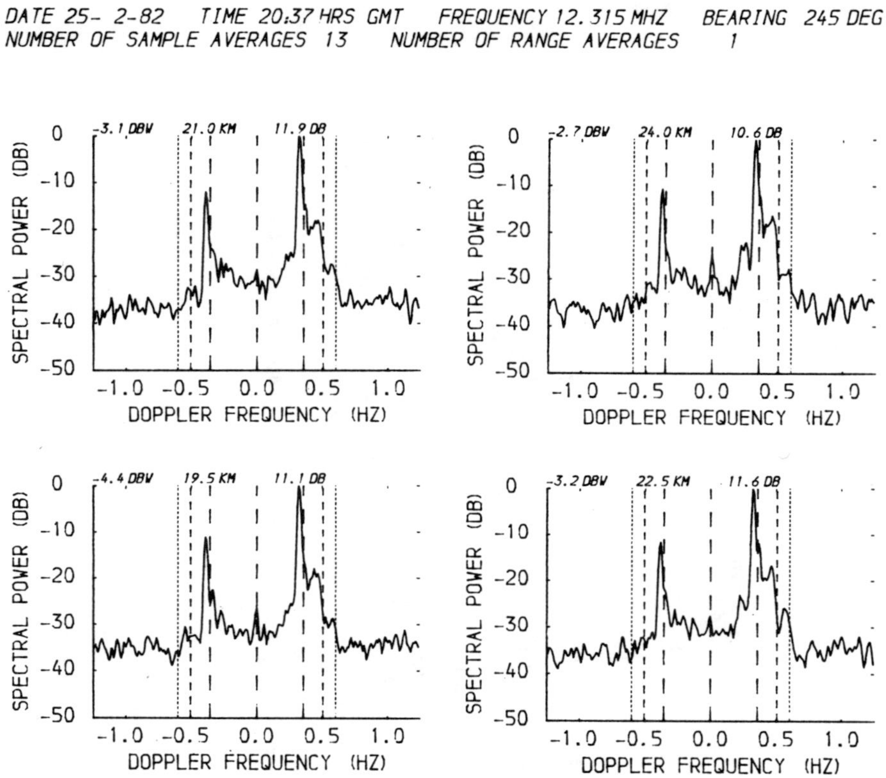

Figure 3.5 12.315MHz spectra measured with 100kHz swept bandwidth hence 1.5km resolution.

separated by the sweep repetition rate (see figure 4.1). This process is used online with a spectrum analyser at East Blockhouse to provide a check on the data quality. At Birmingham we use the second procedure whereby each sweep is analysed separately with an FFT. Each component of the result corresponds to a particular range so that this is equivalent to a range gating procedure in pulse radar. The second FFT is performed for each range for which data is required in exactly the same way as pulse data is processed. Since the range information is determined from the first FFT there is a compromise between range sidelobe levels and reduction in range resolution through time domain weighting. If necessary this can be compensated for by increasing the swept bandwidth and hence the nominal range resolution.

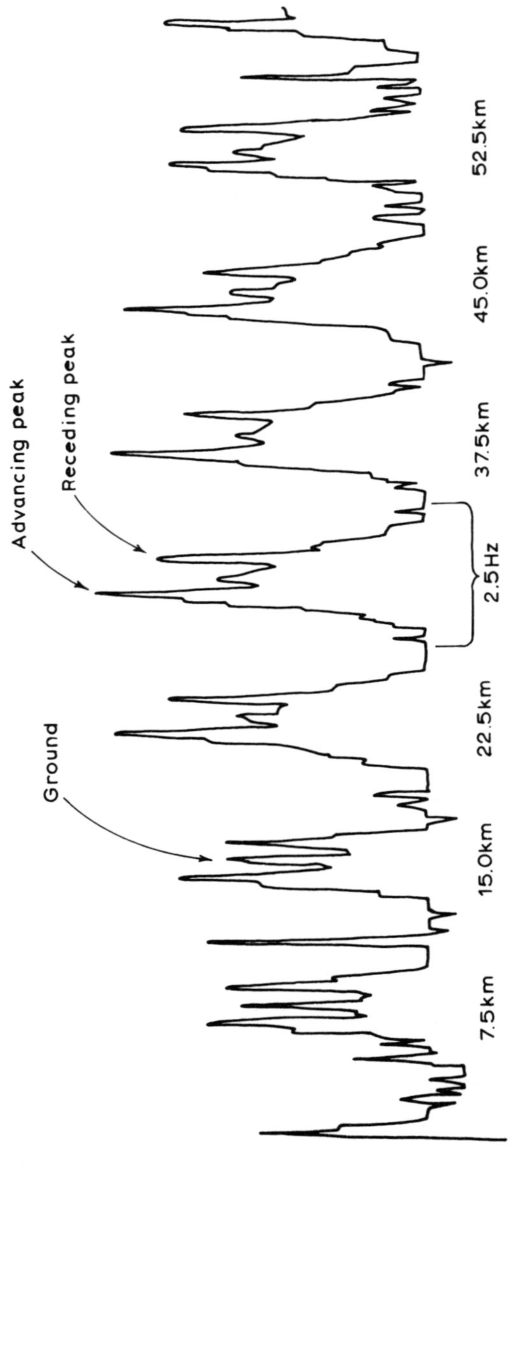

Figure 4.1 9.25MHz spectra analysed on site with a spectrum analyser. Several ranges are displayed simultaneously.

Because the radar is measuring a random process it is necessary to average several spectra to estimate the ensemble average. There are conflicting requirements here of Doppler resolution which requires long records and the need to obtain several spectra before meteorological and oceanographic conditions change the statistical characteristics of the sea surface. We have been collecting data sets of 20 mins to 1 hour in length in order to determine the optimum choice of record length and number of averages. We use a Blackman-Harris window (11) for the second FFT which allows us to carry out the spectral analysis with data sets that overlap by as much as 75% thus increasing the number of averages in a finite time. With 30 minute data sets we have been using integration times of 3 - 7 minutes in length and thus obtain a minimum of 13 spectra to average.

A further check on the processing parameters will be available when we do a detailed comparison between our oceanographic measurements using data from both ground and skywave radars and ground truth measurements.

5. OCEANOGRAPHIC PARAMETER EXTRACTION

5.1 Currents

The measurement of currents with HF radar is very straightforward and has been described by Shearman (1). Figure 5.1 shows a Doppler spectrum with a current displacement in this case due mainly to the tide. The situation is complicated when there is a lot of current variability in a radar resolution cell in which case several peaks can sometimes be seen (see figure 5.2). A number of groups are working on current measurement and a program is underway at the moment to produce a transportable current measuring system in this country.

5.2 Wind direction

This is an indirect measurement. It depends on the assumption that waves generated by the wind have a symmetrical directional distribution about the wind direction. It is of course necessary that the Bragg resonant wave is generated by the wind. A directional distribution of the form

$\cos^s \frac{(\theta - \theta_w)}{2}$ is assumed where θ_w is the wind direction and s is a spreading parameter usually assigned the value 4.

The advancing and receding Bragg waves this have amplitude proportional to $\cos^s \frac{(\theta_a - \theta_w)}{2}$ and $\cos^s \frac{(\theta_a + \pi - \theta_w)}{2}$ respectively and θ_w can be determined by looking at the ratio of their amplitudes. A

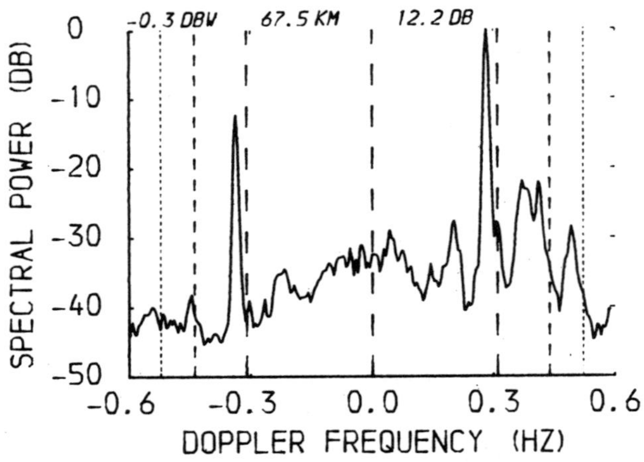

Figure 5.1 Current displacement indicating a current component away from the radar.

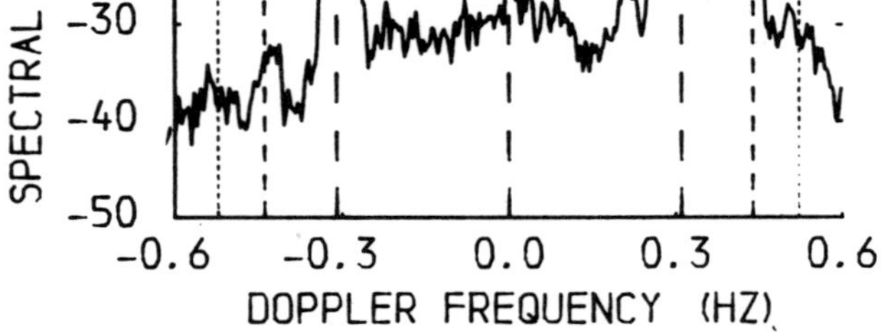

Figure 5.2 Multiple first order peaks indicating a complex current structure in this case probably due to sub-cell scale islands.

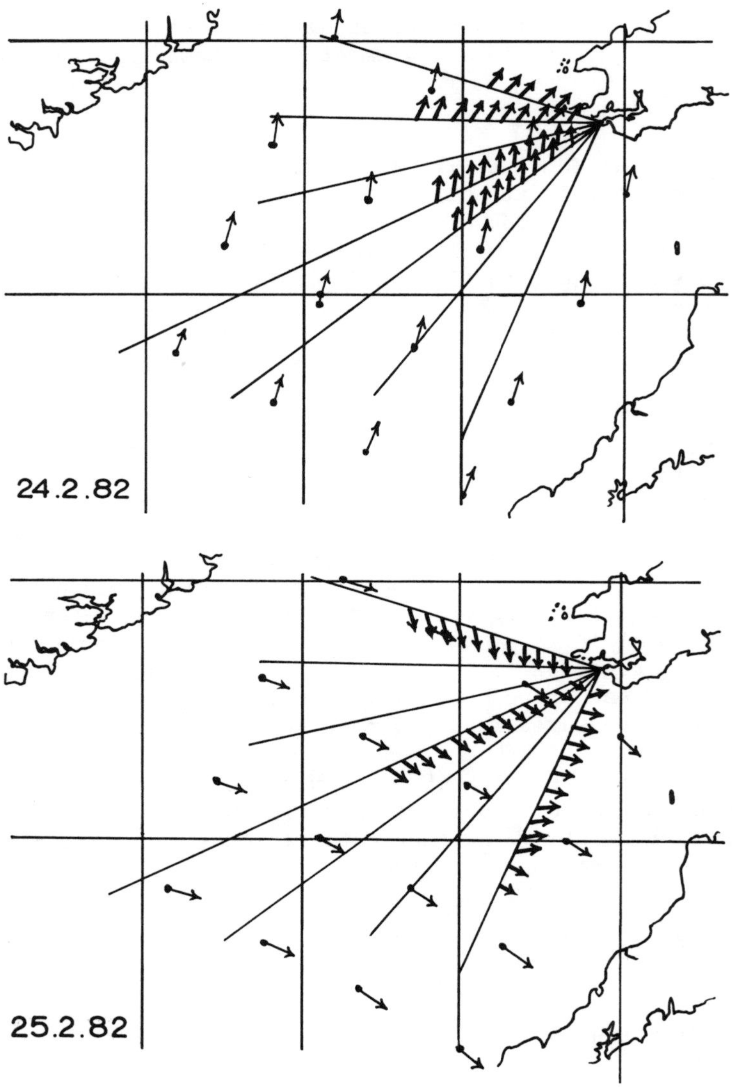

Figure 5.3 Radar-deduced wind directions (heavy arrows) compared with meteorological office analysed winds. The discrepancies in the lower picture are due to the multiple peak structure on this bearing.

left-right ambiguity with respect to the radar look direction exists and must be resolved by continuity arguments. Examples of ground-wave wind direction measurements are shown in figure 5.3.

5.3 Waves

One technique for determining the wave directional spectrum, S($\underline{k}$), would be to look at the first order returns for a range of frequencies and radar look directions. A less cumbersome and less time consuming procedure would be to use the second order returns and hope to solve equation (2) for S($\underline{k}$). The first order return can then be used as a calibration.

Unfortunately, equation (2) is a nonlinear Fredholm integral equation of the first kind for which no standard inversion procedures are available. In fact this type of equation, even if it were linear, is very difficult to solve in that it is ill-posed which means that small errors in the Doppler spectrum can lead to large errors in the inverted wave spectrum. Some success has been achieved by approximating the equation and thus finding solutions for a rather restricted set of conditions. Another approach is to develop empirical expressions describing the relationship between wave parameters and second order structure. Another technique is to assume a model for S($\underline{k}$) the parameters of which are determined by integrating equation (2) and fitting the integral to the measured spectrum. These approaches will be discussed and compared using Doppler spectra measured with ground-wave radar.

Linearisation. Barrick (5) used a series of approximations and transformations to reduce the integral equation to a closed form expression relating the backscatter cross-section to the non-directional wave frequency spectrum.

$$S_t(|\omega - \omega_B|) = \frac{4\,\sigma^{(2)}(\omega)/W(\omega/\omega_B)}{k_o^{\,2}\int_o^{\infty}\sigma^{(1)}(\omega)\,d\omega}$$

W(ω) is a weighting function which is the average value of $|\Gamma|^2$ along a constant Doppler frequency contour.

In developing this expression he linearised the integral equation. The linearisation implies that one of the interacting wave vectors is approximately the Bragg wave vector and the other is approximately constant in amplitude around an integration contour. Reference to figure 2.2 will make clear the limited validity of this approximation at the frequencies that we have been working with so far. We have found that although it gives a reasonable picture of the energy distribution in the frequency spectrum the amplitudes are too large and hence the waveheights predicted by integrating the expression are too large. Barrick (6) has suggested techniques for improving the accuracy which involve making use of systematic differences between buoy measured and radar-deduced waveheights and mean periods. As the radar frequency increases this approximation is appropriate for a larger range of wave frequencies. However, Lipa et al (12) have indicated that measurement becomes more difficult because the first order line tends to broaden and at sufficiently high wind speeds the second order structure becomes absorbed in the first order return.

It turns out that for the case of monochromatic, unidirectional swell, well separated in frequency from the wind-wave spectrum, the linearised equation has a simple solution given by Lipa & Barrick (13). They showed that four sharp spikes appear in the spectrum at frequencies given by:

$$\omega_{m,m'} = m\omega_s + m'\left[\omega_B^4 + 2m\,\omega_s^2\,\omega_B^2\cos\Theta_s + \omega_s^4\right]^{1/4}$$

ω_s, Θ_s are the frequency and propagation direction of the swell. The amplitude of the swell can be found by dividing the amplitude of the peaks by the amplitude of the first order lines. Figure 5.4 is a skywave measured Doppler spectrum with four such peaks indicated. The amplitude, ω_s and Θ_s of swell found for this case showed good agreement with buoy measurements in the area. This technique has been used successfully by a number of people.

Numerical techniques developed by Lipa (14) have been used by Barrick and Lipa (15) to look for more general solutions to the linearised equations. They make use of a fourier series decomposition of the ocean wave spectrum the coefficients of which are the same as those measured by a pitch and roll buoy. Examples of the applications of this technique can be found in (12).

Non-linear Inversion. Lipa (14) made the first attempt to use numerical techniques that have appeared in the literature to deal with ill-posed problems, to solve the full inversion problem. She split the integration region into two parts. In one, both interacting wave vectors were assumed to be in the equilibrium regime with amplitudes proportional to k^{-4} and the equation is solved iteratively for the directional distribution. In the other region one wave is still assumed to be in the equilibrium regime and the equation is solved for the amplitude of the longer wave components. All wave components are assumed to have the directional distribution determined using the first region.

The problem with this method is that long wave components do not in general have the same directional distribution as short wave components. This is particularly true if the long wave components are swell rather than wind waves.

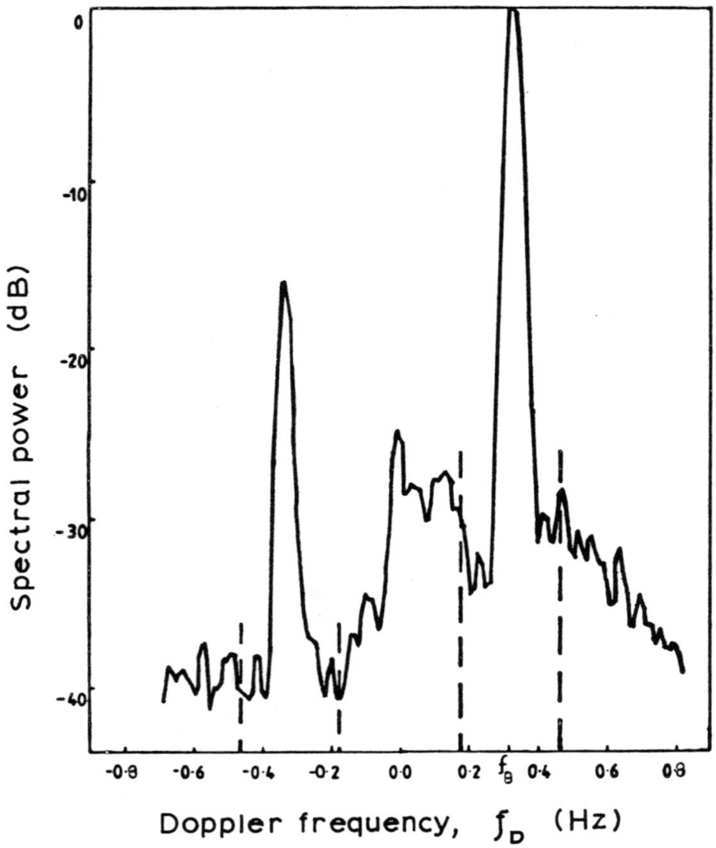

Figure 5.4 A skywave measured Doppler spectrum with four peaks (between the dashed lines) interpreted as swell.

Model Fitting. Model fitting techniques can be used for the windwave part of the spectra assuming JONSWAP or Pierson-Moskowitz type wave spectra. Maresca & Georges (16) have presented the result of fitting a JONSWAP spectrum to skywave data and show good agreement with buoy measurements. They have also suggested a simple relationship to determine waveheight, h as a function of the second order return. They looked for a relationship of the form:

$$k_o h = a R^b$$

where R is an estimate of the ratio of the power in the second order part of the spectrum to that in the first. The constants a, b are determined as a best fit to calculated values of h and R using equation (2) and various model windwave spectra. They found that $a = 0.8$ and $b = 0.6$ gave the best fit and have used this method with some success with skywave data.

Forget et al (17) have looked for characteristic frequencies in the Doppler spectrum which allow the determination of a long-wave cut-off frequency and a dominant wave frequency in the ocean wave spectrum. They have suggested techniques for extracting these quantities and for using the dominant wave frequency to determine waveheight.

All these techniques have been developed with the assumption that the ocean wavefield is locally wind driven. The presence of swell, especially it it cannot be easily identified as suggested above, is likely to reduce the accuracy of the methods.

A more general longwave technique has been developed by Lipa & Barrick (18) which assumes a parametrisation of the long ocean wave spectrum:

$$S(\underline{k}) = a(k) \cos^{s(k)} \frac{(\theta - \theta(k))}{2}$$

where a(k), s(k) and θ(k) are determined by carrying out the integration and fitting the solution in a least squares sense to the measured Doppler spectrum. This technique is designed to apply in the linearisation region but the contribution from the shorter wave is generalised slightly to include the wavenumber variation along a constant Doppler contour assuming also that it is short enough to be in the equilibrium regime thus:

$$S(\underline{k}') = \left(\frac{2k_o}{k'}\right)^4 S(-2\underline{k}_o)$$

The technique has been generalised further at Birmingham by allowing the shorter wavenumber direction to vary along a constant Doppler contour thus:

$$S(\underline{k}') = \frac{\alpha}{k'^4} \cos^4 \frac{(\theta - \theta_w)}{2}$$

In this case the first order return can still be used to determine α and the first order ratio must be used to determine wind direction.

In applying these techniques the assumption is made that the same values of a(k), s(k), θ(k) apply for all values of k along a constant Doppler contour. This assumption becomes increasingly inaccurate as k increases. As a first attempt to compensate for this we have applied the technique to constant Doppler contours close enough in frequency for the corresponding wavenumber bands to overlap and have then applied a four point running average to the solution.

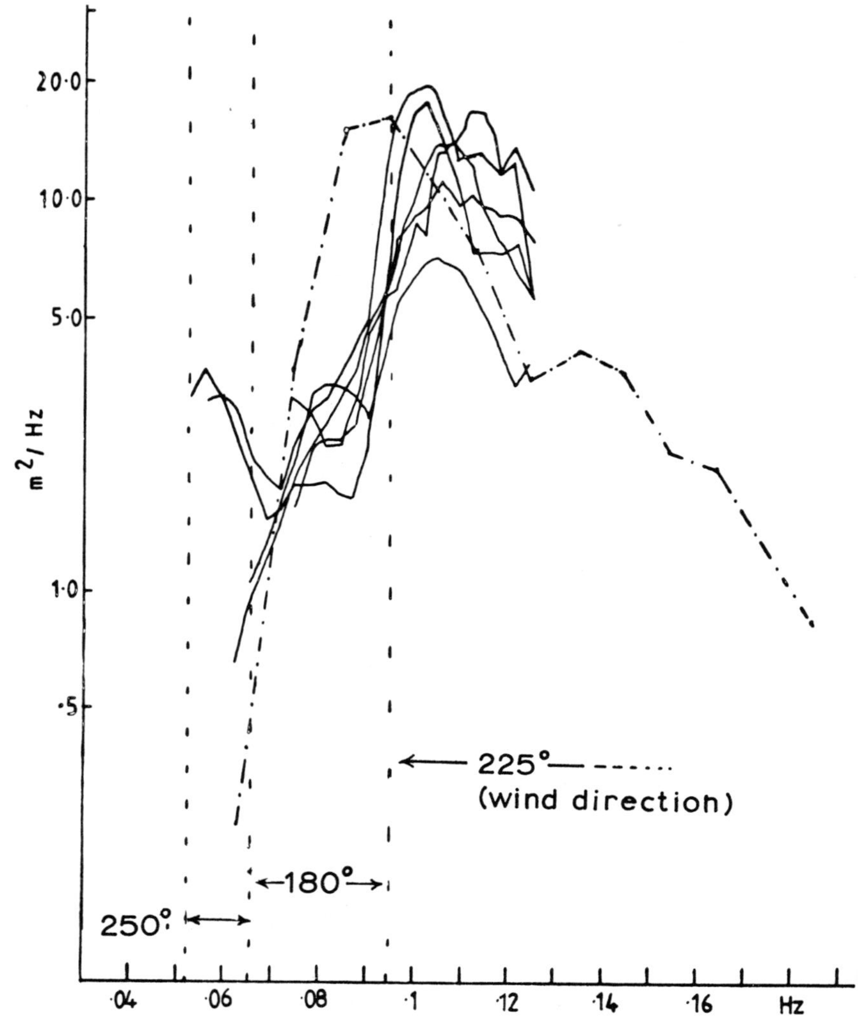

Figure 5.5 Wave frequency spectra
—— measured by 9.25MHz radar.
-·-·- measured with a waverider buoy at the Isles of Scilly.
Also shown are the principal propagation directions (relative to the radar beam) measured by the radar.

Experimental Results. In February 1982 we carried out our first successful trials with the FMCW radar. The data discussed below was collected soon after the passing of a vigorous cold front after several days of calm conditions under the influence of a ridge of high pressure.

Wave frequency spectra and principal propagation direction obtained using the technique described above are shown in figure 5.5. They are compared with buoy measurements from the Isles of Scilly. Unfortunately local buoy measurements were not available. The maximum amplitudes are similar but the frequency of this maximum is lower in the buoy measurement. However, the radar measurements which are from six consecutive cells (22.5→60km offshore) are reasonably consistent in both amplitude and direction. Such consistency was not obtained using the Lipa & Barrick method.

As a check on the solution for one range cell, the inverted longwave spectrum has been combined with a Pierson-Moskowitz wind wave spectrum appropriate for the measured wind speed and direction and substituted into the full non-linear equation. The resulting Doppler spectrum is the one shown in figure 2.1(a) where it is compared with the corresponding measured spectrum. The general pattern shows good agreement. The noticeable difference in the vicinity of the $+\sqrt{2} \times \omega_B$ singularity can be attributed to two possible mechanisms. The inversion is only carried out to $1.4 \times \omega_B$ so that longwave components contributing to larger Doppler frequencies are not determined. Alternatively, the wind speed used could be an underestimate. With a higher wind speed the Pierson-Moskowitz spectrum would have longer wave components.

Table 1 compares waveheight measurements using a number of the techniques described above, conventional measurements and results from the Meteorological Office wave model.

A number of outstanding theoretical problems are being studied using the full non-linear simulation program. Of these the effects of finite beamwidth and the presence of currents are thought to be particularly important. In addition we are working on more general inversion techniques that will allow the determination of the full wave spectrum.

ACKNOWLEDGEMENTS

The experimental work described here would not have been carried out without the dedicated work of many individuals both within this department and at the field sites. In particular I would like to acknowledge the work of G.D. Burrows who both built the FMCW radar and collected the data and of S. Theodoridis and M.D. Moorhead who developed the software for the FMCW analysis.

TABLE 1

Waveheights, Hs.

SOURCE		Hs
Isles of Scilly	waverider	3.49m
	Met Office model	2.9 m
Radar grid	Met Office model	2.47m
	St.Gowan lightship	2.4 m
9.25 MHz	Barrick (1977)	3.8 m
	Maresca and Georges	3.4 m
	Barrick and Lipa (longwaves)	3.1 m
	Birmingham (longwaves)	2.4 m
12.315 MHz	Birmingham (longwaves)	2.7 m

REFERENCES

(1) Shearman,E.D.R.: 1981, in "Remote Sensing in Meteorology, Oceanography and Hydrology", ed. A.P.Cracknell, pp. 312-334.
(2) Barrick,D.E.: 1971, Radio Science 6, pp. 517-533.
(3) Barrick,D.E.: 1972, IEEE Transactions on Antennas and Propagation, AP-20, pp. 2-10.
(4) Barrick,D.E.: 1972, in "Remote Sensing of the Troposphere", ed. V.E.Derr, chap. 12.
(5) Barrick,D.E.: 1977, Remote Sensing of Environment 6, pp. 201-227.
(6) Barrick,D.E.: 1977, Radio Science 12, pp. 415-424.

(7) Barrick,D.E., and Weber,B.L.: 1977, Journal of Physical Oceanography 7, pp. 11-21.
(8) Georges,T.M., Maresca,J.W., Carlson,C.T., Riley,J.P., Jones,R.M., Westover,D.E.: 1981, NOAA Technical Memorandum ERL WPL - 73.
(9) Shearman,E.D.R., and Unsal,R.R.: 1980, IEE Conference Publication 188, pp. 103-107.
(10) Shearman,E.D.R., Wyatt,L.R., Burrows,G.D., Moorhead,M.D., Bagwell,D.J., and Sandham,W.A.: 1982, IEE Conference Publication 216.
(11) Harris,F.J.: 1978, Proceedings of the IEEE 66, pp. 425-436.
(12) Lipa,B.J., Barrick,D.E., and Maresca,J.W.: 1981, Journal of Geophysical Research 86, pp. 4089-4102.
(13) Lipa,B.J. and Barrick,D.E.: 1980, Radio Science 15, pp. 843-853.
(14) Lipa,B.J.: 1977, Radio Science 12, pp. 425-436.
(15) Barrick,D.E., and Lipa,B.J.: 1979, in "Ocean Wave Climate", ed. M.D.Earle and A.Malahoff, pp. 153-201.
(16) Maresca,J.W., and Georges,T.M.: 1980, Journal of Geophysical Research 85, pp. 2759-2771.
(17) Forget,P., Broche,P., De Maistre,J.C., and Fontanel,A.: 1981, Radio Science 16, pp. 917-925.
(18) Lipa,B.J. and Barrick,D.E.: 1982, NOAA Technical Report ERL 420-WPL 56.

CHAPTER 10

REMOTE SENSING VIA METEOR TRAILS

Ray E. Leader, Pres. & Jon B. Jolly, Pres.

Meteor Communications Corp., Jon B. Jolly, Inc.

INTRODUCTION

The technique of using ionized meteor "trails" to reflect (or re-radiate) radio signals in the VHF spectrum has been known for fifty years or more. This phenomenon is referred to in the text as meteor burst communications (MBC). The development of the microprocessor and other solid state devices during the 70's has made the use of these physical phenomena both practical and useful.

After Canada's National Research Council and various American institutions evaluated this communication system's approach during the middle 50's, no further efforts of consequence went forward with the exception of a few U.S. military applications in the 60's and early 70's. In the late 70's environmental data collection applications spurred the development of remote data acquisition platforms, using MBC for the remote telemetry links. Application of the microprocessor has made MBC economically viable for the remote sensing of environmental or other data.

In short, meteor burst communication systems are computers "talking" to each other in short bursts during time periods of between about 50 milliseconds and several seconds. Voice and other continuous link requirements are not practical in MBC.

In operation, the system works as follows: The master or interrogating station transmits a continuous, coded signal, at suitable frequencies. When a meteor trail appears in the proper loca-

A. P. Cracknell (ed.), Remote Sensing Applications in Marine Science and Technology, 207–217.

tion, it reflects that signal to a receiving remote or transponding station. More than a billion meteors large enough to give useable trails enter the earth's atmosphere every day.

The main propagation mode of meteor burst is reradiation of the ionized trails left by micrometeors as they enter the earth's atmosphere, and are vaporized. The remote station decodes the signal, recognizes that it has been addressed, and turns on its transmitter to transmit a signal back along the same path to the master station. (Refer to Figure 1). Information can be sent in either direction until diffusion reduces the electron density in the trail to a value too low to sustain communication. The maximum length of a transmission is on the order of 2000 kilometers. This distance is determined by the height of the meteor's ionized path, relative to the curvature of the earth.

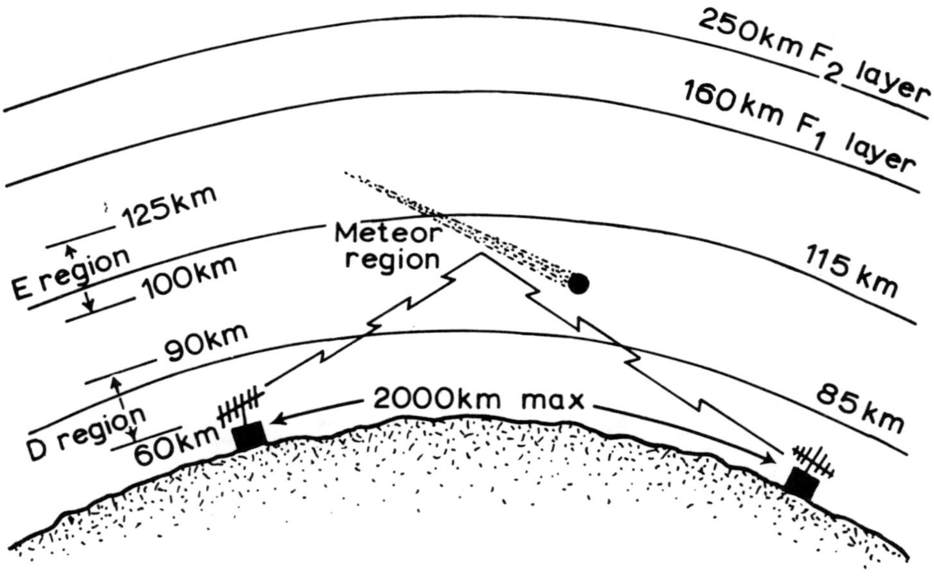

Figure 1. Meteor burst communication system-system geometry.

The meteor trail of interest here will have a useful duration of a few hundred milliseconds, and wait times between such trails, suitably located, can range from seconds to minutes depending on diurnal and seasonal variations, along with system design factors. Hence the transmissions consist of bursts of tens to hundreds of characters, separated by periods of silence. Thus, many links can share a common transmission frequency, providing efficient use of the spectrum.

The above exchanges of information can be in either direction (i.e. full duplex); it can consist of short messages such as sensor readout, coded messages of up to several hundred characters, text messages of a few words, or long messages achieved by splicing together successive bursts. In the latter applications, average data rates of up to about one hundred words per minute may be realized.

A MBC system is ideally suited for operation in harsh environments for the collection of hydrological, meteorological and oceanographic data. Power requirements of the remote units are low and can be met by small solar panels in conjuction with a storage battery.

FUNDAMENTALS

Meteors are defined as extraterrestrial objects that travel in elliptical orbits around the sun, and at some point in these orbits enter the earth's atmosphere.

As the meteor enters the upper atmosphere it is traveling at speeds of 10 to 75 km/second, possessing a large amount of kinetic energy. As it begins colliding with air molecules, much of this kinetic energy is converted into heat which vaporizes atoms from the surface of the parent meteor. These vaporized atoms, which are restricted by air molecules as they progress further into the atmosphere. The result is the transformation of kinetic energy into the energy of ionization which strips electrons from the atoms leaving a trail of positive charged ions and free electrons. It is the electrons that reflect or re-radiate radio waves.

The ionization is distributed in the form of a long, thin paraboloid with the particle at the head. The electron line density (electrons/meter) is proportional to the mass of the meteor and ranges from 10^{10} electrons/meter to about 10^{18} electrons/meter. Typical trails are 25 km long and have an initial radius of about 1 meter.

There are two general classes of meteors; shower meteors and sporadic meteors. Shower meteors can be visually spectacular and are well known; however, they account for only a tiny fraction of the meteors present. The vast majority of meteors used in radio work are sporadic and they move in random orbits around the sun. They are called sporadic because their radiants and times of occurrence are random and cannot be predicted as shower meteors can. It is the intersection of the meteor's orbits with the earth's orbit that causes the meteors to enter the earth's atmosphere. The intersection of these orbits are not uniformally distributed but are concentrated so as to produce a maximum of intersections in August and a minimum in February, with about a 3:1 variation. This is illustrated in Figure 2.

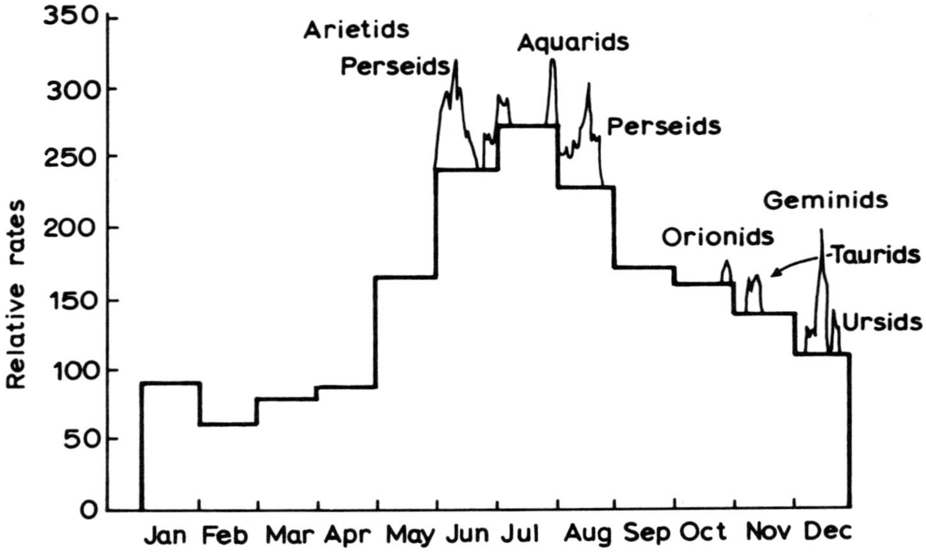

Figure 2. Seasonal variation of meteor rates.

The rate of incidence of sporadic meteors also is dependent on the time of day, with the morning hours being more active. On the morning side of the earth, meteors are swept up by gravity and, in addition, by the forward motion of the earth in its orbit around the sun. On the evening side, the meteors reaching the atmosphere are those which overtake it. A daily variation of about 4:1 occurs.

Practically, this means that, from a location in the northern hemisphere, the variation of meteor incidence on a July morning, to an evening in February, can be as much as 12:1. In the southern hemisphere, these months would be reversed, indicating that the annual variations are a function of the earth's tilt.

SIGNAL CHARACTERISTICS

Radio signals received on a meteor-burst link are of two forms, the most prevalent of which is the underdense trail re-radiation. In this case, the trail is characterized by a relatively low electron line density ($9<10^{14}$ electrons/meter). An underdense trail does not reflect energy; indeed, the radio wave passes through the trail exciting individual electrons. These excited electrons act as small dipoles, re-radiating the signal at an angle equal but opposite to the incident angle of the trail and radio signal.

Signals from such underdense trails rise to an initial peak value in microseconds and have useful durations from perhaps milliseconds to as long as a few seconds. Figure 3 is a typical example of the distribution of the burst time durations at 50 MHz. The decay in signal strength is due to destructive phase interference caused by the radial expansion of diffusion of the trail's electrons.

Meteor trials with electron line densities greater than 10^{14} electrons/meter actually reflect the signal and are called overdense. Here the line density is so great that radio signals cannot penetrate them and are reflected instead of being re-radiated. There are no distinctive signal patterns associated with overdense trails except that the signals usually reach higher amplitudes than underdense meteors and last much longer. Signal fading sometimes occurs due to reflections from two parts of the trial constructively and destructively interfering. Also, ionospheric winds blow the trails around causing occasional fading. As overdense signal reflections are unprdictable in nature, no useful theory has been developed to describe them.

Other means of communication over a MBC system are Sporadic "E" and airplane reflections at distances less than 150 miles. The airplane reflections usually last many seconds and are characterized by a considerable fading. The Sporadic "E" layer, which provides a reflecting surface, is a low loss continuous communication path. Sporadic "E", in the mid-latitudes, is more prevalent during the summer months of June and July, and has been observed to occur for periods of time up to several hours.

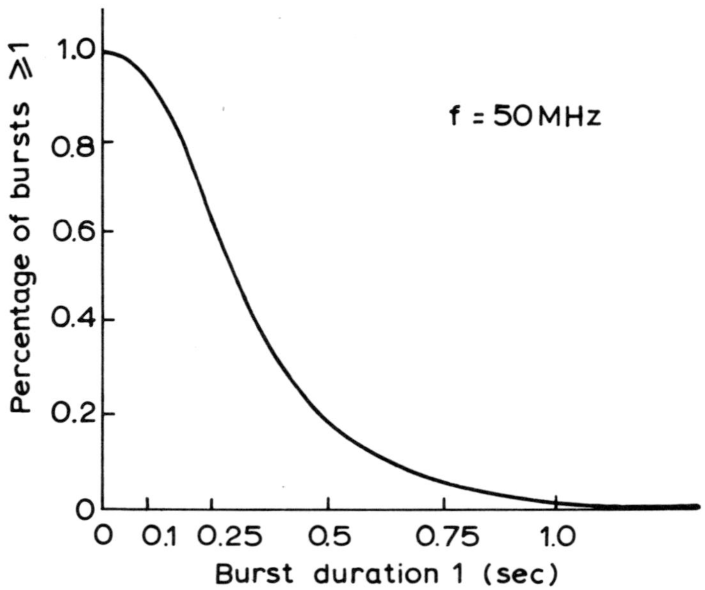

Figure 3. Cumulative distribution of burst duration.

PERFORMANCE REQUIREMENTS

The performance of a meteor burst communications link may be defined as the time required to transfer a message between two stations at a specified reliability. The following system parameters that will influence this "waiting time" are operating frequency, data rate, transmitter power, antenna gain and receiver threshold level, as follows:

Operating Frequency - Meteor trails will re-radiate or reflect VHF signals optimally in the 30 to 200 MHz frequency range. However, since the reflected signal amplitude is proportional to $1/f^3$ and its time duration is proportional to $1/f^2$, the message waiting time increases and data throughout reduces sharply at the higher VHF frequencies. Frequencies in the 30 to 50 MHz range are most practical. A lower limit of 30 MHz exists because of ionospheric F layer reflection, or other scattering phenomena. Antenna size is another practical factor in operating at lower frequencies.

Data Rate - Data rates from 2 Kbps to 4 Kbps are practical, but rates below 1 Kbps are not as useful because of the time required for overhead functions such as receiver start up, bit and word synchronization. RF bandwidth restrictions may limit the data rate. In coherent PSK systems in the U.S. the bit rate is required by the Federal Communications Commission to be less than one half the authorized bandwidth.

Transmit Power - The higher the transmit power the lower the message waiting time; 1000 watts is considered a practical limit.

Antenna Gain - Increasing the antenna gain will decrease the message waiting time; however, this power gain is offset by the loss in common sky area due to narrower beamwidth and the subsequent loss in the number of mutually useable meteors. The practical limits of antenna gain is a function of operating range and is in the region of 16 dBi at short range (400-600 miles) and 21-24 dBi at long range (600-1200 miles).

Receiver Threshold - Receiver Threshold is defined as the signal power required to achieve a particular bit error rate (BER) and is a function of the type of modulation used, the level of bit error rate (BER) required, and the receiver noise. The lower the receiver threshold, the lower the waiting time.

The principal noise sources affecting meteor burst systems are man-made, galactic and receiver noise.

A typical receiver threshold for a system operating at 45 MHz, using coherent bi-phase PSK modulation at 4 Kbps would be -114 dBm to obtain a BER of 10^{-3}.

Two parameters affect the average burst time duration. These are operating frequencies and distance between the stations (range). Figure 4 shows the effect of these parameters. In addition to the above parameters, the rate of data handling is influenced by the mode of operation: half duplex, full duplex, and system protocol. When a 30 character message with a standard 1000 watt master station and a 300 watt remote station is transmitted, operating half duplex at a rate of 4 Kbps, the 30 character message waiting time will be about 1.5 minutes at a probability of .95. These rates will vary with system parameters and diurnal and annual variations. The operation of a 100 baud circuit throughout the year is practical using full duplex master stations.

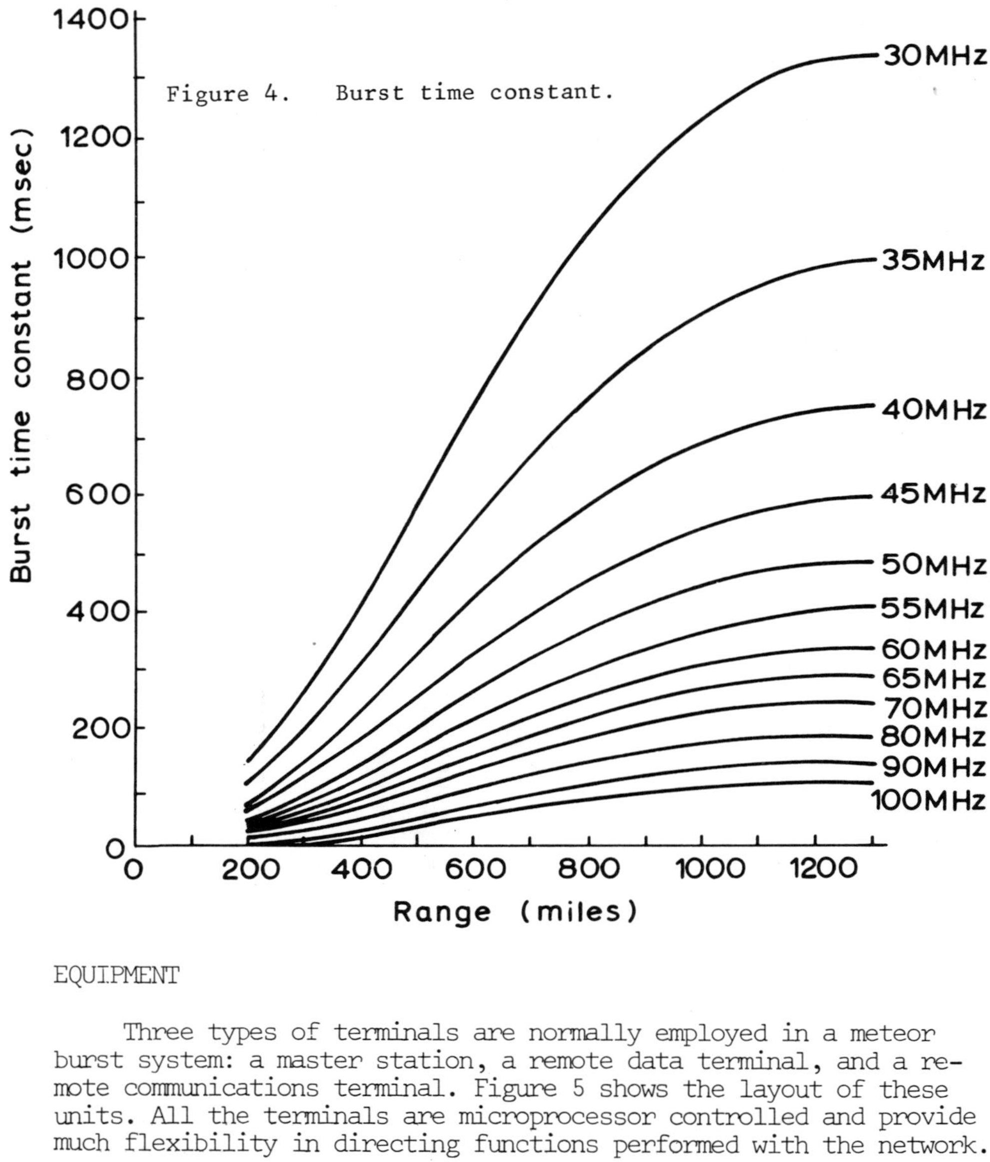

Figure 4. Burst time constant.

EQUIPMENT

Three types of terminals are normally employed in a meteor burst system: a master station, a remote data terminal, and a remote communications terminal. Figure 5 shows the layout of these units. All the terminals are microprocessor controlled and provide much flexibility in directing functions performed with the network.

The master station illuminates the meteor trial within the network with a coded RF signal in an attempt to contact remote units. When a suitably-oriented trail appears, a link is established for communications between the two terminals. The master station can operate half-duplex mode for data acquisition or in a full duplex mode for telecommunications. Also, it can function as a full duplex relay station for extended ranges using data store-and-forward techniques.

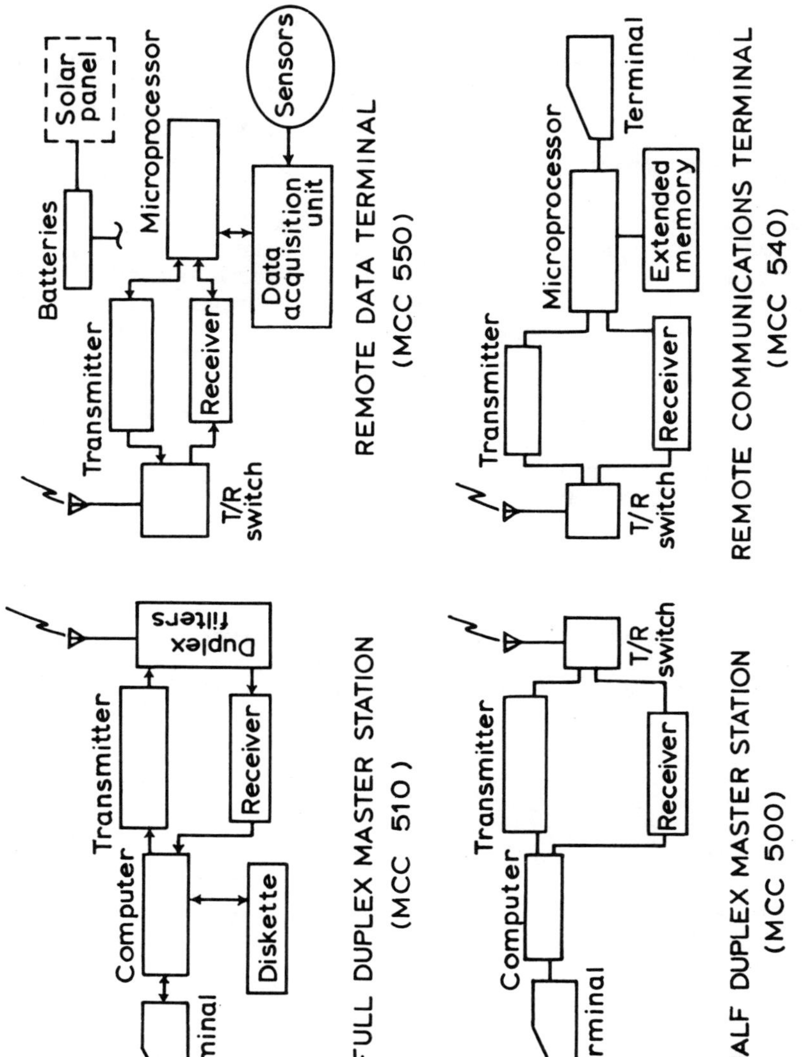

Figure 5. Block diagrams for meteor burst terminal equipment.

The master station controls both the communication protocol and dissemination of the data. This may include a teletype print-out, a CRT display, a mass memory for data logging or into a data modem for communication over telephone or microwave circuits.

Normally, the master station will consist of up to a 1000 watt transmitter, a PSK receiver, duplexers, antenna and a mini-computer/diskette system. The equipment is available rack-mounted or in carrying cases for portable use.

The remote station operates in half-duplex only. Its transmitter remains off until the master station is heard, then responds with a message or data block. Additional equipment required for a remote data site is a dipole antenna, solar panel/battery and the data sensors. Control of the program is in programmable memory, so that modification for custom applications is simple.

Another remote unit is a communications terminal. It has an RS 232 data port for messages up to 6,000 characters in length. The processor than buffers and divides the message into data suitable blocks for transmission over successive meteor trials. It is packaged in a small carrying case for simple deployment and operation.

In addition to these operating terminals, a test terminal is available to verify operation of this remote station before a given site is left unattended. Further, it can be used to inform the master station that the remote is ready to function.

SYSTEM APPLICATIONS

Meteor burst communications systems can be employed effectively for both point-to-point services and multiple station networks for ranges up to about 2000 km as shown. For extended ranges, relay stations may be employed using data store-and-forward techniques.

Communication service can be readily established for a variety of purposes, such as remote oil and mineral exploration activities, airborne platforms, shipboard use and communications with work crews in remote locations. Its rapid deployment capability also makes it ideal for disaster and emergency communications. The equipment is available in physically compact terminals, making it ideal for mobile and buoy applications.

Several major data acquisition networks are in operation today, including the Department of Agriculture's SNOTEL system and the Alaskan Meteor Burst Communication System (AMBCS).

The SNOTEL system contains over 500 remote data sites spread over 11 western states. These sites, unattended and powered by solar panel and battery, are polled daily for accumulated snow, precipitation and temperature data. This data is then used for streamflow forecasting and for making assessments of the water resources in the Western United States. The remote sites are microprocessor controlled and provide data averaging and event reporting. Within the SNOTEL system, seven event reporting sites are on Mt. St. Helens to provide immediate reporting of sudden moisture and temperature changes. The SNOTEL network will be expanded to over 1000 sites in the next few years.

The AMBCS, its master station located in Anchorage, Alaska is owned and operated by five federal agencies. It performs both data acquisition and message communications. These applications include:

air temperature, windspeed and direction, precipitation, snow pack, humidity and river stage

message communications from U.S. Bureau of Land Management remote survey camps

flight weather data from remote air fields

Gulf of Alaska wave data from buoys located near Kodiak

The Federal Emergency Management Agency is procuring portable master stations and a number of remote communications terminals for rapid deployment of emergency communication in disaster areas.

Another application in use is the combination of a Loran C receiver with a meteor burst remote terminal for transmitting position data. Ships have been tracked from Valdez to Seattle and from Seattle to San Diego. An extension of this application is the position monitoring of icebergs and drifting buoys.

SUMMARY

Meteor burst systems today are fully automated and simple to operate. Training is required only for familiarization with the terminal operation. It has proven to be a reliable and dependable communication service without recurring leased circuit costs, and without dependence on the services and support of other agencies or perhaps foreign governments.

REFERENCES

McKinley, D. W. R. Meteor Science and Engineering, McGraw - Hill, 1961

Sugar, G. R. Radio Propagation By Reflection for Meteor Trails, Proc. IEEE, Feb. 1964

CHAPTER 11

THE USE OF AN AERIAL INFRARED SURVEY IN IDENTIFYING ROOF STRUCTURES OF HIGH "U" VALUE IN AN EXTENSIVE BUILDING COMPLEX

J.M. Anderson and S.B. Wilson

Carnegie Laboratory of Physics, The University,
Dundee, Scotland

INTRODUCTION

In the past few years increasing use has been made of aerial infrared surveys by various organisations in an attempt to identify areas of high heat loss where remedial action might prove cost-effective. There is little doubt that the resulting "quick-look" imagery does show up such areas, provided a knowledge of the materials making up the roof structures is available. When the attempt is made to quantify these losses, however, difficulties do arise and it is one of the purposes of this paper to illustrate these difficulties with special reference to a local survey commissioned by the University of Dundee.

DATA ACQUISITION

The techniques involved in gathering the infrared data are now fairly well established. In G.B. the work can be commissioned from Clyde Surveys Ltd. who use a Daedalus Multi Spectral Scanner mounted in an aircraft which overflies the site on an arranged date. The information (relating to the ground covered) is stored on magnetic tape and at the ends of each scan line the detector is exposed to two "black bodies" held at temperatures bracketing the expected range of the subject area. The data is subsequently digitised for ease of handling and now takes the form of a series of lines of data, each piece of data being a number on a grey scale ranging from 0 to 255 and each line beginning and ending with a number on the scale corresponding to a known black body temperature. It is then sufficiently accurate to linearly interpolate the observed data between these known

A. P. Cracknell (ed.), Remote Sensing Applications in Marine Science and Technology, 219–229.

values to obtain an equivalent black body temperature for each element in the scene.

EXTRACTION OF DATA FROM MAGNETIC TAPE

The data recorded during a typical overflight may contain the grey levels of around 5 x 10^6 pixels. Since a typical building could be covered by several hundred pixels techniques need to be developed to locate the section of data required from the magnetic tape containing the overflight information. Essentially what must be done is to obtain a reasonably good visual representation of the total area covered by the overflight from the tape data and then work out the coordinates (in terms of line and column numbers) of the limits of the particular area of interest. This process is made easier if a photographic representation of the data in one whole overflight is available (the so called "quick-look" imagery). An example of such an image is shown in Figure 1. Rough coordinates can be worked out if the pixel size is known and if a large scale map of the area is available. The data in the area of interest can then be extracted using a suitable program. It is then usually advisable to find out the range of the grey levels in the data and to perform a contrast stretch before producing a line printer image. A typical example of the latter is shown in Figure 2. This uses every column and every 4th line to give a very reasonably proportioned representation. It is not necessary to have the "quick-look" image since it is possible to produce a line printer image of the entire overflight although it may be that a contrast stretch is required before a recognisable picture is achieved. An example is given in Figure 3. This uses every 3rd column and every 12th line. The coordinates of the area of interest are then easily obtained and the previously explained procedures can be followed. (Programs capable of carrying out the above processes were adapted from those published by Cracknell, (Ref. 1).)

The detailed line printer image can quite clearly show up features such as areas of high heat loss and it is at this stage that the actual grey level number (every line and column) is extracted for analysis.

ANALYSIS OF DATA

As explained in the earlier text the grey level numbers can be translated into temperatures by using the calibration chart supplied. The temperatures obtained are the equivalent black body temperatures of the surface viewed and will be designated T_{er}, indicating that they are determined by the amount of radiation reaching the scanner. These temperatures do not represent the actual temperatures of the surface which will be designated T_r.

Figure 1. A photographic representation of an entire overflight.

Figure 2. A line printer image of part of an overflight.

The amount of radiation reaching the scanner depends on various factors. These are:

(1) The emitted radiation from the roof surface.

(2) The reflected sky radiation.

(3) The amount of radiation absorbed by the atmosphere between the ground and the aircraft.

(4) The spectral range of the detector.

These factors are now discussed in turn.

The radiation emitted by a surface depends on the temperature of the surface and its emissivity. In this case since only that part of the radiation between 8μm and 14μm is recorded by the scanner the spectral emissivity, ε_s, in this range is the important factor. This is because the variation of emissivity with wavelength must be allowed for. The obvious importance of spectral emissivity is that it is possible for a material of very

Figure 3. A line printer image of an entire overflight.

low spectral emissivity (eg. aluminium sheet has $\varepsilon_s = 0.1$) to appear much colder to the scanner than a surface of high emissivity (eg. water of $\varepsilon_s = 0.96$) even when its true temperature is appreciably higher.

The magnitude of the reflected sky radiation depends on the amount of radiation in the range 8μm - 14μm from the sky reaching the surface and on the reflectivity of the surface. The reflectivity can be taken as $(1 - \varepsilon_s)$. While it is possible to obtain a reasonable estimate of the total radiation flux from the sky, covering all wavelengths, incident on the ground (and this will be discussed later) this is not distributed evenly over the whole spectral range and it is probably more realistic to attempt to measure this quantity on the night of the overflight. This can be attempted by examining data from materials of low emissivity, and therefore high reflectivity. In particular an area of sheet aluminium (covering about 15 pixels) can be laid out on the roof of a building on the theory that this will mostly reflect sky radiation.

Atmospheric absorption can be shown to considerably affect the results. It can be estimated if the temperature and humidity of the atmosphere are known. Using the method outlined by Roberts et al (Ref. 2), but considering only gaseous absorption (ie. no aerosol absorption) and neglecting the effect of nitrogen - H_2O collisions the transmission is then calculated (98% for the night of the overflight). This does, however, assume a uniform distribution of water molecules between the ground and the aircraft and as mentioned, neglects the effects of any possible aerosol contribution. It is reasonable therefore, to proceed without assuming this quantity but to attempt to estimate it from the observed data.

Since the detector only picks up that part of the radiation incident on it in the range 8 - 14μm allowance must be made for this in the analysis. Essentially the Planck radiation formula must be integrated only over this range of wavelengths. A simple way of accommodating this difficulty has been indicated by Goldstein (Ref. 3) who shows that the energy emitted by a surface in the wavelength interval $\lambda_1<\lambda<\lambda_2$ is given by $Q'_{\lambda_1\to\lambda_2} = \varepsilon_s CT_r^n$ where, for the range $8\mu m<\lambda<14\mu m$ and over the temperature range 250K to 290K, n is very nearly 5. C can be calculated by integrating the Planck formula numerically and this turns out to be:

$$C = 7.26 \times 10^{-11} \ Wm^{-2} \ K^{-5}.$$

The detected energy can now be written as:-

$$C\,T_{er}^5 = C\varepsilon_s\,T_r^5 + (1 - \varepsilon_s)Q'_{sky}$$

where Q'_{sky} is that part of the total sky flux lying between the wavelength limits of the detector. This equation must now be modified to allow for the absorption by the atmosphere.

$$C\,T_{er}^{5} = FC\varepsilon_s\,T_r^{5} + F(1-\varepsilon)Q'_{sky} \qquad \text{(a)}$$

where F is a fraction probably near 0.98 which represents the transmission of the atmosphere.

Use may now be made of the ground truth data taken for various surfaces of known emissivity concurrently with the over-flight. The quantities not known accurately in equation (a) are the fractional transmittance F and the spectral sky flux Q'_{sky}. Figure 4 shows a collection of graphs (referring to the Dundee overflight) for each of the surface materials, aluminium, oxidised copper, concrete, quarry dust and water. In each case T_{er}, T_r and ε_s were inserted in equation (a), F was varied and Q'_{sky} calculated. The graphs can be seen to cross in a region where $F = 0.98$ and $Q'_{sky} = 115\ Wm^{-2}$.

Having established these values it is now possible to use equation (a) to calculate unknown values of T_r from the grey level data provided the emissivity of the surface is known. Alter-natively if further ground truth measurements are available the emissivities of the corresponding materials can be estimated.

An example of the latter concerned a roof covered in felt, the outer surface of which consisted of crushed green slate. The mean values of T_{er} and T_r were 277K and 281K respectively and using the estimated values of F and Q'_{sky} led to a value of $\varepsilon_s = 0.53$, much lower than would be expected for plain roofing felt or felt covered in granite chips.

It has now been established that, given sufficient ground truth measurements, the true surface temperatures can be determin-ed for all surfaces of known emissivity.

ESTIMATION OF HEAT LOSS FROM FLAT ROOF SURFACES

In order that the total radiative heat loss from a roof surface can be calculated it is necessary to obtain an estimate of the total radiative flux arriving on the roof structure from its surroundings. Assuming for simplicity that only the sky contrib-utes this can be calculated from a knowledge of the air temperature on the night of the overflight. Swinbank (Ref. 4) shows that this can be estimated as $Q_{sky} = 5.31 \times 10^{-13}\ T^6\ Wm^{-2}$ which for the night in question when the air temperature was $9.5^{o}C$ results in a value of $270\ Wm^{-2}$. The total radiative transfer between the flat roof and its surroundings is then given by:-

$$\Delta Q = \varepsilon(\sigma T_r^{4} - 270)Wm^{-2}$$

where ε is the total emissivity of the roof surface and σ is

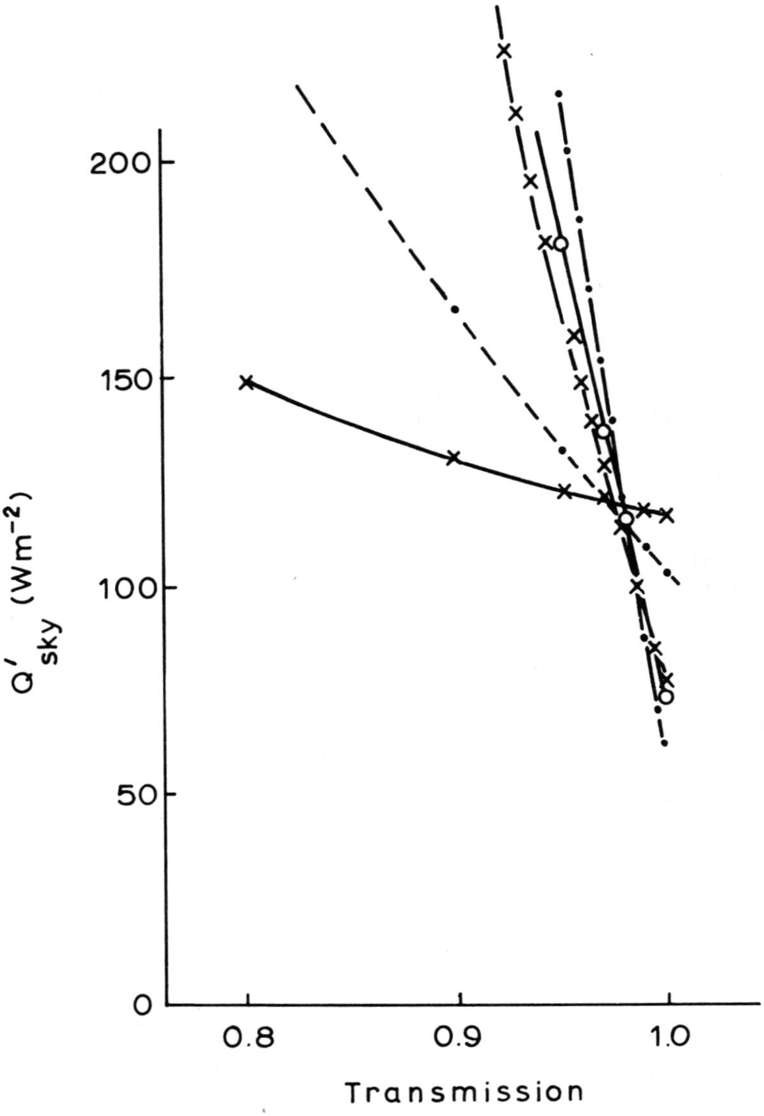

Figure 4. Dependence of Q'_{sky} on transmission for several surfaces.

– – – –	Carnegie Roof - Oxidized copper
————	Tower Building - Aluminium
–x–x–x–	Tower Building - Concrete
–•–•–•–	Frankland Pond - Water
–o–o–o–	Hawkhill Car Park - Quarry Dust

Stefan's constant (5.67×10^{-8} Wm^{-2} K^{-4}). As an example, picked from the data of the present survey, a roof surface of crushed slate where the spectral emissivity was measured at $\varepsilon_s = 0.53$ had a value of $T_r = 8.1^oC$. Assuming $\varepsilon_s = \varepsilon$ gives $\Delta Q = 44.8$ Wm^{-2}.

Yet to be taken into account is the effect of convective heat transfer between the roof surface and the adjacent air. Assuming turbulent forced convection flow over a smooth flat surface Goldstein (Ref. 2) shows that the heat transfer coefficient would be given by:

$$h = \frac{3.91\ u^{4/5}\ x^{-1/5}}{1-0.148\ (ux)^{-1/10}}\ Wm^{-2}\ K^{-1}$$

where u is the wind velocity in ms^{-1} and x is the distance in m from the leading edge of the flat surface. In fact this reduces to:

$$h = 4.37\ u^{4/5}\ x^{-1/5}\ Wm^{-2}\ K^{-1}$$

for typical values. This should of course be evaluated over the width of the roof which means multiplying the expression by the factor 1.25 and replacing x by L, the width of the roof, giving:

$$h' = 5.46\ u^{4/5}\ L^{-1/5}\ Wm^{-2}\ K^{-1}.$$

Taking as an example the building quoted earlier where

$$u = 2.5\ ms^{-1} \quad L = 15\ m,$$

and the roof surface is rough and at a lower temperature than that of the air,

$$h' = 6.6\ Wm^{-2}\ K^{-1}.$$

Following Goldstein's suggestion this is doubled because the surface is rough and reduced by a factor to allow for the fact that the surface is colder than the surrounding air. Allowing for this the value arrived at is h' lying between 4.5 and 13.5 Wm^{-2} $^oK^{-1}$. This gives a convective heat gain lying between 6.3 Wm^{-2} and 18.9 Wm^{-2}. Taking the largest extreme gives a net loss of energy for this roof structure as 44.8 - 18.9 = 25.9 Wm^{-2}. The "U" value would then be 2.1 Wm^{-2} K^{-1}. This value is much higher than one would expect indicating that the convective heat gain on the night has been underestimated or that the roof surface was not at its equilibrium temperature during the overflight and still retained some solar gain. In fact the measurements taken on the night go against the likelihood of the latter possibility and it is more probable that proper procedures for calculating the convective heat loss or gain are not yet available. This view is also held by Bowman who conducted controlled experiments on flat roof heat losses (Ref. 5).

More realistically differences in "U" value can be calculated.

With the present data a good example is illustrated by an obvious anomally which is evident in the different apparent temperatures of a "high" and "low" section of one of the main teaching blocks. The average temperature measured during the flight were:

$$T_{er}^{high} = 278.8K$$

$$T_{er}^{low} = 280.1K$$

Since no ground truth data were available the true roof temperatures are then calculated using $\varepsilon_s = 0.96$ (the value for granite chippings) giving:-

$$T_r^{high} = 280.1K \text{ and } T_r^{low} = 281.4K.$$

This then gives the difference in radiative heat transfer as

$$\varepsilon_s \sigma (T_r^{low\,4} - T_r^{high\,4}) = 6.5 \text{ Wm}^{-2}.$$

Assuming all other factors are equal this gives a difference in "U" value of 0.54.

CONCLUSION

It has been established that if the emissivity of the material making up the flat surface is known the correct surface temperature can be derived from the remotely sensed data, provided atmospheric absorbance is taken into account. Accurate absolute measurement of "U" values are considered, however, to be unattainable at present from the survey data because of the uncertainties in the values calculated for convective heat loss. Differences in "U" value are, however, attainable.

REFERENCES

(1) A.P. Cracknell (Ed.), Computer Programs for Image Processing of Remote Sensing Data, University of Dundee, 1982.

(2) R.E. Roberts, J.E.A. Selby, and L.M. Bibermann: "Infrared Continuum Absorption by Atmospheric Water Vapour in the 8-12μm Window", Applied Optics, 15, 1976.

(3) R.J. Goldstein: "Applications of Aerial Infrared Thermography to the Measurement of Building Heat Loss", ASHRAE Transactions, 84, 1978.
(4) W.C. Swinbank: "Long Wave Radiation from Clear Skies", Quat. Jour. Roy. Met. Soc., 89, 1963, pp. 339-348 and 90, 1964, p. 488.
(5) R.L. Bowman et al.: "Feasibility of Determining Flat Roof Heat Losses Using Aerial Thermography", NASA, TM79152, 1979.

CHAPTER 12

SEA SURFACE TEMPERATURES FROM INFRARED MEASUREMENTS.

S.M. Singh

Carnegie Laboratory of Physics, Dundee University,
DUNDEE DD1 4HN, Scotland, U.K.

and

D.E. Warren,
Department of Electrical Engineering and Electronics,
Dundee University, DUNDEE DD1 4HN, Scotland, U.K.

ABSTRACT

The physical principles underlying the transmission of infrared radiation through the Earth's atmosphere are discussed. The inflight calibration procedure for the Advanced Very High Resolution Radiometer is described. A procedure for computing the brightness temperatures and for comparing these temperatures with simultaneous in-situ sea-surface temperatures is presented.

Various schemes for correcting brightness temperatures to reduce the atmospheric contributions are compared. The multi-spectral correction technique is described in some detail as it is the most reliable strategy that can be used at present. The possibility of using observations along two atmospheric paths is mentioned in connection with the Along Track Scanning Radiometer proposed for the European Remote Sensing satellite, ERS-1.

1. INTRODUCTION

Modern oceanography was given birth by the expedition of H.M.S. Challenger during 1872-76, an excellent account of which can be found in reference (1). About 54 oceanographic expeditions followed the H.M.S. Challenger (2) until 1960. One of many

A. P. Cracknell (ed.), Remote Sensing Applications in Marine Science and Technology, 231–262.

objectives of these expeditions had been measuring and mapping the distribution of seawater temperature on a global scale; this included mapping of the sea-surface temperature. Such oceanographic expeditions provide reliable data but these expeditions are slow and expensive.

The idea of remote measurement of the surface temperature of an object stems from Planck's radiation law which states that the spectral radiant exitance $M(\kappa,T)$ of a perfect emitter (blackbody) at an absolute temperature T within an infinitesimal wavenumber interval from κ to $\kappa+d\kappa$ has the distribution of the form

$$M(\kappa,T)d\kappa = c_1\kappa^3 d\kappa \;/\; \{\exp(c_2\kappa/T)-1\} \tag{1}$$

where κ is in cm^{-1} and c_1 and c_2 are known as the first and second radiation constants, respectively,

$$c_1 = 3.74151 \times 10^{-8}\ W/(m^2\ cm^{-4})$$

$$c_2 = 1.43879\ \ cm\ K.$$

Another expression for Planck's law which is convenient in practice is obtained by dividing Eqn.(1) by π. If we denote this quantity by $B(\kappa,T)$ then

$$B(\kappa,T) = M(\kappa,T) \;/\; \pi \tag{2}$$

is termed the spectral radiance of the source and has the units of $W/(m^2\ sr\ cm^{-1})$. For a real object having emittance ε, the spectral radiance is given by $\varepsilon B(\kappa,T)$. Measurement of this quantity in the thermal infra-red region enabled Hoffman (3) to detect a low flying aircraft and his findings were openly published in 1919. Similar applications of Planck's radiation law were sought through the Second World War (4). Application of Planck's radiation law for remote measurement of the sea-surface temperature (SST) using air/satellite-borne infra-red (IR) detectors is well known today.

For purposes of identifying and mapping features of the sea-surface temperature, infrared images need to be rectified and properly registered to be useful. The temperature differences can be estimated to accuracies better than 1 deg K. The brightness temperatures calculated from the satellite data are found to be, in general, lower than simultaneous _in-situ_ data (5-7). There are many sources of errors which contribute to this discrepancy (5). The primary contaminator of the satellite data is the atmosphere. To achieve better accuracy in remotely measured SST, it is necessary to compensate for the effect of the atmosphere. In the following sections we shall see that the atmosphere is rather complex and the removal of atmospheric effects is difficult.

Perhaps the primary motivation for requiring accurate measurements of SST with the kind of coverage and resolution provided by satellites arises from efforts to improve the accuracy and reliability of weather forecasting. Maintaining ocean weather ships on station to make routine measurements is expensive and as a result the density of the network is very low. In the last thirty years there have been, at most, nine weather ships to make observations over the entire North Atlantic. This is less than one observing station for every ten degree latitude-longitude square. Apart from the SST and the routine surface meteorological observations each weather ship makes two upper-air observations every day. The temperature and humidity profiles produced from these observations describe the vertical structure of the atmosphere and are the raw data from which the input data for numerical weather prediction models is produced.

Temperature and humidity sounding from satellites is possible and although the vertical resolution is not as great as that which can be obtained from radio-sondes the horizontal coverage is significantly better. However the data from a satellite sounding instrument (infra-red radiances in perhaps twenty different wave-length intervals) do not necessarily define a unique temperature profile. The accuracy of soundings derived from this type of data is increased and the retrieval of soundings is made considerably easier if the sea-surface temperature is known beforehand.

One of the important factors in understanding the ocean weather is the interactions which take place at the air-sea interface, especially involving those phenomena through which energy transfer between the water and the air takes place, e.g. condensation, precipitation, evaporation, radiation, etc. A successful weather prediction model should include all the energy transfer mechanisms (8) on a global scale. The point pertinent to our discussion is that some of these energy transfer mechanisms demand a knowledge of the sea-surface temperature on a global scale - a quantity which may be obtained from the satellite data.

It is perhaps worth pointing out that among the observational criteria specified for the Global Atmospheric Research Programme was the collection of a reliable set of sea-surface temperature measurements every 100km over the entire globe to an accuracy of 0.2deg K with a frequency at each point of one measurement every twelve hours. Measurements of this accuracy have still to be achieved from satellite data.

The infra-red data from satellites can be used to detect upwelling areas because the deeper, cooler water is forced to the surface. Therefore the sea-surface water temperature is much below the normal temperature for that latitude. The reasons for ones interest in detecting upwelling areas are well known. For example, the upwelling areas have high productivity because upwelling brings into the euphotic zone nutrient-rich deep water on which fish feed. It is chlorophyll-a in the plankton which shows up in the green channel of the Coastal Zone Colour Scanner (CZCS) on board the NIMBUS-7 satellite. Unusually high chlorophyll-a concentrations have been observed during upwelling (9-11). Although it is difficult to obtain a quantitative correlation between upwelling and marine productivity, a qualitative relation seems apparent because about half of the world's fish supply is thought to be produced in upwelling regions (12). Therefore, the detection of upwelling areas may help in the conservation and exploitation of marine life.

The sea level decreases in the upwelling areas because the upwelled water is denser than the surrounding water. It is these sea-surface slopes which, together with wind stress on the surface water develop geostrophic currents (13,14). Upwelling also affects the climate off a coast. The cold upwelled water cools the air close to the water surface thereby increasing the relative humidity. The result is mist, fog or clouds. Thermal infra-red data are also useful for monitoring eddies, warm or cold currents and the movement of thermal fronts. The position of fronts is thought to influence the movement of fish because the colder side of the front corresponds to upwelling (15) and hence a nutrient rich zone.

There have been some indications that thermal infra-red data might be useful for detecting large oil slicks on sea-water (16, 17). The basic physics underlying this idea is contained in a relation similar to Stefan-Boltzmann law, i.e.

$$B(\kappa,T_s) = \varepsilon a' T_s^{b'} \tag{3}$$

where T_S denotes the temperature of the surface in question and a' and b' depend on the wavenumber and width of the spectral window (ref.18, chapter 9). The thermal infra-red emittance of water is higher than that of a mixed oil and water surface (19). It is easy to see from Eqn.(3) that the oil slick may have warmer, similar or cooler spectral signatures compared with the spectral signature of the surrounding water. These three possibilities have been encountered in practice. For example, both warmer and colder signatures were observed by Estes and Senger (16). The oil spillage in the Gulf of Mexico from June 3, 1979 was not detectable (20) in the thermal infra-red imagery from the Advanced Very High Resolution Radiometer (AVHRR).

With these motivations, in what follows we shall discuss the measurement of sea-surface temperature using the thermal infra-red data from satellites and the difficulties underlying such measurements.

2. THE ATMOSPHERE

The atmosphere is complex. The pressure and temperature conditions change rapidly and this causes atmospheric circulations and hence redistribution of atmospheric constituents. Although water vapour, carbon dioxide and ozone are the primary atmospheric absorbers and emitters in the thermal infrared, further complications arise due to the fact that the amount of water vapour is a highly variable quantity. To achieve better accuracy of the sea-surface temperatures deduced from the remotely sensed data, the amounts of these atmospheric constituents within the instantaneous field of view of the remote sensor must be known. This relies on a knowledge of the distribution of these constituents within the atmosphere.

In order to facilitate computations for the atmospheric transmittance and for the emission from the atmospheric constituents we simplify the atmosphere by assuming that the optical properties of the atmosphere vary only in the vertical direction, i.e. the distribution of atmospheric constituents in the horizontal direction is assumed to be homogeneous. Such an atmosphere is known as a plane-parallel atmosphere (ref.18, chapter 8).

2.1 Distribution of Atmospheric Absorbers

We will now briefly describe the main features of the distribution of water vapour, carbon dioxide and ozone in the atmosphere and the degree to which these distributions can be expected to vary from day to day. Figure 1 shows two temperature and moisture soundings which are representative of conditions in mid-latitudes. The temperature profiles define the depth of the troposphere, the lowest layer of the atmosphere in which the temperature decreases with height. Above this layer the temperature increases or remains constant and this suppresses vertical motion and restricts the transport of material upwards from the troposphere. The upper boundary of the troposphere is typically at a height of about 10km to 20km. It is lower in winter and at high latitudes and higher during summer and nearer to the equator.

The two soundings shown in figure 1 demonstrate that most of the water vapour in the atmosphere is confined to the troposphere. The concentrations of moisture in the stratosphere are five orders of magnitude smaller than those lower down. Even within the troposphere the water vapour distribution is weighted towards the lower levels. This is fortunate. It means that most of the absorption and emission by water vapour takes place at temperatures and pressures which are close to the surface values.

In mid-latitudes the relative humidity near to the surface is fairly constant at around 60% to 70%. Therefore we might expect

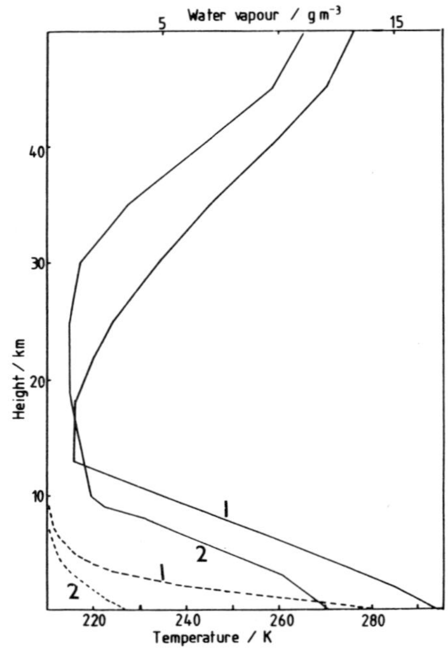

Figure 1. Standard atmosphere temperature and water vapour concentration profiles, mid-latitude summer, 1, and winter, 2.

there to be more water vapour in the atmosphere during the summer when the temperature is higher. This is indeed the case. The water vapour density near to the surface in the summer profile is nearly three times greater than the corresponding winter value. The total quantity of water vapour contained in a vertical column of these two atmospheres are 1.0g cm^{-2} in the winter profile and 3.0g cm^{-2} in the summer one. Again because of the increase of temperature the total water vapour content tends to be greater nearer to the equator. The tropical atmosphere is likely to contain up to 5g cm^{-2}.

As an illustration of the horizontal gradients of water vapour content which can occur figure 2 shows the values calculated at the upper air stations in the British Isles for midday on 15th May 1980. The total water vapour content doubles over a distance of the order of 1000km from the north-east of the chart to the south-west.

Carbon dioxide constitutes 0.033% of the atmosphere by volume.

It is normally considered to be uniformly distributed throughout the atmosphere and to remain constant from day to day.

Ozone is created by several photochemical reactions which occur in the upper atmosphere and involve ultra-violet radiation and oxygen. The reaction dynamics lead to a maximum in the ozone concentration profile at a height between 20km and 30km. Since the reactions depend on the presence of sunlight there is a marked diurnal variation in the amount of ozone in a vertical column from 2mg cm^{-2} at night to 3mg cm^{-2} in the daytime.

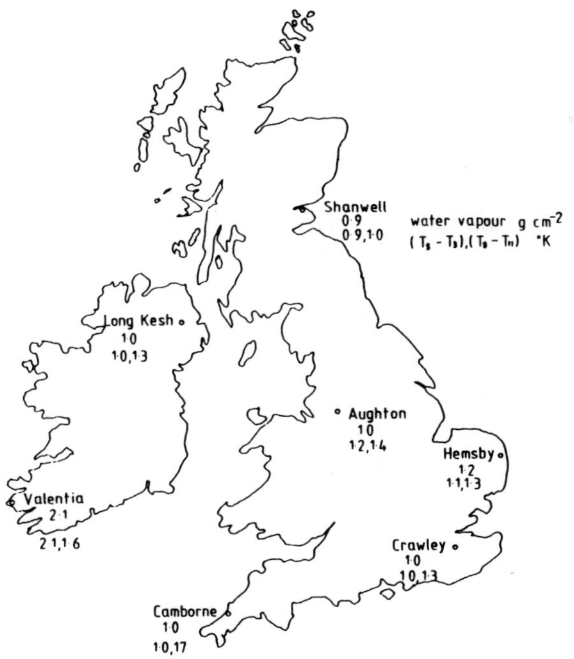

Figure 2. Total water vapour content and atmospheric corrections in the 3.7μm and 11μm channels calculated from sounding data of 1200 GMT, 15 May 1980.

2.2 Atmospheric Transmittance

The transmission of electromagnetic radiation through the atmosphere is determined by the amount of absorption and scattering of the radiation by the atmospheric constituents. However, in the thermal infrared region of interest to us the scattering effects can be ignored for clear atmospheric conditions (i.e. without clouds, haze, smog, fog or smoke) because the wavelength of the radiation (say between 3 and 100μm) is much larger than the sizes of either the molecules (which are of the order of 0.0001μm) or the aerosol particles (the radii of aerosol particles fall in the range from about 0.01μm up to about 10μm). The number of aerosol particles whose radii lie within the radius interval r and r + δr is approximately inversely proportional to the fourth power of the radius r (ref.21, p.185,ff.) and in clear air the majority of the aerosols are of sizes at the lower end of the range, up to about 0.1μm.

Let $\rho_a(z)$ denote the volume density of the absorbing matter at an altitude z. If a beam of radiation is traversing in the direction Θ with the vertical then it is easy to see that the transmittance $\tau(\kappa,\Theta;z,z_1)$ of the atmospheric layer lying between altitudes z and z_1 for a monochromatic radiation of wavenumber κ is given by

$$\tau(\kappa,\Theta;z,z_1) = \exp\{-\sec(\Theta) \int_z^{z_1} k(\kappa)\rho_a(z')dz'\} \qquad (4)$$

where $k(\kappa)$ is the absorption coefficient which depends on the width and strength of the absorption lines. The natural width of a spectral line is very small, of the order of 10^{-11} to 10^{-9} cm^{-1} but in practice the width is much larger than the natural width.

Two mechanisms broaden the absorption (emission) lines of the atmospheric constituents. The random motion of the molecules introduces Doppler shifts which result in spreading of the frequencies absorbed or emitted. Since the velocities of gaseous molecules are proportional to the square root of the absolute temperature, the Doppler broadening is strongly temperature dependent. At 300K the Doppler width of a water vapour line at 1493 cm^{-1} (6.7μm) is about 0.003 cm^{-1} which is large compared to the natural width.

Although Doppler broadening is the dominant mechanism in the higher parts of the atmosphere, below about 50km the broadening of the spectral lines is due to collisions. The half width of a collision broadened spectral line can be about 0.1 cm^{-1} (ref.22, chapter 4). The collision broadening of a spectral line results in the absorption being a function of both temperature and pressure. However, in the Earth's atmosphere below 50km the pressure varies over five orders of magnitude from 0.1 mb to about 1000 mb at the surface, whereas the temperature is confined roughly between 200K and 300K. The temperature dependence of the collision broadening seems to be weak (ref.22, chapter 4) while the pressure dependence is complicated (23). Nevertheless, one can conclude that near the centre of an absorption line the absorption coefficient is inversely proportional to the pressure but in the far wings of the absorption line the proportionality is direct. This results in the spectral line being progressively smeared out as the pressure increases, with absorption at wavenumbers remote from the line centre becoming increasingly more important.

There is no well-defined mathematical relation which will predict temperature and pressure at a given altitude. Nevertheless, as was pointed out in section 1, it is possible to measure the vertical profiles of pressure P, humidity H and temperature T. In practice, therefore, when the PHT profiles are available, Eqn.(4) becomes useful if it can be regarded as a function of the pressure instead of the variable z. To achieve this we assume that the atmospheric gases obey the ideal gas law. If r(p) denotes the mass mixing ratio of an absorber at pressure level p, i.e.

$$\rho_a(p) = r(p)\rho(p) \tag{5}$$

where $\rho(p)$ is the volume density of atmospheric gases, then Eqn.(4) can be written as

$$\tau(\kappa,\theta;p,p_1) = \exp \{\frac{\sec(\theta)}{g} \int_p^{p_1} r(p')k(\kappa)dp'\} \tag{6}$$

where p and p_1 represent the pressures at altitudes z and z_1, respectively, and g is the acceleration due to gravity which we have considered as a constant with respect to the altitude.

2.3 Atmospheric Windows

Approximately 13% of the energy emitted at infra-red wavelengths at the surface of the Earth escapes from the top of the atmosphere to space. The rest is absorbed by the atmosphere. However, although the average transmittance of a vertical column of the atmosphere is so low, the variation with wavenumber is relatively large. There are three spectral intervals in the thermal infra-red in which the vertical transmittance may be as high as 90%. Within these windows it is possible to "see" the surface from a satellite in Earth orbit. Of these three windows, (see figure 3), from 3μm to 5μm, from 7μm to 8μm and from 9.5μm to 13μm, the first is the most nearly transparent.

Unfortunately, as we shall see in Section 3.6, the 3.7μm window has proved to be of less practical value than the others because the reflected solar radiance within this window is large enough to swamp the thermal emission from the surface.

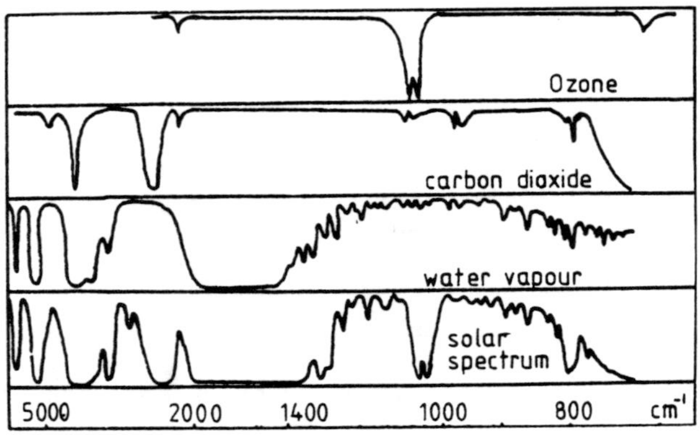

Figure 3. The absorption spectra of ozone, carbon dioxide, water vapour and of all atmospheric constituents.

2.4 The Continuum Absorption

Although the transmittance within the infrared windows is high there is still some absorption which cannot be ignored. There are no absorption lines within the windows; the absorption spectrum in these regions is a smooth continuum. Some of this is undoubtedly due to absorption in the far wings of the strong spectral features on either side of the windows, for example by the water vapour bands at 6.7μm and the carbon dioxide absorption beyond 13μm. However line absorption has been found to account for only a small part of the continuum. The remaining "anomalous" component is thought to be due to associations of pairs of water vapour molecules, known as "dimers" (24). The evidence for water vapour dimers being responsible for the anomalous absorption is for the most part indirect. The absorption coefficient has been found to depend on the water vapour partial pressure, e mb, and the atmospheric pressure p in the following fashion (24).

$$k(\kappa) = k_1(\kappa)p + k_2(\kappa)e \qquad (7)$$

where the first term is due to collision broadening and the second term, which was suggested by Bignell (24), represents the absorption coefficient due to the water dimer. Since the population of the water dimer should rise with the water vapour content, the second term should become increasingly important with increasing humidity. This in turn implies that in very moist atmospheres, at low latitudes, the transmittance, even in the windows, may not be sufficient to permit accurate sensing of the surface in the infra-red. The fact that the spectrum of the dimer would be expected to consist of very broad lines and the agreement between the temperature dependence of the coefficient $k_2(\kappa)$ and that expected for the dimer lend further support to this hypothesis.

2.5 Relative Importance of Main Absorbers

The attenuation caused by each of the three main absorbers in the 3.7μm and 11μm channels of the AVHRR has been calculated, using an atmospheric transmission model, and expressed as a temperature difference (25). The results shown in table 1 distinguish between the effect of the water vapour line absorption and the continuum. The continuum contribution is much more important at the longer wavelength and it increases dramatically in the moister atmosphere.

Table 1. Calculated attenuation in 3.7μm channel (top three lines) and 11μm channel (bottom three lines) of the AVHRR.

	Water vap. content $g\ cm^{-2}$	Water lines K	Water continuum K	Carbon-dioxide K	Ozone K
Winter	1.0	0.4	0.1	0.4	0.0
Summer	3.0	1.0	0.3	0.4	0.0
Tropical	4.3	1.3	0.4	0.5	0.0
Winter	1.0	0.2	0.2	0.4	0.0
Summer	3.0	0.6	1.6	0.3	0.0
Tropical	4.3	0.9	3.4	0.3	0.0

2.6 Atmospheric Emission

In a non-scattering atmosphere, the radiation propagating in the direction θ at an altitude z_1 includes the transmitted radiation and the radiation emitted by the atmospheric gases. In the case of remote sensing of the Earth's surface temperature one is interested only in that component of the radiation which originates from the surface of the Earth. This radiation is indistinguishable from the radiation emitted by the atmospheric gases. Therefore, one has to devise a way of evaluating the atmospheric contribution. To achieve this we assume that local thermodynamic equilibrium, LTE, exists and that the local temperature can be defined at each point in the atmosphere (26). The assumption of LTE is a good approximation for the lower atmosphere but does not apply to the upper atmosphere (ref.22, chapters 2 and 5). Since the primary absorber (water vapour, Table 1) lies in the lower atmosphere, see Fig.1, we may assume a case of LTE. If the density of water vapour content is assumed to decay exponentially then the scale height for water vapour is found to be approximately 2.5km (ref.21, p.183).

Let us imagine an infinitesimally thin slab of the atmosphere of thickness dz lying between z and z + dz, where the pressure conditions are p and p + dp, respectively. This slab of atmosphere will emit radiation in all possible directions. Let T(p) represent the temperature of that surface of the slab where pressure is p. Then the assumption of LTE enables Kirchhoff's law to express the emission coefficient in terms of the absorption coefficient. It is now easy to show that the amount of spectral radiance of the slab in the direction θ with the vertical is $r(p)k(\kappa)\sec(\theta)B(\kappa,T(p))dp/g$. As this radiance propagates upward, it is attenuated. If $dL_a(\kappa)$ is the fraction reaching the altitude z_1 where pressure is p_1 then we see that

$$dL_a(\kappa) = r(p)k(\kappa)\sec(\theta)B(\kappa,T(p))\tau(\kappa,\theta;p,p_1)dp/g. \quad (8)$$

If we differentiate Eqn.(6) then we see that the relation (8) can also be written as

$$dL_a(\kappa) = B(\kappa,T(p))d\tau(\kappa,\theta;p,p_1) \quad (9)$$

with

$$d\tau(\kappa,\theta;p,p_1) = \sec(\theta)r(p)k(\kappa)\tau(\kappa,\theta;p,p_1)\,dp/g. \quad (10)$$

3. SATELLITE-RECORDED AND SATELLITE-RECEIVED RADIANCES

The satellite data are available in a digital form and therefore the radiance computed from the digital data involves some error which will be discussed later. For the time being if we ignore this error then the radiance $L(\kappa)$ received by a satellite borne detector and the corresponding radiance $L^*(\kappa)$ which is actually recorded are related through

$$L^*(\kappa) = \int_{\kappa_1}^{\kappa_2} L(\kappa')\,\hat{\phi}(\kappa')d\kappa' \quad (11)$$

where $\hat{\phi}(\kappa)$ is the detector response function which is normalized to unity, i.e.

$$\int_{\kappa_1}^{\kappa_2} \hat{\phi}(\kappa)d\kappa = 1. \quad (12)$$

The limits of integration are such that the response of the detector outside these limits is zero. The satellite received radiance $L(\kappa)$ may be written as

$$L(\kappa) = L_e(\kappa) + L_{sc}(\kappa) + L_r(\kappa) + L_a(\kappa) \quad (13)$$

where the subscripts e, sc, r, and a refer to the respective contributions to $L(\kappa)$ from emission from the surface of the Earth, scattering processes by the atmospheric constituents, reflection at the sea-surface and emission from the atmospheric gases. It is understood that all of these components of $L(\kappa)$ are directed towards the sensor.

As indicated earlier, we may assume a non-scattering atmosphere so that $L_{sc}(\kappa) = 0$. The reflected component $L_r(\kappa)$ has two sources of origin: First, solar radiation is reflected at the sea-surface towards the sensor. Secondly, the downwelling atmospheric emission is reflected at the sea-surface towards the remote sensor. Let $L_{sr}(\kappa)$ and $L_{ar}(\kappa)$ denote the satellite received radiances due to these two causes, respectively. Then,

$$L_r(\kappa) = L_{sr}(\kappa) + L_{ar}(\kappa). \qquad (14)$$

Once the satellite-received radiance which is given by Eqs.(13) and (14) is known, the satellite-sensed radiance can be evaluated using Eqn.(11). Now we show a way of evaluating various contributions to $L(\kappa)$ for a non-scattering atmosphere.

3.1 The Contribution From Sea-Surface Emission.

The radiance of the sea-surface with surface temperature T_s is given by $\varepsilon B(\kappa,T_s)$. In general, the emittance ε of an object is a function of wavenumber and temperature. For example, the emittance of gases varies very rapidly with wavenumber in the neighbourhood of the absorption (emission) lines. But, for sea-water ε may be treated as constant with respect to κ and T_s. The emittance of sea-water may also depend on the state of the sea-surface. However, Sidran (5) points out that it is independent of the roughness of the sea-surface up to wind speeds of 14 ms^{-1}. If we ignore the presence of any material which is not part of sea-water, e.g. oil pollution, industrial waste, etc., then ε may be regarded as a constant. Let p_o be the atmospheric pressure at the sea-surface. By definition, the pressure on the top of the atmosphere is zero. The atmospheric transmittance of the entire atmosphere in a specified direction θ is obtained from Eqn.(6) by setting $p = p_o$ and $p_1 = 0$. Thus, the radiance reaching the detector from the view angle θ is

$$L_e(\kappa) = \varepsilon B(\kappa,T_s)\tau(\kappa,\theta;p_o,0). \qquad (15)$$

If the detectors are not tilted (in the case of the Coastal Zone Colour Scanner (CZCS) the detectors can be tilted so as to look either ahead or behind the spacecraft) then the view angle can be computed using the expression

$$\theta = (N-N_o)\delta\theta \qquad (16)$$

where N is the pixel number along a given scan line from which the radiance $L_e(\kappa)$ is being evaluated, N_o is the nadir pixel number and $\delta\theta$ is the angular separation of adjacent pixels along a given scan line. As an example, for TIROS-N/NOAA satellite series, $N_o = 1024$ and $\delta\theta$ is calculated using the fact that the maximum cross-track scan angle from nadir is $\pm$ 55.4 degrees, i.e.

$$\delta\theta = 55.4/1024 \text{ (degrees)}. \tag{17}$$

3.2 Contribution From Solar Radiation Reflected at the Sea-Surface

The change in the solar reflected radiance due to wind roughened sea-surface by wind speeds up to 14m s^{-1} introduces an error of less than 0.04K in the sea-surface temperatures retrieved from the satellite data (5). Therefore, we assume that the sea-surface behaves as a perfectly diffuse (Lambertian) surface. Such an assumption cannot be made when the spatial resolution of the detector is of the order of a few wavelengths of the ocean waves unless the scanning rate is so slow that the signal output from the detector is averaged over many waves.

Let $E(d,\kappa)$ denote the solar irradiance on the top of the atmosphere at wavenumber κ per unit area perpendicular to the incident direction on Julian day d. If θ_s is the solar zenith angle then the amount reaching unit area of sea-surface is $E(d,\kappa)\cos(\theta_s)\tau(\kappa,\theta_s;0,p_o)$. Since the water is an opaque object for thermal infrared radiation, the incident radiation must be either absorbed or reflected. Thus, a fraction $(1-\varepsilon)$ of the solar irradiance is reflected. The fraction of this reflected radiation which propagates towards the remote sensor is further attenuated along its upward path. Since the transmittance for the reflected radiation is $\tau(\kappa,\theta;p_o,0)$, the quantity $L_{sr}(\kappa)$ is given by

$$L_{sr}(\kappa) = \frac{1}{\pi}(1-\varepsilon)E(d,\kappa)\cos(\theta_s)\tau(\kappa,\theta_s;0,p_o)\tau(\kappa,\theta;p_o,0). \tag{18}$$

The orbit of the Earth around the Sun is elliptic with eccentricity $e' = 0.0167$. The Earth is at perihelion on about January 3 and is at aphelion on about July 5. This implies that the solar irradiance on the top of the atmosphere changes slightly from day-to-day. If $\bar{E}(\kappa)$ denotes the solar irradiance for the mean Sun-Earth distance, then $E(d,\kappa)$ may be evaluated from

$$E(d,\kappa) = \bar{E}(\kappa)\ \{1 + e'\cos(2\pi(d-3)/365)\}^2. \tag{19}$$

The values of $\bar{E}(\kappa)$ may be taken from the measured values on the top of the atmosphere (27,28) or may be evaluated using the following expression

$$\bar{E}(\kappa) = M(\kappa,T)(R_s/D)^2 \tag{20}$$

where $R_S = 695.3 \times 10^6$m is the radius of the Sun, $D = 149.6 \times 10^9$m is the mean Sun-Earth distance, $M(\kappa, T)$ is given by Eqn.(1) with T as the effective temperature of the Sun.

Experimental data for the solar irradiances (27,28) indicate that the Sun does not behave as a perfect blackbody and that the emittance of the Sun is wavelength-dependent. If one compares the experimental data with the blackbody radiation then one finds that the effective temperature of the Sun is wavelength dependent. For example, in the 3.7μm region of the solar spectrum, the Sun behaves as a blackbody radiator at a temperature of 6000K whereas in the 11μm region the effective temperature is 5000K (ref.18, chapter 3).

3.3 Contribution From Atmospheric Emission Reflected at the Sea-Surface.

The downwelling atmospheric emission is obtained by setting $p_1 = p_o$ in Eqn.(9) and integrating it from the top of the atmosphere to sea-level. If the sea-surface can be assumed to be a plane reflector (5,6,25), then only that fraction of the downwelling atmospheric emission for which the angle of incidence is equal to the view angle of the remote sensor contributes to the satellite-sensed signal. The radiation emitted downward by the top atmsopheric layer is attenuated by the entire atmosphere so that the lower limit of integration is $\tau(\kappa,\theta;0,p_o)$. The radiation originating from the bottom layer of the atmosphere is not attenuated and hence the upper limit of integration is 1. As in the previous section, a fraction $(1-\varepsilon)$ of this downwelling atmospheric emission is reflected at the sea-surface and a fraction $\tau(\kappa,\theta;p_o,0)$ of this reflected radiation reaches the top of the atmosphere. Thus, the contribution $L_{ar}(\kappa)$ is given by

$$L_{ar}(\kappa) = (1-\varepsilon)\tau(\kappa,\theta;p_o,0) \int_{\tau(\kappa,\theta;0,p_o)}^{1} B(\kappa,T(p))d\tau(\kappa,\theta;p,p_o). \qquad (21)$$

When regarded as a function of pressure p the above relation can be written as

$$L_{ar}(\kappa) = (1-\varepsilon)\tau(\kappa,\theta;p_o,0) \int_{o}^{p_o} B(\kappa,T(p))\frac{d\tau}{dp}(\kappa,\theta;p,p_o)dp \qquad (22)$$

where we have used the identity

$$d\tau(\kappa,\theta;p,p_o) = \frac{d\tau}{dp}(\kappa,\theta;p,p_o)dp. \qquad (23)$$

We have utilized the relations (21-23) in this work. However, it is worth pointing out that the sea-surface may also approximate to a Lambertian surface. In such a case, $L_{ar}(\kappa)$ is given by

$$L_{ar}(\kappa) = \frac{1}{\pi}(1-\varepsilon)\tau(\kappa,\theta;p_o,0)E_a(\kappa) \qquad (24)$$

where $E_a(\kappa)$ is the irradiance of a unit area of sea-surface due to atmospheric emission,

$$E_a(\kappa) = \iint B(\kappa, T(p))\, d\tau(\kappa, \theta'; p, p_o) \cos(\theta')\, d\Omega'. \tag{25}$$

3.4 Contribution From Atmospheric Emission Towards The Sensor

The contribution $L_a(\kappa)$ in Eqn.(13) is obtained by integrating Eqn.(9) from the sea-level to the top of the atmosphere and by noting that $p_1 = 0$. Thus,

$$L_a(\kappa) = \int_{\tau(\kappa, \theta; p_o, 0)}^{1} B(\kappa, T(p))\, d\tau(\kappa, \theta; p, 0). \tag{26}$$

Or, in terms of pressure variable, we have

$$L_a(\kappa) = \int_{p_o}^{0} B(\kappa, T(p)) \frac{d\tau(\kappa, \theta; p, 0)}{dp}\, dp. \tag{27}$$

The total radiance reaching the remote sensor is obtained by substituting Eqs.(14,15,18,22 and 27) into Eqn.(13),

$$\begin{aligned} L(\kappa) &= \varepsilon\, B(\kappa, T_s) \tau(\kappa, \theta; p_o, 0) + \\ &+ \frac{1}{\pi}(1-\varepsilon)\, E(d, \kappa) \cos(\theta_s)\, \tau(\kappa, \theta_s; 0, p_o) \tau(\kappa, \theta; p_o, 0) + \\ &+ (1-\varepsilon)\, \tau(\kappa, \theta; p_o, 0) \int_{0}^{p_o} B(\kappa, T(p)) \frac{d\tau(\kappa, \theta; p, p_o)}{dp}\, dp + \\ &+ \int_{p_o}^{0} B(\kappa, T(p)) \frac{d\tau(\kappa, \theta; p, 0)}{dp}\, dp. \end{aligned} \tag{28}$$

The radiance sensed (recorded) by the detector $L^*(\kappa)$ is obtained by substituting Eqn.(28) in Eqn.(11), provided that all the quantities on the r.h.s. of Eqn.(28) are known.

3.5 Inflight Calibration and Satellite-Recorded Radiance

In order to compensate for changes in the response characteristics of the detectors and the electronics of satellite radiometers some form of inflight calibration is incorporated into most of the instruments currently in use. The calibration is performed by directing the radiometer towards sources of radiation whose radiance is known. One convenient standard source is deep space which has zero radiance at infrared wavelengths. The other calibration points are provided by blackbody targets carried within the radiometer. The blackbodies are either kept at a known temperature or the temperature is monitored.

The infrared radiometers carried by the GOES satellites are not dynamically calibrated. Those flown on METEOSAT and NIMBUS-7 (the Coastal Zone Colour Scanner) use a single point calibration from on-board blackbody sources. The Advanced Very High Resolution Radiometer (AVHRR) is calibrated using a two-point calibration, the two sources of radiance being an on-board blackbody target and deep space.

Performing inflight calibration of the detectors enables one to compute the satellite-recorded radiance $L^*(\kappa)$. Within the digitization approximation, evaluation of $L^*(\kappa)$ from the digital data $DN(\kappa)$ assumes a linear relation of the form

$$L^*(\kappa) = G(\kappa)\ DN(\kappa) + I(\kappa) \qquad (29)$$

where κ signifies that the various quantities refer to spectral channel centered at wavenumber κ, $G(\kappa)$ is the slope and $I(\kappa)$ is the intercept.

The inflight calibration procedure for the CZCS is described in ref.29. Here, we shall present an example of the two point calibration of the AVHRR instrument.

The blackbody target temperature is monitored by four platinum resistance thermometers (PRTs). The signal output from these PRTs is digitized and stored in the data stream. Using these data the blackbody target temperature is computed using a predetermined algorithm (30). If $\overline{T}$ denotes the mean target temperature then the radiance $L(\kappa,\overline{T})$ detected by a detector is computed from

$$L(\kappa,\overline{T}) = \sum_{i=1}^{n} B(\kappa_i,\overline{T})\ \hat{\phi}(\kappa_i)\Delta\kappa_i \qquad (30)$$

where B is the Planck function, $\hat{\phi}(\kappa_i)$ is the normalized spectral response function of the detector, Eqn.(12), n denotes the number of discrete wavenumbers within the spectral window at which the response of the detector was measured during the preflight investigation, $\Delta\kappa_i$ is the width of the ith interval within which the response function was measured.

NESS (30) supplies 60 values of the normalized response function with specified values of $\Delta\kappa_i$. Using these data, the standard radiance $L(\kappa,\overline{T})$ can be evaluated from Eqn.(30).

Strictly speaking, the response of a detector may not be a linear function of the radiance as was assumed in Eqn.(29) and this is the case for the detectors used in the 11-12μm region in the AVHRR. Therefore, the slopes and intercepts must be corrected for this non-linearity. To a first approximation, this is achieved by adding a correction term to the deep space radiance (30, Table 2). Let $L(\kappa,ds)$ denote the deep space radiance including the correction term and if $X(\kappa,\bar{T})$ and $X(\kappa,ds)$ represent the digital numbers (which are averaged over many samples) corresponding to the radiances from the on-board target and deep space respectively, then Eqn.(29) yields

$$G(\kappa) = \frac{L(\kappa,ds) - L(\kappa,\bar{T})}{X(\kappa,ds) - X(\kappa,\bar{T})} \tag{31}$$

and

$$I(\kappa) = L(\kappa,ds) - G(\kappa)\, X(\kappa,ds). \tag{32}$$

The satellite-recorded radiance from the direction of a given picture element is now given by Eqn.(29). For the purpose of later reference, the values of slopes and intercepts for a scene of our interest are given in Table.2.

Table 2

TIROS-N, Orbit No. 4544, 31-8-79, 15:48 GMT
Solar zenith angle = 52.6°
View angle = 28.6°
Seawater temperature = 290K
1 TRU = 1 mW/(m^2 sr cm^{-1})

Wavelength (μm)	3.7	11
Deep Space Radiance (TRU)	0.0	-1.151
Slope (TRU/Count)	-1.619×10^{-3}	-0.2007
Intercept (TRU)	1.5834	198.6006
DN (Counts)	705	495
$L^*(\kappa)$(TRU)	0.3643	99.2393
Digitization error (TRU)	$\pm 0.8095 \times 10^{-3}$	± 0.1004

3.6 Estimation of Satellite-Recorded Radiance Components

The relations (11,13 and 14) indicate that for a non-scattering atmosphere, there are four terms contributing to the satellite-recorded radiance. To see the relative importance of each of these terms we consider a typical example of an AVHRR sub-scene for which the in-situ sea-surface temperature measurements exist. By comparing the sea-surface temperature maps with the AVHRR data, a typical picture element was selected. The measured sea-surface temperature within this particular pixel was identified to be 290K. For the AVHRR channels 3 and 4, the relevant inflight calibration constants, the satellite data for the selected pixel and the satellite-recorded radiances from this pixel (computed from Eqn.29) are listed in Table 2.

Since the satellite data are in a digitized form, there is some loss of the actual information sensed by the sensor. It is easy to see that this digitization error is $\pm$ half the slope value. These values are also included in Table 2 because while comparing the magnitudes of various terms in the radiative transfer equation (28) these terms should also be compared with the digitization error. The reason for this is that if certain contributions are smaller than the error involved due to digitization then there does not seem to be any good reason for including such small contributions in the radiative transfer equation.

For simplicity we substitute Eqs.(13) and (14) in Eqn.(11) and write it as

$$L^*(\kappa) = L^*_e(\kappa)+L^*_{sr}(\kappa)+L^*_a(\kappa)+L^*_{ar}(\kappa) \quad (33)$$

where, we recall that the meaning of '*' is that these are satellite-recorded radiances. The radiances $L^*_e(\kappa)$ and $L^*_{sr}(\kappa)$ can be computed from Eqs.(15),(18) and (11) provided that the atmospheric transmittance is known. $L^*(\kappa)$ is given by Eqn.(29). Thus, the total atmospheric contribution, from Eqn.(33), is

$$L^*_a(\kappa)+L^*_{ar}(\kappa) = L^*(\kappa)-L^*_e(\kappa)-L^*_{sr}(\kappa). \quad (34)$$

In order to estimate the relative strength of $L^*_a(\kappa)$ and $L^*_{ar}(\kappa)$ we make a crude assumption, namely,

$$L^*_{ar}(\kappa) \simeq (1-\varepsilon)\tau(\kappa,\theta;p_o,0)L^*_a(\kappa). \quad (35)$$

In this example, the atmospheric transmittance is not known. For this reason the various contributions have been evaluated as a function of the atmospheric transmittance. These results are shown in figure 4 for the 3.7μm channel and in figure 5 for the 11μm channel of the AVHRR. From these results and the data in Table 2 we may draw the following conclusions:

Concerning the 3.7μm Spectral Channel. (a) The contribution from the solar radiation reflected at the sea-surface is comparable to the contribution from the sea-surface emission and, therefore, the solar contribution must be removed from the day time satellite data. (b) The negative values of atmospheric contributions imply too large a value of atmospheric transmittance. This means that the contributions from the sea-surface and solar radiation have been over estimated. The physically meaningful atmospheric contributions should be positive. The zero atmsopheric contribution would imply a unit transmittance, but figure 4 shows that it corresponds to vertical transmittance of about 0.9. The amount of negative atmospheric contribution is equivalent to a

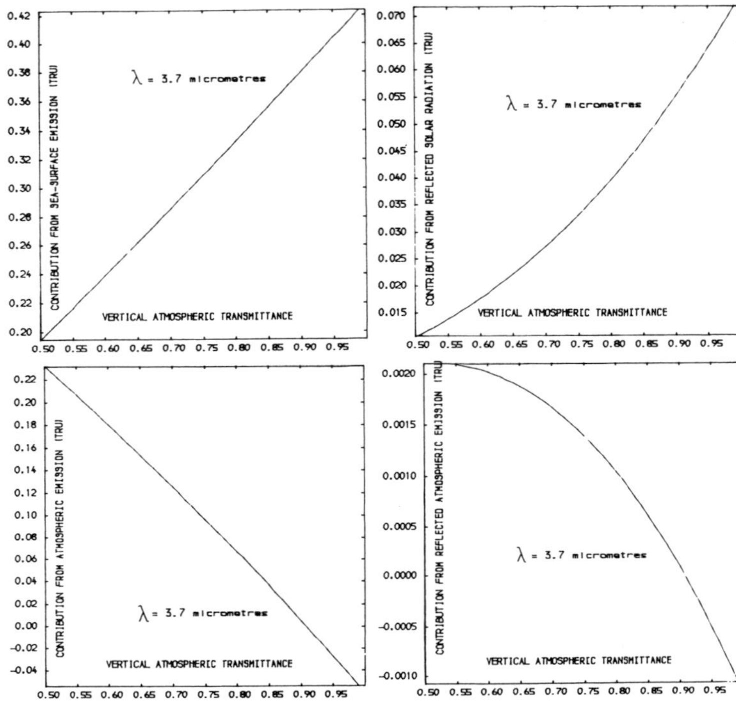

Figure 4. Various components of the satellite-recorded radiance in the 3.7μm channel of AVHRR.

digital number of about 31 counts. Thus, the digital number shown in Table 2 is too large by about 31 counts. And (c) further examination of the raw data in the vicinity of the picture element under consideration indicated that there is an alternate pixel striping (there was no striping in the 11μm channel). Alternate pixel values differed by about 50 counts but the count of 705 shown in Table 2 is the lower value. Therefore, there seems to be some other problem associated with this spectral channel which is responsible for the shift in the count values by about 31 counts.

Concerning the 11μm Spectral Channel. For physically meaningful atmospheric transmittances, (a) the contribution from the solar radiation reflected at the sea-surface is about ten times smaller than the digitization error. Therefore, for all practical purposes, the solar contribution can safely be ignored. (b) The atmospheric contributions constitute a sizeable fraction of the contribution from the sea-surface emission. The atmospheric

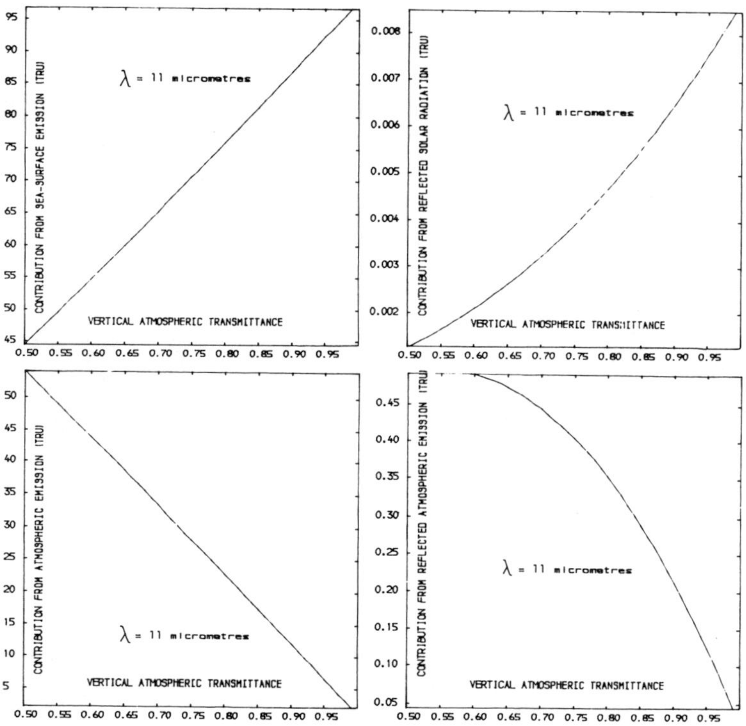

Figure 5. Various components of the satellite-recorded radiance in the 11μm channel of AVHRR.

contributions are also large compared to the error due to digitization. Therefore, the atmospheric components should be removed from the satellite data.

Because of the problem of negative atmospheric contributions in the 3.7μm spectral channel, it is not possible to make a definite statement about the relative effect of the atmospheric contribution in the two spectral channels considered, but it has been claimed (25) that the atmsopheric effects in the 3.7μm channel are less serious than in the 11μm channel.

4. THE BRIGHTNESS TEMPERATURE

Let T_b denote the temperature of a blackbody the radiance of which is equal to the radiance received by the detector. The radiance $L^*(\kappa)$ detected by the detector is related to this blackbody radiance $B(\kappa,T_b)$ through Eqn.(11) with $L(\kappa')$ replaced by

$B(\kappa', T_b)$. Thus, in practice we have

$$L^*(\kappa) = \sum_{i=1}^{n} B(\kappa_i, T_b)\hat{\phi}(\kappa_i)\Delta\kappa_i \tag{36}$$

where we recall that κ refers to the spectral channel centred around wavenumber κ. In principle, the relation (36) can be solved for the temperature T_b which is usually called the brightness temperature. But, in our case Eqn.(36) cannot be solved for T_b exactly because the left hand side quantity is an integrated quantity over the spectral band width. Therefore, we seek an approximate solution for T_b as follows.

We select a range of T_b which is appropriate to the sea-surface temperatures encountered. Since the response function of the detector is known, the value of $L^*(\kappa)$ can be computed for various values of T_b within the selected range of T_b. To find a relation between T_b and $L^*(\kappa)$ we note that T_b appears in the exponential form through the Planck function. Therefore, a physically meaningful relation should have the following form

$$\ln(L^*(\kappa)) = \alpha + \frac{\beta}{T_b} \tag{37}$$

where α and β are parameters which depend on the selected range of T_b and the absolute value of T_b. These parameters can be determined by a least squares analysis. In Table 3 we have shown the results obtained from Eqn.(37) for three different temperature ranges. In obtaining these results, the radiance was expressed in terms of a convenient unit which we refer to as the thermal radiance unit (TRU) and which is defined as

$$1 \text{ TRU} = 1 \text{ mW}/(\text{m}^2 \text{ sr cm}^{-1}). \tag{38}$$

A high correlation coefficient (better than -0.999998) was obtained for all the temperature ranges shown in Table 3. The data in column five of this table corresponding to a given temperature range (column two) is the maximum absolute difference between T_b and the recomputed value of T_b using the parameters α and β . The worst result of $|\Delta T|_{max}$ = 0.2K is obtained for the 11μm channel and for a temperature range from 200 to 310K. This range of temperatures may probably be suitable when one is interested in the temperatures of sea and clouds. For smaller temperature ranges, the value of $|\Delta T|_{max}$ is smaller. In conclusion, therefore, the relation (37) with known parameters α and β serves as a convenient relation for computing the brightness temperature T_b from the satellite-sensed radiance $L^*(\kappa)$.

Perhaps this is a good place to estimate the error in the brightness temperature due to the digitization error. It is easy to see from Eqn.(37) that this type of error in T_b is given by

$$\Delta T = \frac{T_b^2}{\beta} \frac{\Delta L}{e^{(\alpha + \beta/T_b)}} \quad (39)$$

where ΔL is the digitization error. Typical values of ΔL are given in Table 2. Using the data from Tables 2 and 3 in Eqs.(37) and (39) for 11μm channel and for the temperature range 270 to 300K we obtain $\Delta T = \pm 0.06$K. Comparison of this error

Table 3. Results obtained from a least squares analysis using Eqn.(37) for 3.7 and 11μm channels and for various temperature ranges.

Wavelength μm	Temperature range (K)	α	$-\beta$ (K)	$\lvert\Delta T\rvert_{max}$ (K)
	200 - 310	12.631	3117.08	0.06
3.7	250 - 310	13.060	3587.10	0.01
	270 - 300	13.282	3862.09	0.002
	200 - 310	9.126	1312.52	0.20
11	250 - 310	9.145	1317.85	0.06
	270 - 300	9.149	1319.16	0.01

with the value of $\lvert\Delta T\rvert_{max}$ in Table 3 corresponding to the same temperature range indicates that the relation (37) is a fairly good approximation for computing brightness temperatures.

5. ATMOSPHERIC CORRECTION STRATEGIES

Given an array of satellite brightness temperatures which are uncontaminated by values from pixels containing cloud or troublesome land surfaces, but which have been properly rectified and transformed onto a convenient map projection there are a number of alternative techniques by which the effect of the atmospheric contribution can be estimated.

5.1 Calibration of an Individual Image

In principle it ought to be possible to calibrate an individual image or portion of an image from a reasonable number of simultaneous in-situ SST measurements. One of the factors which limits the accuracy of this technique is the scale of the horizontal variations of the atmospheric correction which is largely determined by the absorption by water vapour, see Table 1. This may be judged by examining the horizontal distribution of the water vapour content of a vertical column. For instance, the distribution described in figure 1 leads to differences between satellite brightness temperature and the in-situ measurement increasing by a factor of four from 0.5K to 2.0K, from the drier region in the north-east to the moister area in the south-west. In the remainder of this section we present an illustration which describes this calibration procedure.

About 50 sea-surface temperature maps for the south western coast of Portugal were produced (31) by correlating ship data with airborne infrared sensor data at the University of Lisbon. The accuracy of these temperature maps was only 0.5K. These temperature data were provided to us by Professor Fiuza (31) in the hope that we might be able to validate the AVHRR data. The aircraft was flown at an altitude of about 150m so as to minimize the atmospheric contributions. Platt and Troup (32) have also carried out similar experiments. Their results indicated that the sea-surface temperatures and the brightness temperatures measured from such a low altitude differed by about 0.5K whereas the estimated error in the brightness temperature was ± 0.4K.

Out of these 50 data sets, 18 sets were for relatively cloud-free days. Only five of these data sets have been compared with the brightness temperature obtained from the TIROS-N AVHRR data for the 11μm channel. Since the temperature appears through the Planck function, the brightness temperature T_b and the sea-surface temperature T_s should have a relation of the form

$$B(\kappa,T_b) = \varepsilon\tau^* B(\kappa,T_s) \quad (40)$$

where ε is the emittance of water and τ^* is the effective transmittance of the atmosphere within the instantaneous field of view of the sensor. For the temperature range of our interest and for the infrared wavelengths $\exp(c_2\kappa/T) \gg 1$, where T stands for either T_b or T_s and c_2 is given below Eqn.(1). Therefore, Eqn.(40) simplifies to

$$\frac{1}{T_b} = a_1 + \frac{1}{T_s} \quad (41)$$

with

$$a_1 = -\ln(\varepsilon\tau^*)/c_2\kappa. \quad (42)$$

The relation (41) suggests that if a suitable number of T_s and the corresponding brightness temperatures T_b are available then one can relate T_s and T_b using a relation of the form

$$\frac{1}{T_b} = a_1 + \frac{b_1}{T_s} \quad (43)$$

where a_1 and b_1 are parameters to be determined by a least squares analysis. By comparing Eqn.(43) with Eqn.(41) we may think that the value of b_1 should be unity if Eqn.(41) is a fairly good approximation of Eqn.(40). In practice, however, (i) the quantities T_s and T_b are not error free, (ii) T_s is usually a point measurement whereas T_b is an average over the instantaneous field of view of the detector, (iii) T_b is the sea-surface (skin) parameter whereas T_s is a bulk property, (iv) the difference between the skin and bulk temperature could be different from place-to-place, (v) the satellite data are recorded fairly quickly compared to the time required for collecting a reasonable set of in-situ data from a ship. Therefore, the atmospheric conditions may change during the in-situ SST measurements. Because of reasons such as these, we have included in Eqn.(43) the parameter b_1 whose value could be different from unity.

Table 4. Results from a least squares analysis using equation (43).

Date	15/6/79	27/7/79	17/8/79	24/8/79	31/8/80
Number of data (n)	7	7	7	5	5
View angle θ (degrees)	40.0	10.7	35.0	5.2	28.6
Correlation Coefficient R	0.996	0.998	0.997	0.999	0.998
$a_1 \times 10^{-4} (K^{-1})$	-5.47	1.73	4.24	5.31	4.17
b_1	1.16	0.95	0.87	0.85	0.88
$\sqrt{\Sigma(T_s-T_b)^2/n}$ (K)	0.6	0.1	0.8	0.4	0.3

In Table 4 we have included the results obtained from a least squares analysis using Eqn.(43). However, these results are obtained for a small number of data values shown in column two. This limitation is due to a small variation in the measured sea-surface temperatures and the poor accuracy (0.5 deg K) of sea-surface temperature measurements. In line three we have listed the view angles for the centre of the scene of our interest; they have been computed using Eqn.(16). The view angles are included because the brightness temperature depends on the atmospheric transmittance and the atmospheric transmittance depends on the view angle as indicated by Eqn.(4). Therefore, even if the atmospheric conditions were unaltered from day to day (which is highly unlikely) one cannot expect a unique relation of the form of Eqn.(43), i.e. with unique values of a_1 and b_1, to represent the relationship between T_b and T_s without expecting poorer correlation coefficients than those shown in Table 4 for the individual dates. The variation in the value of a_1 from day to day indicates that the effective transmittance τ^* is varying from day to day. As expected, the value of b_1 is not unity but deviates from unity by up to 16%. The entries in the last line of Table 4 are the rms differences between T_b and T_s. The smallness of these values indicate dry atmospheric conditions. Unfortunately, the temperature, pressure and humidity profiles were not recorded for all of the dates listed in Table 4. The only profile which was recorded was for 15th June 1979 and indicates dry atmospheric conditions.

5.2 Calibration Over a Short Period

A succession of images of the same area made over a few days can be calibrated in a similar way although a similar constraint applies. Calibrations made in this way are unlikely to be of any use for images of other places or even of the same area at a different time. They should best be regarded as means of interpolating the SST between conventional measurements.

5.3 Explicit Calculations of the Atmospheric Correction

If a transmittance function is available and if temperature and humidity profiles for the appropriate area can be obtained then it is possible to compute the atmospheric correction, the accuracy of which depends on the accuracy of the transmittance function and on the model of the radiative transfer equation. Various transmittance models have been explored in the literature (for example, see references 5, 6, 24, 32, 33). Although the solar contribution term in Eqn.(28) can be ignored for the 11μm channel, the reflection term arising from atmospheric emission has been ignored by some investigators (6, 34). The error introduced by neglecting this term is much larger than the digitization error (Eqn.(39) and ref.25).

Obtaining the sounding data is likely to present less of a problem. Although the Ocean Weather Ship network is very sparse, so that radio-sonde ascents are not available over large areas of the ocean, these data can be obtained from the operational temperature and humidity sounders flown on the NOAA satellites. Until recently (17th November 1981) the data from these instruments was used to compute the atmospheric correction to obtain the GOSSTCOMP(Global Ocean Sea-Surface Temperature Compilation) sea-surface temperatures. The rms difference between the GOSSTCOMP SST corrected this way and the simultaneous <u>in-situ</u> data is found to be about 1.8 deg K (35).

The European Space Agency (ESA) uses a similar method to obtain sea-surface temperatures from the METEOSAT infra-red channel. The necessary sounding data is taken, not from measurements, but from a numerical weather prediction model operated by the United States Weather Service.

We have used the atmospheric correction program (34) for the data of 15th June 1979 discussed in the previous section. The correction was found to be about 0.5 deg K which is in agreement with the rms difference between the brightness temperature and *in-situ* SST, Table 4.

5.4 Multi-spectral Correction

There is a strong incentive to develop a method of obtaining corrected SSTs from data collected entirely by a satellite. Only if this is possible can the advantages of the greatly extended coverage and frequency of observation be realised.

The multi-spectral technique was first suggested by Anding and Kauth (36). They proposed making radiance measurements in two wave bands, one between 7μm and 9μm and the other, on the other side of the ozone band, between 10μm and 12μm. They argued that since the same physical processes are responsible for absorption in both of these wave bands the effect in one ought to be proportional to that in the other. Therefore one measurement in each wavelength interval could be used to eliminate the atmospheric effect.

A more formal justification of the technique can be obtained by judicious manipulation of the radiative transfer equation (28). Ignoring the solar contributions, this relation can be rewritten as

$$B(\kappa,T_b) = B(\kappa,T_s) - \int_{po}^{o} [B(\kappa,T_s)-B(\kappa,T(p))]\frac{d\tau}{dp}(\kappa,\theta;p,o)dp$$

$$-(1-\varepsilon)\tau(\kappa,\theta;p_o,0)[B(\kappa,T_s)-\int_{o}^{po} B(\kappa,T(p))\frac{d\tau}{dp}(\kappa,\theta;p,p_o)dp \qquad (44)$$

where the second and third terms are the contributions due to direct and reflected atmospheric emission, respectively. If we assume that (a) T_b and $T(p)$ are close to the sea-surface temperature T_s then the Planck function can be expanded about T_s. Retaining only the first order term, we have

$$B(\kappa,T) = B(\kappa,T_s) + (T-T_s)\frac{\partial B}{\partial T}\bigg|_{T_s} \qquad (45)$$

where T stands for either T_b or $T(p)$. To proceed further we need two more assumptions: (b) we include the effect of water vapour only and (c) the transmittance is a linear function of the water vapour content. These three assumptions enable one to express the direct atmospheric emission effect as (25)

$$\Delta T_a(\kappa) = k(\kappa)f(T(p),m(p)) \tag{46}$$

where the absorption coefficient is a function of a wavenumber and water vapour content m(p). The function f(T(p),m(p)) is independent of wavenumber and depends only on the atmospheric temperature and pressure profile.

If the reflection term in Eqn.(44) is ignored for the moment then to correct the brightness temperatures obtained from two of the infra-red channels of the AVHRR we need only know the values of $k(\kappa)$. Then the function f(T(p),m(p)) can be eliminated and a corrected SST obtained. In practice, the corrected SST is obtained using a linear relation of the form

$$T_s = e_o + e_1 T_b(\kappa_1) + e_2 T_b(\kappa_2) \tag{47}$$

where κ_1 and κ_2 refer to the two spectral channels and the parameters e_o, e_1 and e_2 are determined by a multivariate regression analysis. The effects of other absorbers and of the reflection term are, therefore, automatically included.

Several simulations of the multispectral correction have been carried out in order to assess its likely accuracy. The simulations depends heavily on the model of the atmospheric transmittance they employ. Certain features, such as the effect of aerosols, are difficult to model and the results of such simulations seem to be optimistic when contrasted with the results obtained in practice.

5.5 Multispectral Correction in Practice

It was hoped that the two infra-red channels of the AVHRRs flown on TIROS-N and NOAA-6 would provide data from which the SST could be obtained to an accuracy of better than 1 deg K. Unfortunately, although computer simulations predicted that this would be possible it was found that during daylight hours there is a sufficiently large reflected solar component in the 3.7μm channel to swamp the SST information completely. Data from this channel are useful during the night, although even then it does seem to be very sensitive to the presence of solid aerosol and haze or mist in the lower atmosphere. Later versions of the AVHRR incorporate a third infrared channel between 11μm and 12μm which should overcome this difficulty. In fact since November 1981 the SSTs published by NOAA have been corrected by a multispectral technique, using the 11 mμ and 12μm data.

5.6 Multipath Correction

An alternative to the multispectral technique, which still permits the satellite to gather all the data required is the multipath correction. This requires a radiometer capable of looking in two directions at once. Such an instrument is planned for the European Satellite ERS-1. Figure 6 shows schematically how the instrument would operate. A modified version of the AVHRR would be flown scanning the Earth's surface across its own track vertically downwards and looking forwards at, say, 30 degrees to the vertical. Each point on the surface will then be scanned twice within two minutes. The paths through the troposphere will be sufficiently close together for differences in the temperature and humidity to be small enough to be ignored. The atmospheric correction for the vertical and the slant path will be simply related to the difference in the two path lengths.

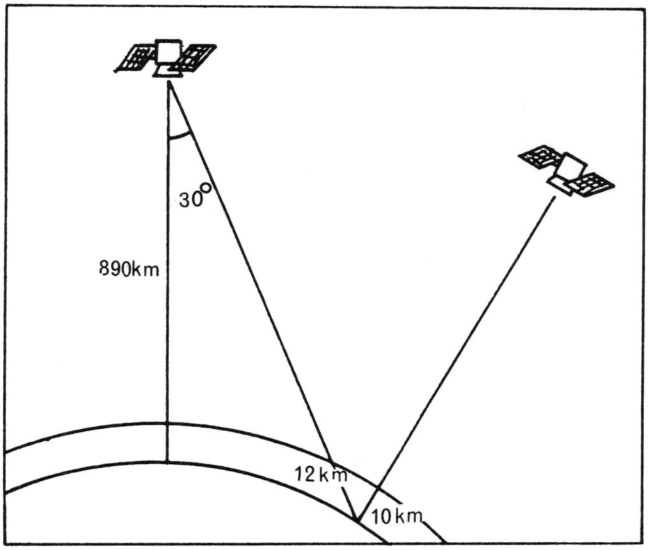

Figure 6. Geometry showing two angle scanning instrument on board European satellite ERS-1.

ACKNOWLEDGMENTS

The sea-surface temperature maps were provided to us by Professor Armando F.G. Fiuza of the University of Lisbon, Portugal. This work was partly supported by a grant from the Science and Engineering Research Council.

REFERENCES

1. Deacon, M., 1972, in "Oceanography, Contemporary Readings in Ocean Sciences", (R.G. Pirie, Ed.), Oxford University Press, pp. 24-44.

2. Gerard, R., 1966, in "The Encyclopedia of Oceanography", Vol.1, (R.W. Fairbridge, Ed.), Reinhold Pub. Co., New York, pp.602-10.

3. Hoffmann, S.O., 1919, Phys. Rev. 14, pp.163-66.

4. Arnquist, W.N., 1959, Proc. of the IRE, 47, pp.1420-30.

5. Sidran, M., 1980, Rem. Sens. Environ., 10, pp.101-14.

6. Prabhakara, C., Dalu, G. and Kunde, V.G., 1974, J. Geophy. Res., 79, pp.5039-44.

7. Rao, P.K., Smith, W.L. and Koffler, R., 1972, Monthly Weather Review, 100, pp.10-4.

8. Simpson, R.H., 1972, in "Oceanography, Contemporary Readings in Ocean Sciences", (R.G. Pirie, Ed.), Oxford University Press, pp.171-76.

9. Anderson, G.C., 1964, Limnol. Oceanogr., 9, pp.284-302.

10. Curl, H. and Small, L.F., 1965, Limnol. Oceanogr., 10, pp.R67-73.

11. Smith, R.L., 1972, in "Oceanography, Contemporary Readings in Ocean Sciences", (R.G. Pirie, Ed.), Oxford University Press, pp.126-47.

12. Ryther, J.H., 1969, Science, 166, pp.72-6.

13. Knauss, J.A., 1978, "Introduction to Physical Oceanography", Prentice-Hall, Inc., Englewood Cliffs, New Jersey.

14. LaFond, E.C., 1966, in "The Encyclopedia of Oceanography", Vol.1, (R.W. Fairbridge, Ed.), Reinhold Pub. Co., New York, pp.957-59.

15. Simpson, J.H., 1981, in "Remote Sensing in Meteorology, Oceanography and Hydrology", (A.P. Cracknell, Ed.), Ellis Horwood Ltd., Chichester, pp.295-311.

16. Estes, J.E. and Senger, L.W., 1972, Rem. Sens. Environ., 2, pp.141-63.

17. "Manual of Remote Sensing", 1975, Vol.II, (L.W. Bowden, Ed.), American Society of Photogrammetry, Falls Church, Virginia, pp.1570-74.

18. Slater, P.N., 1980, "Remote Sensing, Optics and Optical Systems", Addison-Wesley Pub. Co.

19. Buettner, J.K. and Kern, C.D., 1965, J. Geophys. Res., 70, pp.1329-37.

20. Hayes, R.M., 1980, "Proc. of the Fourteenth Intern. Symp. on Rem. Sens. of Environ.", Environmental Res. Inst. of Mich., Ann Arbor, Michigan, pp.1187-96.

21. "Manual of Remote Sensing", 1975, Vol.I, (F.J. Janza, Ed.), Am. Soc. of Photogrammetry, Falls Church, Virginia, U.S.A.

22. Houghton, J.T., 1977, "The Physics of Atmospheres", Cambridge University Press.

23. Gross, E.P., 1955, Phys. Rev., 97, pp.395-403.

24. Bignell, K.J., 1970, J. Roy. Met. Soc., 96, pp.390-403.

25. Phulpin, T. and Deschamps, P.Y., 1980, "Proc. of the Sixth Annual Conference of the Remote Sensing Society", (A.P. Cracknell, Ed.), pp.47-58.

26. Chandraseker, S., 1960, "Radiative Transfer", Dover Publication, New York, Chapter 1.

27. "American Institute of Physics Handbook", 1972, (D.E. Gray, Ed.), Third Ed., p.6-216.

28. Thekaekara, M.P., Kruger, R. and Duncan C.H., 1969, Appl. Opt., 8, pp.1713-32.

29. Cracknell, A.P. and Singh, S.M., 1981, "Proc. of an EARSeL-ESA Symp., Voss, Norway", pp.157-67.

30. "NOAA/NESS Technical Memorandum, NESS 107, 1979, Data Extraction and Calibration of TIROS-N/NOAA Radiometers", Washington, D.C.

31. Fiuza, A.F.G., 1981, Report on one year of sea-surface temperature measurements between Cape Raso and Vila Nova de Milfontes, Oceanography group, University of Lisbon, Portugal.

32. Platt, C.M.R. and Troup, A.J., 1973, Rem. Sens. of Environ., 2, pp.243-7.

33. Davis, P.A. and Viezee, W., 1964, J. Geophys. Res., 69, pp.3785-94.

34. Weinreb, M.P. and Hill, M.L., 1980, NOAA Technical Report NESS 80, Washington, D.C.

35. McClain, E.P., 1980, "Proc. of COSPAR/SCOR/IUCRM Symp. on Oceanography from Space", May 1980, Venice.

36. Anding, D., and Kauth, R., 1969, Report 2676-1-P, Willow Run Laboratories, Inst. of Sc. and Tech., University of Michigan, Ann Arbor, Michigan.

CHAPTER 13

REMOTE SENSING OF BIOLOGICAL SUBSTANCES

Charles S. Yentsch

Bigelow Laboratory for Ocean Sciences
W. Boothbay Harbor, Maine 04575 U.S.A.

ABSTRACT

The time and space problems so traditional to the study of ocean processes can be partially resolved using high altitude colorimetry and fluorescence. The primary goal of remote sensing has been to estimate the amount of photosynthetic pigment - chlorophyll. The remote systems for sensing chlorophyll depend on the manner in which phytoplankton chlorophyll influences the color of water and/or light emission (fluorescence) from the pigment itself.

The obvious justification for the measurement of phytoplankton chlorophyll is that these organisms constitute the first link in the food chain. However, phytoplankton patterns of distribution in time and space provide information as to how oceanographic processes regulate primary production.

INTRODUCTION

In 1969, a group of meteorologists and oceanographers was assembled by the U.S. National Academy of Sciences to discuss the potential uses (research) for studying the oceans. The group to which I was assigned was chaired by Dr. Gifford Ewing, a research oceanographer who had pioneered the study of features on the oceans surface using aircraft. This group had wide interdisciplinary coverage; frequently one found the discussion flowing from water color to beach erosion. Giff Ewing assigned me the task of searching out the research needs which could be

A. P. Cracknell (ed.), Remote Sensing Applications in Marine Science and Technology, 263–297.

satisfied or supplemented by viewing ocean color from space. I approached this problem by assembling some prestigious oceanographers and asking if, in their opinion, we had the means for studying the ocean; were observations made from ships giving the correct picture of the oceans? Was it possible that the picture (from ships) was distorting our mental picture and perhaps misdirecting our science. All of the consultants agreed that the time and space problems in oceanography were staggering and therefore, the experimental approach as common to other disciplines, was difficult to design and interpret. Most felt that they could not ascertain how far their conception of "how the ocean works" - was in error, but they all admitted that their interpretations were influenced by sampling strategy which could influence one's thinking about how things work.

The point of telling this story is to emphasize the need for remote sensing. I have been asked "suppose you have a global picture of chlorophyll patterns, what will you do with it?" The answer is that we already believe we have the pattern and know it's causes - a knowledge obtained by years of activity on the high seas. But, is it the correct pattern? Has the enormity of the oceans and the temporal scale of the oceans and the temporal scale of time fooled us?....This is where the satellite information can play a major role, by showing that the patterns of chlorophyll in water masses support or reject our hypotheses on how the fertility of the oceans functions and is distributed.

The substance of this lecture series, because of the prejudices of the speaker, is strongly influenced by phytoplankton dynamics which presents a procedural problem. I could, for example, discuss spatial and temporal changes in phytoplankton abundance and attempt to demonstrate the role remote sensing might have in these studies. On another course, I could show how remote sensing of phytoplankton abundance is possible and the physical-chemical basis which allows us to see organic signals on earth from space.

I have chosen the latter approach for two reasons. First, I believe anyone involved in trying to understand the patterns of ocean color and/or fluorescence should know something of the cellular processes behind these. Secondly, the application of these remote tools, colorimetry and fluorescence, are just in the formative stages. Many of us are trying to bridge the gap of total visualization as opposed to a scheme produced by putting together scalar bits. With this in mind, it seems logical that some statement - the importance of phytoplankton has to be made. This then, must be followed by how phytoplankton produces signatures that can be detected remotely, and concluding with some of the potential applications of the present remote systems.

WHY MEASURE PHYTOPLANKTON?

In the early history of this planet, it is believed that the atmosphere and the oceans contained very little oxygen. However, the oceans had salts and water necessary for life, and as geological time passed, they accumulated a weak broth of organic matter. It is believed that the first living cells arose from this broth. In doing so, they faced a paradox. With no other sources of energy, the the organic broth would soon be used up - the early cells were living on borrowed time. Fortunately, by consuming organic broth (an anaerobic fermentive process), carbon dioxide was produced. The early cells evolved a process of using carbon dioxide to produce organic material. The energy needed was sunlight, and the process, as we know it, is called photosynthesis. The meaning of photosynthesis to the early cells was that they no longer had to live on the organic broth. With photosynthesis they could synthesize and store organic substances.

Since photosynthesis involves the use of light and water as well as carbon dioxide, one of the by-products was oxygen. Its appearance greatly increased the efficiency of decomposing organics and allowed for almost the complete extraction of energy from organic matter - a process we call oxidative respiration. Now the primeal cells could meet their maintenance ration with a small amount of expended energy. Therefore, the importance of oxygen to life as we know it cannot be underestimated; photosynthesis coupled with anaerobic respiration allows organisms to maintain themselves, but photosynthesis coupled with aerobic (oxygenated) respiration allow cells enough surplus to multiply and grow, utilizing less energy.

Another benefit to life occurred with the appearance of oxygen. In the atmosphere a layer of ozone formed which shielded the surface of the earth from harmful ultraviolet radiation. This allowed organisms to develop complex genetic material without the ultraviolet light destroying nucleic acids and proteins. This also allowed organisms to live on the land surface of the planet and not use the water submersion as a protective means from ultraviolet light. In summary, the two principal benefits of photosynthesis are: (1) the production of organic material and oxygen from light energy and resulting aerobic oxidation and growth, and (2) the production of the ozone which shielded cells from harmful wavelengths, thereby genetic diversity.

We cannot conceive of life on earth without plants. Annually, the earth produces about 150 billion tons of carbon, about one half of this occurs in the ocean. Photosynthesis is characterized by the uptake of simple inorganic substances which are converted to complex organic compounds. It is the affinity of these compounds for oxygen which provides the energy for the rest of life in the sea.

In the course of evolution of photosynthesis, organisms needed to develop antennae for capturing sunlight. These appeared as pigmented proteins which function in a manner similar to colored dyes. The pigmented proteins fill the visible window of the sunlight's radiation (Fig. 1). Since none of these proteins would survive in the presence of ultraviolet radiation, a step in their evolution must have involved partial submergence of the cells using the water as a protective shield before the ozone was established. But the cells could not have resided too deep in the water column, because water itself absorbs the energy especially at the long wave lengths of visible light. Thus, the optimal choice for the pigmented antennae had to be somewhere between the ultraviolet and the near infrared, that is the visible window of sunlight. In the course of their evolution, the algae have virtually filled the

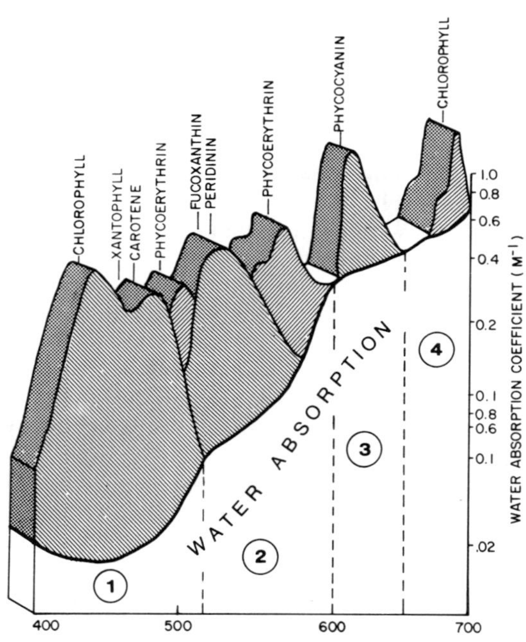

Figure 1. The absorption of light by different algal pigments and the "windows of clarity" in water absorption. The spectra for the pigments approximate those measured in vivo. For the sake of this illustration, fucoxanthin and peridinin absorption are superimposed (43).

visible window with their antennae, establishing firmly their potential to utilize practically all wavelengths in this region for photosynthesis. Yet, even the casual observer recognizes that this visible window of ocean waters is never completely filled - one may not even realize that there are plants there at all, since the ocean is neither a broth nor a soup of algae. But why is it not?

To ask the above question introduces the reasons behind the measurement of phytoplankton and their activities. To complete the introduction we must at a very elementary level explore some of the major causes for the dominance of certain species of algae - and the present day biochemical structure of the oceans. Remembering that the primeal oceans are thought to have had little oxygen, the production of oxygen by photosynthesis presented problems to bacterial procaryotes. The physiology of these problems is not well understood but are manifested in the following way: In some of the ancient procaryotes (evolutionary old) which are still around today, their respiratory rate is accelerated by light and oxygen. It is as if some of these primitive algae cannot keep photosynthesis from reversing itself. As fact as the organic compounds are made in light, they are consumed.

With the presence of oxygen the nitrogen metabolism changed which also posed a problem for the procaryotes. In the anaerobic oceans the nitrogen was in all probability in the reduced form as ammonia. This form could be readily utilized by the procaryotes. However, in the presence of oxygen the form of nitrogen became oxidized to nitrate (NO_3). Thus, the early organisms had to develop a mechanism for reducing nitrate to ammonia for their proten synthesis.

To cope with these problems a group developed we now call eucaryotes. These organisms have a highly organized nucleus containing genetic information which can combat these problems, and they soon began to dominate the seas. Organisms, such as the diatoms, have a very low maintenance respiration and hence active photosynthesis easily produces new growth. They can efficiently convert nitrate to ammonia for protein synthesis. It was, so to speak, a new ball game when the eucaryotes appeared. With adequate light and nutrients they could quickly outcompete the procaryotes. But, the question remains: Why is the ocean not filled with these organisms? The answer to the question is at the crux of why we measure phytoplankton. It depends not only upon the understanding of the physiology of the phytoplankton but knowledge of the dynamics of the water mass movements in the oceans.

In a general sense we can classify the modern oceans as an unbalanced medium for phytoplankton growth. By this I mean that the light energy necessary for photosynthesis penetrates to a depth above the reservoir of nutrients which lies in deep waters. Thus, the upper photosynthetic layers (euphotic) generally contain too little plant nutrients, such as nitrate-nitrogen, which is needed to support vigorous growth. To put it the other way, euphotic waters have limiting concentrations of nutrients because in the processes of photosynthesis, nutrients have been incorporated into new cells faster than they can be resupplied. This situation is characteristic for ocean water masses having a thermocline, that is, two layers at different density. The top buoyant layer is warmed by sunlight and the water is of low density. The bottom layer consists of cold high density water. The nutrient-light imbalance is corrected when two layers mix, and the occasions for mixing may occur on time scales of days to months, depending on the source of energy needed to mix the layers. These sources of energy are generally wind, currents, and tides, in combination with bottom friction. The close coupling between vertical mixing of the two layers and the variability of energy input, i.e., changing winds, tides, currents, etc., means that the patterns of phytoplankton growth will reflect the variability of the energy inputs in both time and space, which in the latter case may involve hundreds of kilometers. Therefore, the prediction of events of phytoplankton growth in today's oceans depends upon having considerable climatological and hydrographic information. Some of the mixing-growth events can be numerically modeled to a reasonable degree of accuracy. Other events, for example, those induced by large-scale oceanic circulation, are much less easily modeled. However, in either case, measurements for confirmation or disputing the models require large-scale synoptic measurements. This is why satellite sensing systems are so attractive.

But still you might ask why must we be able to predict the changes in phytoplankton in time and space? At the onset of this discussion it was clear that the appearance of phytoplankton in early times played a unique role in developing life as we know it today. In my opinion, it follows that by understanding the mechanisms behind this primary formation of organic material we have taken a major step toward the understanding of the other forms of life in the sea. This in the most practical sense aids in man's attempt to use and manage the sea and its resources -- a problem most are aware of. But in the more academic sense we have to understand existing biochemical systems and what are the constraints of the present environment on these systems so we can speculate as to the mechanisms operating in the primitive oceans. Some of those primitive systems still exist in the sea, and it is against this background that a sensible means can be sought to explore the origin of life. It is also against this background that remote sensors are developed. By sensing growth processes on earth the same sensors can be put into space for the search of life on other worlds.

FACTORS AFFECTING WATER COLOR

General

Most of the light entering the sea never leaves it. The amount of light leaving the sea amounts to about 3-5% of that entering it. Thus, the fate of the solar energy is to be eventually absorbed. Critical to this discussion are the absorbing components in natural waters and how they compete for spectral energy. The equations shown in Fig. 2 show the budget in a general fashion, absorbing and scattering components which attenuate the light; the light energy absorbed by water is converted to heat. The light energy absorbed by particles can be used in photochemical processes by algal particles. There are other light absorbing particles, organic and inorganic, whose photochemical properties are poorly understood. All of the absorbing particles scatter light and can function as black bodies which liberate heat. The algal particles fluoresce light of specific wavelengths. Other particles also fluoresce, the nature of these particles and their fluorescence is just now becoming understood. Other substances which absorb light in natural waters are like dissolved dyes. The general term given these is dissolved yellow substances, yellow u.v. absorbing material or Gelbstoff.

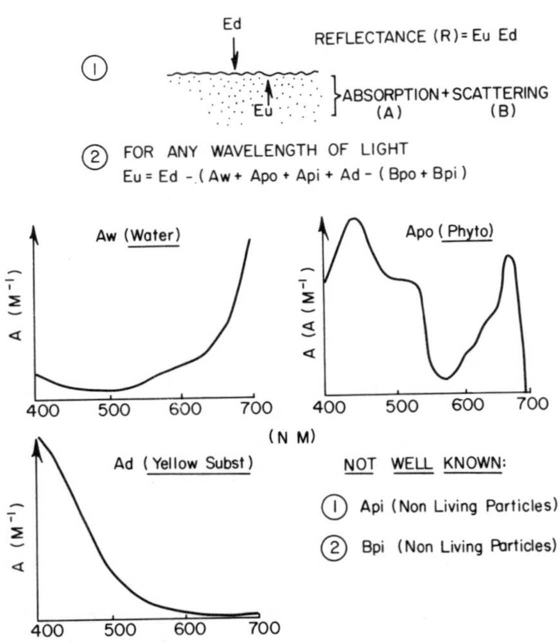

Figure 2. General scheme for partitioning of light in natural waters (43).

A case can be made for an apparent relationship between the absorption by major algal pigments and the major windows of clarity in the water spectrum (Fig. 1). The major "clear window" for water lies between 425-450 nm, is filled with the Soret absorption bands of chlorophylls a, b, c, and carotenoid bands from all groups of algae found in the marine environment. The window of clarity between 525-575 nm is filled with absorption by carotenoids (fucoxanthin and peridinin) of diatoms and dinoflagellates and the phycoerythrins of red and blue-green algae (= cyanobacteria). The third conspicuous window between 525-650 nm is filled by another blue-green algae with chromoprotein-phycocyanin. The fourth window between 665-680 nm, positioned next to a strong water absorption band, is filled with the long wavelength absorption bands of chlorophyll.

Phytoplankton Attenuation

Most of the information in which the visible light spectrum is attenuated by phytoplankton comes from attempts to measure light absorption by algal cells. These, for the most part, are comparative measurements of light passing through solutions of suspended cells and solutions free of cells. The interpretation of these results in terms of cellular absorption (or for that matter attenuation) is complicated by front surface scattering (phase change) between the liquid and the cell wall and by the distribution of pigments and their physical-chemical state. For example, if all of the light E emerging from the cell suspension is measured, then

$$(1) \quad E = T + S$$

where T is that transmitted and S that scattered. Treatment of these data in terms of Beers Law,

$$(2) \quad E_z = E_o e^{-k \cdot c \cdot z},$$

makes it apparent that because the incident light, E_o, has been scattered, the length of the light path is unknown and not easily measured. Furthermore, the concentration of absorbing molecules c is variable; some of the incident light passes unabsorbed while other portions are absorbed. The net result in the absorption coefficient, k, is uncertain.

In natural populations, most of the information on attenuation of light by phytoplankton has come from comparisons of chlorophyll content and the total diffuse attenuation coefficient; in 1956, G.A. Riley pioneered the use of partitioning the diffuse attenuation coefficient into the components responsible for the total attenuation, namely water, phytoplankton and other attenuating substances. Since then, a number of workers have used this approach to estimate the amount of light being absorbed by phytoplankton in a water column and to estimate the maximum theoretical yield of primary production (2, 31, 38). In partitioning, one argues that the diffuse attenuation coefficient for irradiance (K_t) is equal to

$$(3) \quad K_t = K_w + k_c \cdot C + K_x.$$

In this case, K_w is the diffuse attenuation per meter for pure water, k_c is the specific attenuation for phytoplankton expressed as milligrams of chlorophyll over a meter light path, C is the concentration of chlorophyll in milligrams per cubic meter and K_x, the coefficient. Following Beer's Law,

$$(4) \quad E_z = E_o e^{-k + k_c \cdot C + K_x},$$

one measures irradiance at the surface (E_o) and then at depth (E_z) for the calculation of K_t. Thus, in equation (1), K_t is known; K_w is assumed to be that for the clearest waters. The phytoplankton attenuation is estimated from the measurement of chlorophyll and the remainder ($K_t - K_w + k_c \cdot C$) is K_x, i.e. attenuation by substances other than phytoplankton or water.

There is a considerable amount of data in the literature which has been treated by this partitioning technique; however, by far the most extensive data set treated in this fashion was collected by Scripps Institute and analyzed by Smith and Baker (27, 28) (Fig. 3). The term attenuation length (K_t^{-1}) in

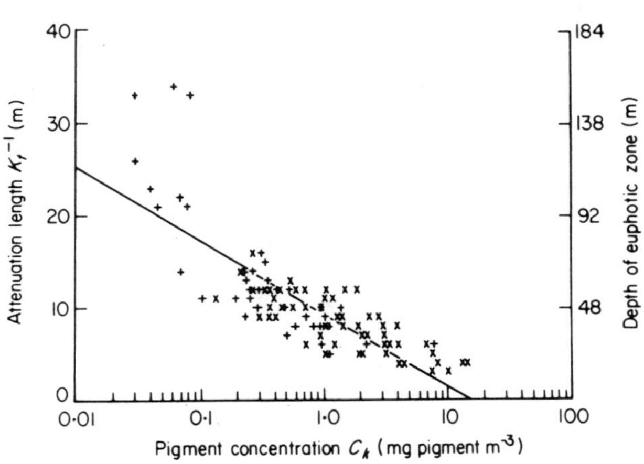

Figure 3. Attenuation length, K_t^{-1}(m), versus log average "chlorophyll-like" pigment concentration C_k(mg pigments m^{-3}) in the depth zone to one attenuation length. Solid line is a least squares fit to the data given by $K_t^{-1} = 9.45 - 7.90 \log C_k$ (N=104, r = 0.806, $P_c < 0.01$). (27, 28).

meters is used by these workers instead of K_t, as it is a more useful parameter in studies of remote sensing; K_t^{-1} refers to the depth at which 90% of the irradiance is reflected out of the water column. Predictably the phytoplankton diffuse attenuation coefficient should be related to K_t in linear fashion, as other data sets have shown the K_t:chlorophyll relationship in non-linear at low concentrations of chlorophyll in very clear waters. Smith and Baker reason that the non-linearity is due to biogenic substances which are proportionately higher in poor waters than those waters rich in phytoplankton. This reasoning is based on the data and observations by Hobson (9) for particulate organic carbon concentrations in many areas of the oceans. These workers show that when the particulate carbon in eutrophic and oligotrophic waters are compared, more detritus per unit phytoplankton is in the latter. These data in Fig. 5 can be fitted by two functions: the oligotrophic is dominated by the detrital component; the other, relatively free from detritus, is dominated by phytoplankton.

Smith and Baker used this slope as a first approximation of the diffuse attenuation coefficient for phytoplankton (k_c = 0.016) and toward that it agreed well with those by other workers. This agreement, coupled with the detrital carbon observations, tend to support the idea that the attenuation by phytoplankton in natural waters does indeed follow Beer's Law; that is, phytoplankton chlorophyll would relate in linear fashion to K_t if no detritus were present, or if the proportions of the two were constant. Throughout the literature, there is a number of comparisons between K_t and chlorophyll a. These all warrant individual study and are dependent on geographical and seasonal variations. A hypothetical relationship between the diffuse phytoplankton attenuation and water (Fig. 4) shows that the two contribute equally to the total attenuation when the biomass of chlorophyll reaches a level of 2.0 μg/ℓ. Below this, water dominates; above this value, phytoplankton is dominant. If one assumes that the chlorophyll content in coastal waters is frequently higher than 2.0 μg/ℓ, then the K_t of coastal waters is dominated by phytoplankton if no other attenuators are present. In oligotrophic waters, where chlorophyll content is less than 2.0 μg/ℓ, the water absorption dominates the K_t value.

The value K_c represents an average value for attenuation in white light. The spectral analyses of oceanic particulate matter (Fig. 5) show that in the euphotic zone, the red absorption band at 675-680 nm is conspicuous. However, the other chlorophyll and accessory pigment bands absorbing in the blue-green region are not conspicuous. They are obscured by the "detritus" in the water. The red band of chlorophyll a decreases in intensity with increasing depth in the euphotic zone and becomes undetectable at about 100 m. This is the approximate depth of the euphotic zone in the clear waters of the Sargasso Sea. With

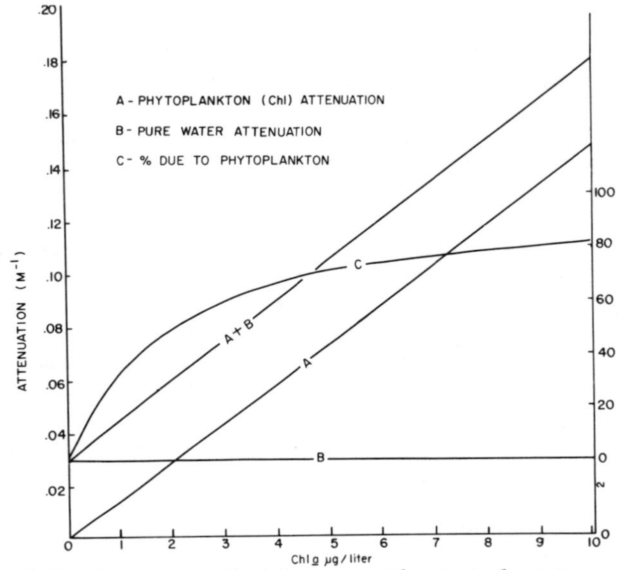

Figure 4. Relative contributions to the total attenuation by phytoplankton (measured as chlorophyll and water absorption)(43).

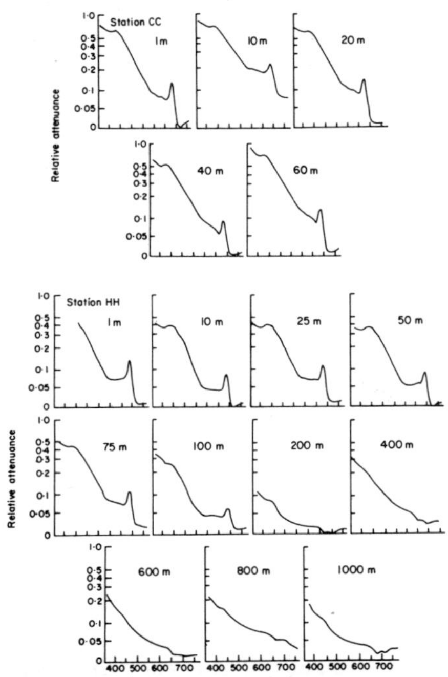

Figure 5. Relative attenuance of particulate matter collected from sea water at station CC $71^{\circ}35'$W, $40^{\circ}31'$N depth 73 meters and at station HH $70^{\circ}20'$W, $38^{\circ}46'$N depth 2,830 meters (36).

increasing depth, there is also a decrease in the short wavelength absorption; this rate decrease with depth is much slower than that for the chlorophyll a band at 675-680 nm.

One concludes that the overall effect of the combination of phytoplankton absorption and that by detritus is to give the particulate matter an "apparent absorption" at short wavelengths, which is much greater than would be predicted by phytoplankton alone (23). Thus, at short wavelengths, the detritus and phytoplankton compete for light energy. If the assumption that the short wavelength absorption is due to organic detritus, chances are that these substances are primarily absorbers of ultraviolet light. The absorption of light by detritus alone is unknown. One speculation is that it resembles the spectra obtained in deep water below the euphotic zone. One way to test this is to subtract "true phytoplankton absorption" from curves obtained in natural waters (Fig. 6). The difference between the two is assumed to reflect the absorption by detritus and agrees rather well with spectra measured below the euphotic zone. Another complexity is the fact that the spectra of phytoplankton changes as a function of environmental stress. In some species, nitrogen deficiency causes the disappearance of chlorophyll, as opposed to carotenoids. The overall spectral effect is a high blue:red ratio (23, 36).

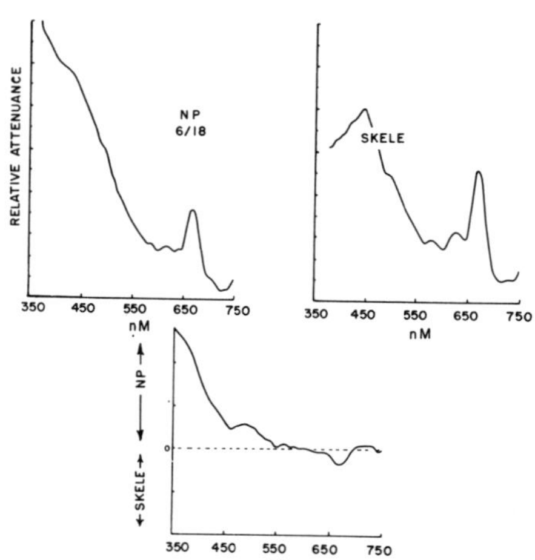

Figure 6. The spectral difference between pure phytoplankton light attenuation (SKELE = Skeletonema costatum) and attenuation of light by particulate matter of natural populations (43).

ABSORPTION OF DISSOLVED YELLOW SUBSTANCES

Many areas of the world's oceans, especially the coastal waters, are discolored by a dye-like material called "Gelbstoff," a term adopted by early German oceanographers (14, 15). In a general sense, Gelbstoff appears to be u.v. absorbing compounds which overlap the visible portion of the spectrum (Fig. 7). It is also weakly fluorescent; when excited at 350-390 nm, it fluoresces in the blue-green (4, 33). Some workers have noted rather sharp spectral bands in the u.v. region (5, 37) which apparently are due to the presence of heterocyclic compounds, but these bands are not always apparent. As to the source of yellow substances, they have been isolated from algae and are abundant in many salt marshes, mangrove swamps and estuaries. The manner or rate of release of yellow substances is not well understood. Dissolved yellow substances are rapidly released in water from brown seaweed and phytoplankton. The rate of release is accelerated in the light; however, it does appear at a substantial rate in darkness. Associated with this "light effect" is a rapid decrease (decomposition) of chloroplastic pigments which suggested to Yentsch and Reichert (37) that the yellow material is a

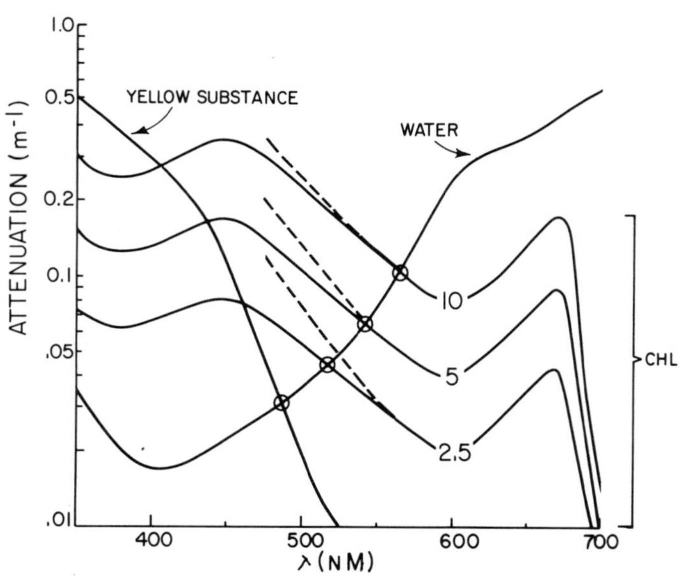

Figure 7. Comparative attenuation by phytoplankton and yellow substances. Numbers are chlorophyll concentrations in μg/ℓ. Circles indicate the wavelength of maximum light transmission in the water. Dotted lines are the attenuation when phytoplankton and yellow substances are added. Values for attenuation by clearest water from Tyler and Smith (34) (43).

part of the pigment-protein complex in the intact cell. The proteins of this complex are believed to be composed of the amino acids, tyrosine and/or tryptohane.

Yellow substance competes with short wavelength light absorption by phytoplankton -- as such they alter water color. Figure 3 compares the attenuation by dissolved yellow substances with that by different concentrations of phytoplankton. The attenuation coefficient at 350 nm for yellow substance is 0.50 M^{-1} and is about average for coastal waters. The diffuse attenuation coefficient is taken from Smith and Baker and the curve for clear water from Tyler and Smith. The circles shown in this figure at the point where the attenuating substance (either phytoplankton or yellow material) intersect with the attenuation of light by water. The wavelength indicated by the circled intersection are those of maximum transmission. The dotted lines indicate the attenuation due to the combined effects of phytoplankton and yellow substance. Figure 7 demonstrates that any short wavelength absorbing substances has the capacity of changing water color. In terms of attempting to implicate water color change due to the abundance of phytoplankton chlorophyll, serious problems can occur when the chlorophyll content is relatively low as compared to yellow substance. At high chlorophyll concentrations (i.e. those approaching 10 $\mu g/\ell^{-1}$) the concentrations of yellow substances is not a serious interference.

Oceanographers have recognized the correspondence between the amount of yellow substances and terrigenous discharge (1, 10, 11, 14), the surface salinity, and yellow substance. The distribution of Gelbstoff in oceanic and coastal waters can be directly related to the freshwater inflows. For example, the Amazon discharge discolors the adjacent ocean many miles to sea (39). The color gradient decreases moving seaward from the river mouth. One can rightly question the "conservative" nature of this substance: Why are the oceans not yellow? Although the evidence is sparse, the fate of yellow substance (disappearance from the water column) appears to be the result of adsorption on particles which sink, and photooxidation and slow microbial oxidation (18).

PHYTOPLANKTON FLUORESCENCE

The amount of visible light upwelled from ocean waters is in the range of 3-5% of that incident at the surface. Most of this light is backscattered from particles and molecules - a small portion is due to fluorescence from dissolved and particulate organic substances. Phytoplankton ecologists using sensitive fluorometers routinely measure phytoplankton fluorescence to estimate phytoplankton biomass. From microscopic observations one would predict that the majority of fluorescence would be derived from light absorbed by marine phytoplankton. Figure 8 shows the fluorescence characteristics of known algal groups - all of these have representative species in the marine environment. At the top of the figure is the action spectrum for exciting a fluorescence whose spectral emission is shown in the lower half of the figure. Note that, in general, the excitation spectra are broader than the emission spectra. Also note that with the exception of the blue bioluminescence protein B, the lowest peak emission wavelength is 575-580 nm, while the longest is at 685 nm. The spectra are numbered to identify the particular action spectrum associated with a particulate emission. Scheme 1a is the signature for green algae and green plants in general. This signature differs from that of diatoms and dinoflagellates (1) only in that the latter absorb light at longer wavelengths (530-550 nm); this long wavelength absorption is associated with the accessory carotenoid pigment fucoxanthin in diatoms and peridinin in dinoflagellates. Schemes 2 and 3 implicate the phycoerythrin and phycocyanin pigments which are found in cyanobacteria, cryptomonads and red algae. Curve B shows the excitation and emission spectra for the blue fluorescent protein associated with bioluminescence in marine dinoflagellates. It is spectrally identical to the bioluminescence of dinoflagellates (32).

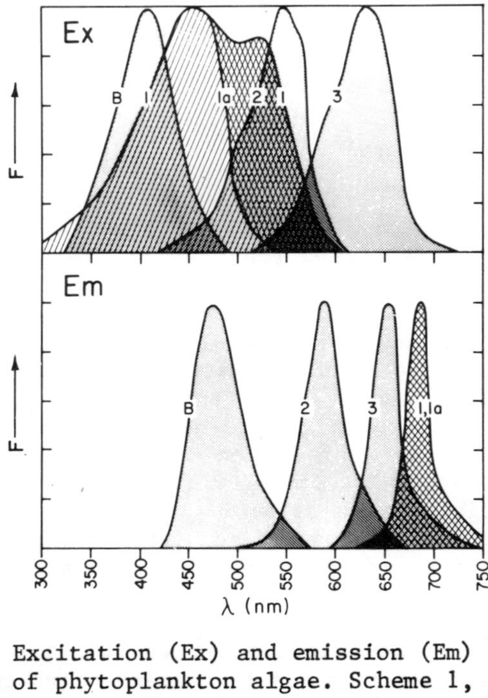

Figure 8. Excitation (Ex) and emission (Em) spectra for different groups of phytoplankton algae. Scheme 1, green algae; scheme 1a, diatoms and dinoflagellates; scheme 2, cyanobacteria. (46).

We know from microscopic counts of microorganisms in the sea that the populations that belong to different taxonomic groups are not in fixed proportions everywhere in the oceans. One would anticipate, however, that the dominance of diatoms and dinoflagellates would be apparent in the characteristics of the fluorescence spectra - and they are. For example, note the shoulder at 525-530 nm in the excitation spectrum shown in Fig. 3. This shoulder results from the absorption of light by fucoxanthin and peridinin, which are the accessory pigments which transfer light energy to chlorophyll a at 685 nm. Figure 4 shows spectra selected from different oceanic regions. Note that the 525-530 nm shoulder is apparent in all of these. The intensity of this absorption varies, presumably as a function of the numbers of diatoms and dinoflagellates present. Because these groups are abundant in eutrophic coastal waters, the intensity of absorption at these wavelengths is greater in coastal waters than in the open ocean (42). A strong emission peak at 570-580 nm is present in every measurement. The association of the phycoerythrin emission with small cell size has been confirmed by comparison of fluorescence measurements on material collected by fine plankton nets (30-μm mesh opening) with those made on material retained by 0.45-μm membrane filters. The spectra of organisms retained by the netting is characteristic of diatoms and dinoflagellates. No phycoerythrin is observed. Phycoerythrin is observed in the organisms retained by the 0.45-μm filter. This appears to be due to small (0.5-1.0 μm) coccoid cyanobacteria by oil immersion epifluorescence (35).

BIOLOGICAL REMOTE SENSORS

CZCS colorimetry

At the onset, it was argued that one of the main reasons for going into space was to get a better perspective of large areas of the earth. Of particular interest is the imaging color scanner (CZCS) flown in the NIMBUS satellite. In this imager, the spectral channels have been positioned (Fig. 9A) to sense substances which affect ocean color and surface temperature (12). The NIMBUS satellite is positioned about 600 miles in space in a sun-synchronous orbit. The instantaneous field of view is about 1 km; the total scanning width (swath) is 2000 km. A NASA team of scientists has developed algorithms for correcting for the effects of the atmosphere and for estimating chlorophyll a concentration from color changes in surface waters (Fig. 10).

It was recognized that colorimetric estimates of chlorophyll could be most difficult in coastal waters where the inter-

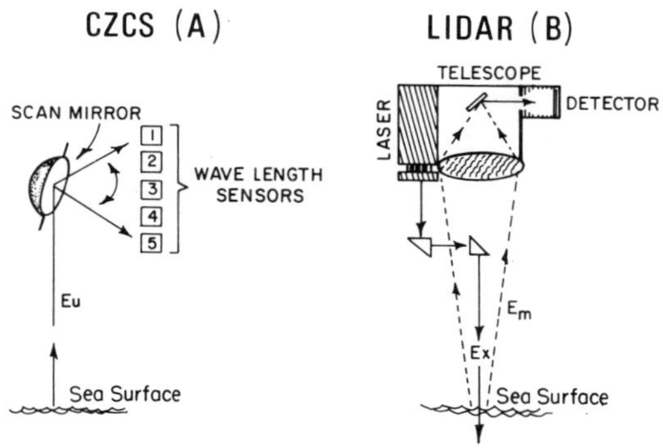

Figure 9. Schematic of NASA's remote sensing system for measuring phytoplankton chlorophyll. A) The NIMBUS-G Coastal Zone Color Scanner. Rotating mirror beams backscatter light across an array of wavelength detectors. B) Laser induced fluorescence system (LIDAR). Laser is the excitation source. Detector views fluorescence through a telescope. (41).

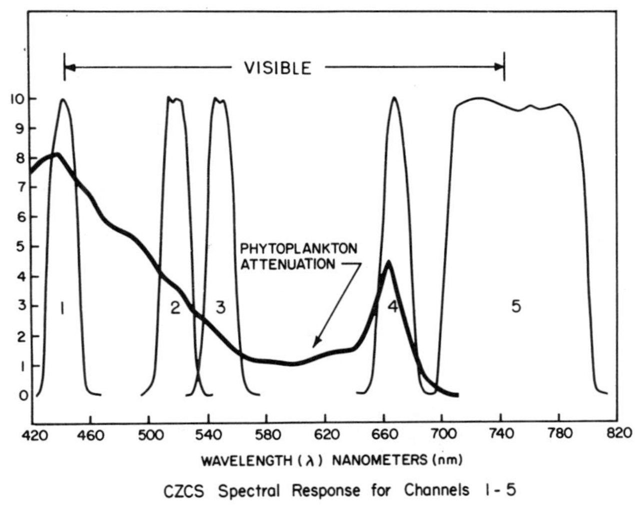

Figure 10. Spectral response for CZCS channels 1-5.

ferences by yellow substances and suspended sediments would be greatest. However, in offshore waters, the colorimetric estimate of chlorophyll from two CZCS channels (L_W 443/L_W 550) correlates with ground truth data better than 0.85, with the worst estimate in error by a factor of two (7). In addition, the Gordon water algorithm has been tested by direct measurements of light absorption in natural populations of phytoplankton (46). These demonstrate that the extracted chlorophyll can be estimated with an error of 15% at the 95% confidence level.

Laser Induced Fluorometers

At the present time, these systems (Fig. 9B) are flown in aircraft at low altitudes. The instrument induces fluorescence by pulsing strong bursts of blue or green light into the water column, and with these units it is possible to resolve patches of a few meters in size.

When considering the problems associated with the use of reflected light for remote sensing, the possibility of using the fluorescence of algae stands out. By fluorescence, one is measuring the wavelengths of light active in photosynthesis. This measurement circumvents the detritus problem by pinpointing the photosynthesis activity unique to algae.

There are other benefits derived from the measurement of fluorescence - particularly the measurement of the excitation spectra for chlorophyll a fluorescence. Excitation peaks correspond to absorption by accessory pigments, hence one can (by identification of accessory pigmentation) indicate the types of organisms present. Studies in our laboratory show that algae with accessory pigmentation such as the carotenoid proteins of brown algae and diatoms (e.g. fucoxanthin) and the chromoproteins of red and blue-green algae (e.g. phycoerythrin) can readily be identified in particulate matter collected from ocean water. This cannot be done by light attenuation because of the interference by detritus.

The disadvantages of measuring fluorescence is that light emitted from fluorescence is a function of a variety of environmental and physiological factors. Hence, the relationship between chlorophyll biomass and fluorescence is subject to error. However, this is not to suggest that these "errors" cannot be capitalized upon.

RESEARCH APPLICATION FOR REMOTE SENSING

Areal Partitioning

As stated in the introduction, one of the reasons for the use of satellite remote sensing was to test hypotheses generated by observations made from ships. The following is an example of a situation where the satellite information confirms and adds to the hypotheses.

In 1928, H.B. Bigelow made a number of hydrographic stations throughout the Gulf of Maine and Georges Bank. From these he computed the density and then estimated the stability $(\sigma t_0 - \sigma t_{40})$. From these data Bigelow outlined what he believed to be areas which were strongly mixed vertically by tidal currents as opposed to those which are not (Fig. 11). Bigelow observed that the mixed areas supported much more plankton than the stable water masses; the reasons being that the formation of the seasonal thermocline prevented nutrient-rich water from mixing into the euphotic zone thereby limiting growth. Thus he reasoned that the degree of vertical mixing regulated the production of phytoplankton and thought that the mixing energy was tidally generated.

Garrett et al. (6) have computed the degree to which tidal action mixes the water column using a model developed by Simpson and Hunter (24). This model (Fig. 12) proposes that the difference between mixed and stratified waters is dictated by the ratio of the potential energy required to thoroughly mix the water column to the rate at which energy is dissipated by the flow of a tidal current over the bottom. For areas where heat-buoyancy effects are constant, the relevant parameter separating mixed from stratified waters is Log H/U^3 - the water depth (H) divided by the tidal velocity - fractional component (U^3). The model demonstrates that tidal-mixed areas are centered on Georges Bank and Nantucket Shoals (Fig. 11). Minor tidal mixings occur off Nova Scotia and the Bay of Fundy. For the remainder of the area, tidal velocities are too low and/or the water column too deep for significant mixing.

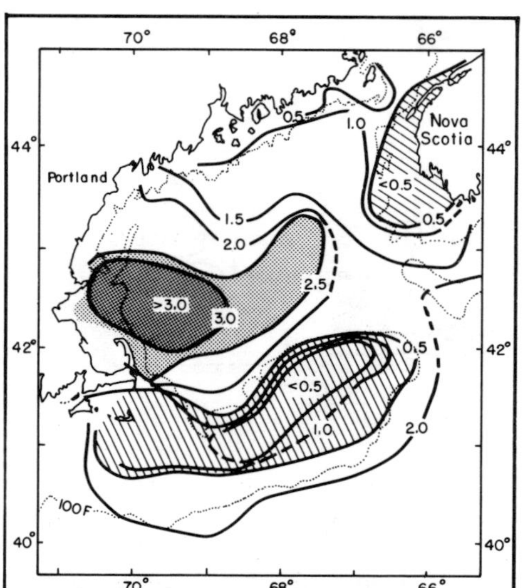

Figure 11. Spatial differences in vertical stability. Contour units are sigma-t difference of surface and 40 meters (45).

Comparison of satellite thermal imagery with the model analysis by Garrett et al. (6) confirms the notion that much of the mixing is tidally driven. Figure 13 shows the thermal structure for the Gulf of Maine region in early summer. The light areas indicate water masses which are cold, i.e. they are mixed vertically with respect to temperature, as opposed to dark areas which are warm and vertically stratified, i.e. warm water overlying cold bottom water. The surface temperatures for the cold mixed waters are about 11-13°C, whereas the warm stratified water is about 16-18°C. The extremely dark filamentous segments appear to be intrusions from the Gulf Stream and/or warm neighboring slope waters - these have surface temperatures over 20°C. The principal mixed regions are Georges Bank, Nantucket Shoals, off Nova Scotia, and the Bay of Fundy. The cold water masses along the Maine coast are identified with the underlying bathymetry known as Jeffreys Bank. The cold regions labelled "coastal" are probably local wind-driven upwelling systems. The tidal model (Fig. 12) predicts that the water column would be mixed at depths much shallower than the depth at which the thermal fronts appear. For example, the frontal edge around Georges Bank is centered around the 60 m isobath. Using the maximum tidal velocities, the destratification depth (Log $H/U^3>2.0$) is about 30-40 m, which is considerably less than that estimated by satellite imagery. This suggests that processes other than tides are important in destratifying the water column.

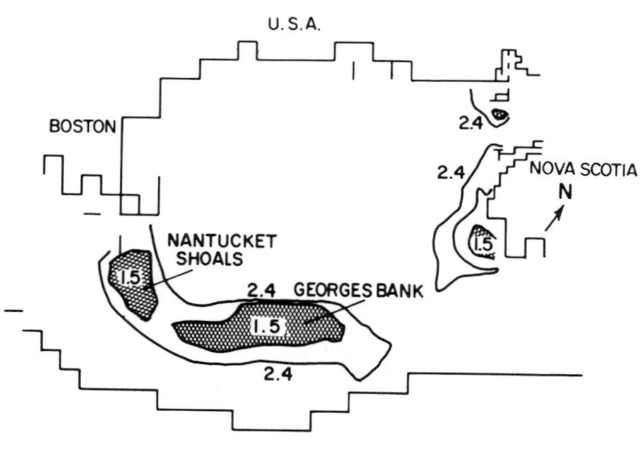

Figure 12. Gulf of Maine and Bay of Fundy. Contours of log (H/D) with depth H in m and dissipation D in Wm^{-2}. Only contours between 1.5 and 2.4 are shown. Hydrographic data for July and August show the areas with log (H/D) less than 1.9 (shaded) to be well mixed (6).

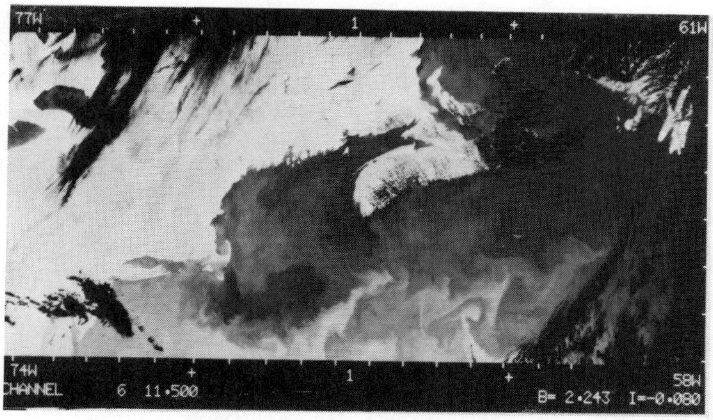

Figure 13. Infrared channel NIMBUS-G (CZCS) June 14, 1979. Dark areas, warm; light areas, cold (45).

Vertical mixing is crucial to phytoplankton growth and distribution. During the months when the water column is being heated, the greater buoyancy of surface waters tends to isolate deeper nutrient-rich water from the euphotic zone. However, in the areas where the water column is mixed, phytoplankton growth thrives on the nutrients which are mixed into the euphotic zone. Thus, during the warm months, the mixed layers potentially maintain high growth as opposed to areas which are strongly stratified. In the context of attempting to establish the magnitude of the productivity, it is important to establish how much of the area is vertically mixed. Figure 14 shows the areal distribution of surface chlorophyll for the Gulf of Maine region in August 1979. The data base used to draw the contours was obtained by measuring fluorescence continuously while underway aboard ship. Discrete station samples have also been included. The shaded areas represent regions where chlorophyll concentration exceeds 2.0 mg/m^3. Note that they occur in regions designated in Fig. 3 as major frontal regions which were identified.

The CZCS image (Fig. 15) shows that high concentrations of phytoplankton chlorophyll are located on Georges Bank, Nantucket Shoals and in the coastal waters off Maine. Low chlorophyll concentrations are in the slope waters off New England. The central basin region in the Gulf of Maine is intermediate between these two extremes. All of these regions correspond to the areas identified by Bigelow as regions of low stability. Conversely, regions of high stability reflect areas of low chlorophyll concentrations. The question asked is, what is the size of the contribution, to the total productivity of the Gulf of Maine, made by the major frontal mixing regimes?

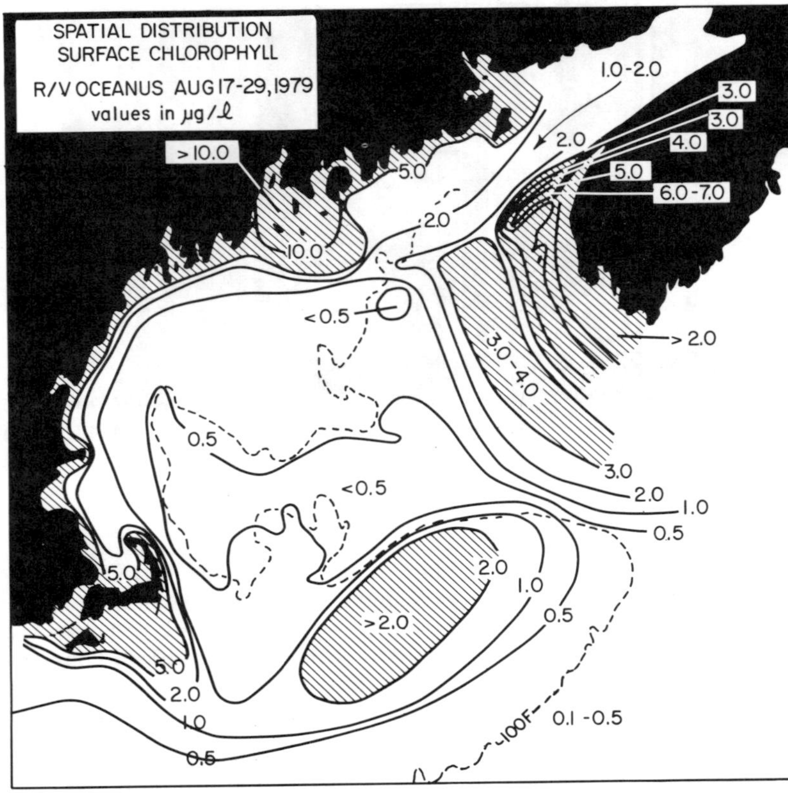

Figure 14. Spatial distribution of surface chlorophyll, August 17-29, 1979 R/V OCEANUS. Expressed as µg/ℓ (45).

If we assume that surface chlorophyll is a reasonable index of water column productivity, the frontal contribution can be approximated as follows. The areas of the mixed regions are measured. The big three are Georges Bank, the areas bounded by the Scotian Front, and Nantucket Shoals. We have estimated the relative contribution of each area to the total area for the Gulf of Maine, 1.40×10^5 km^2, which includes the central portion of the Gulf of Maine and Georges Bank out to the 100 m isobath. Excluded from the total is the Bay of Fundy, St. Mary's Bay and Nantucket Shoals. With these exclusions, the mixed regions defined by thermal fronts occupy about 30% of the total area of the Gulf of Maine. Of this 30%, Georges Bank makes up about one-half and the other areas the other half. On an average, the chlorophyll concentrations of the frontal regions are five times higher than in the non-frontal areas. The non-frontal areas represent 70% of the total, the frontal areas

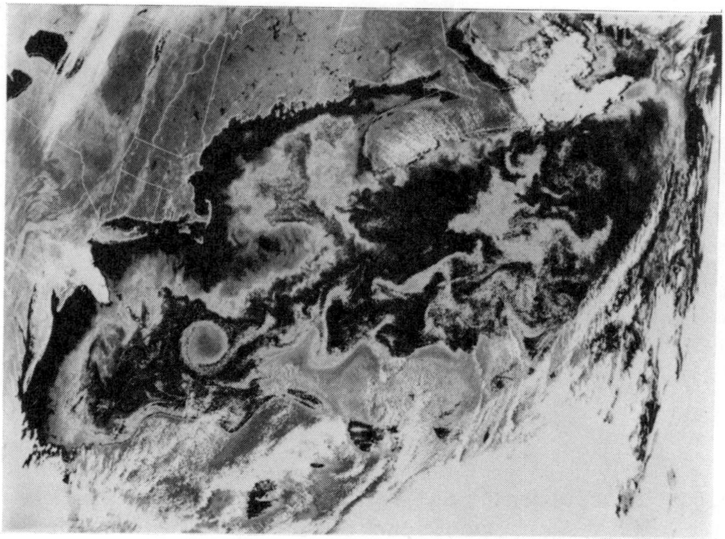

Figure 15. CZCS 3326, 14 June 1979, CZCS colorimetry chlorophyll.

30% of the total. Therefore, the frontal areas, because of their high productivity, are producing 2½ times more than the non-frontal areas. Or to put it another way, the importance of the frontal regions is that the 30% of the area they occupy is producing 2/3 of the total production. This approximation would be for at least five months of the year (May-September).

The manner in which we analyzed these images represents very elemental strategy. Nevertheless, it demonstrates that the level of synopticity is far greater than has been obtained heretofore. From the satellite image the synoptic detail is so great that it challenges our conventional modes of data interpretation. Another way to put it is that one image is almost too much! Recognizing this problem, Smith and Baker (29) and Smith et al. (30) have partitioned productivity and chlorophyll areally using CZCS data. They have accumulated the chlorophyll data from pixels in areas off Southern California covered by satellite overpasses (Fig. 16). Their approach is to assemble the data as a frequency distribution for chlorophyll where 1.0 μg/ℓ = Log C (sat) = 0. This curve can then be compared to the percentage of the total area surveyed by the satellite and cumulative chlorophyll for the area. One can then, for example, state for that particular time (during satellite survey) that the area of the Southern California Bight, 68% of phytoplankton chlorophyll in over 90% of the area was less than 1.0 μg/ℓ.

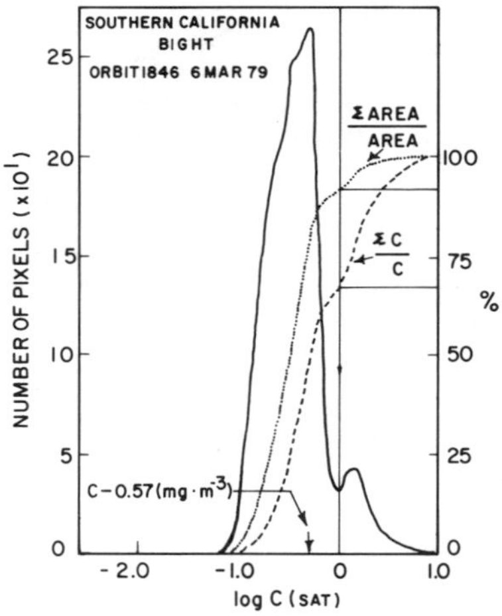

Figure 16. Statistical analysis of chlorophyll image derived from Orbit 1846. Continuous curves: frequency distributions of number of pixels (i.e. area) vs. log C_{13}; dashed curves: cumulative chlorophyll distributions as a function of log C_{13}; dotted curves: corresponding cumulative area distributions as a function of log C_{13}; C: mean chlorophyll concentration for cloud-free area analyzed (29).

Estimates of phytoplankton biomass by satellite chlorophyll is in fair agreement with that measured from ships. This type of areal partitioning can aid in identifying viable regions for the habitats of fishes and marine mammals.

Large Scale Features

The use of satellite colorimetry has provided a means of observing ocean surface patches. Gower et al. (8), using LANDSAT satellite imagery, demonstrated that phytoplankton is distributed in patches formed by eddies 10-100 km in diameter. These high energy features are associated with marked differences in density formed between ocean currents and adjacent gyres. This important observation suggests that the pattern of large-scale plankton patchiness is more complex than would be anticipated from general features of ocean circulation. The identification of patchiness at the 10-100 km scale raises questions about distribution of phytoplankton at much larger scales. The pattern of primary productivity is known to be associated with large-scale changes in the density field of the oceans (40), which in turn, is a reflection of oceanic circulation. Thus, Gower et al. (8) have set the stage for comparison of different size scales of phytoplankton distribution patterns. If mesoscale patterns are dominated by high energy gyres, can the situation at the much larger planetary scales be resolved by colorimetry? Can the geostrophic features associated with large ocean currents be identified by satellite colorimetry?

From NIMBUS-6 on orbit 130, November 2, 1978, the CZCS (13) imaged most of the western Gulf of Mexico (Fig. 17). This was the first image processed by NASA using methods and algorithms (7), which correct for atmospheric effects and result in estimates of the chlorophyll content of the waters. The chlorophyll computations yield values in the central region of the Gulf between 0.05 to 0.10 μg/ℓ. Off the Mississippi Delta, values exceed 10 μg/ℓ. In light of sea truth, measurements made in this region at this time (see Gordon et al., 1980) and what is known from previous studies, these values are considered realistic. The image shows a wide variety of mesoscale features. However, the reader is asked to concentrate on the shape of the general pattern of chlorophyll for the entire region. The chlorophyll pattern on the western side of Florida is much broader than the front on the Atlantic side. The physical dimensions of these fronts correspond to the physical dimensions of the continental shelves on both sides of the Florida peninsula. In the case of the west coast shelf, the distance from shore to the 200 m isobath is over 300 km at latitude 26°N. On the east coast, at the same latitude, it broadens moving north-

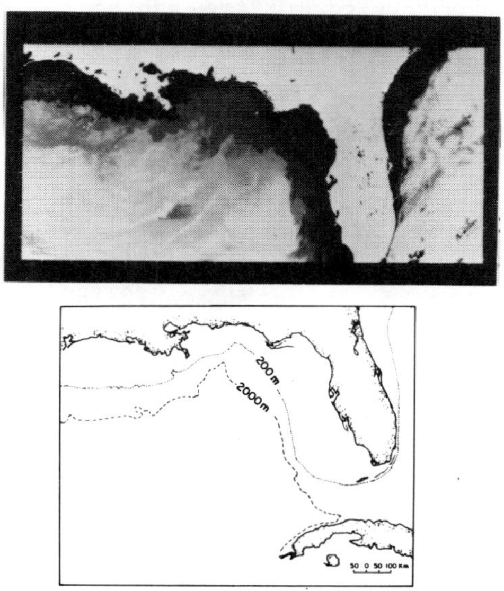

Figure 17. (Upper) Chlorophyll as viewed by NIMBUS 7 (CZCS) orbit 130, November 2, 1980. Lower shows major bathymetric outline in the region (44).

The three major frontal boundaries (A, B, C, D) for chlorophyll which outline the general sequence of decreasing chlorophyll moving seaward from land are outlined in Fig. 18. The frontal positions have been redrawn to correct for distortion caused by orbital configuration. The fronts, excluding small-scale perturbations, can be seen to parallel the isobaths along both coasts of Florida. The more seaward part of the front appears as a boundary which distinguishes oligotrophic water from slope water.

Except for the northern area of the Gulf of Mexico, all of the fronts outlined lie within the 200 m isobath and probably between the 50 and 100 m isobaths. Lacking other geophysical information pertinent to this image, the general position of the fronts can be interpreted to be due to the flow of the Loop Current in the Gulf and the Florida Current on the eastern side of Florida. The dynamic topography in this region shows that channel-shelf constraints markedly influence horizontal velocities and thus the pattern of mass transport (3, 17, 19). For example, in the western Gulf of Mexico, the effect of the Loop Current is seen as forming a high level ridge (Fig. 19) which parallels the Florida escarpment along the west coast shelf. In this case, the sea-level surface slopes about 50 cm from the ridge to the edge of the escarpment at the 1000 m isobath. From the image and bathymetric charts, the positions of the most seaward chlorophyll fronts on either coast begin around vicinity of the 100-200 m isobath and increase in density of color to the 50 m isobath. Proceeding toward the shore, chlorophyll color increases, but at a reduced rate.

Because of the pattern of density across the axis of transport, a close spatial relationship between water color and surface temperature is anticipated. The inclination of isopyc-

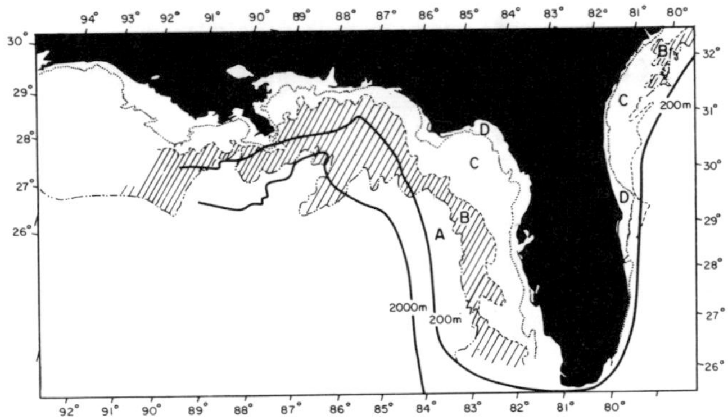

Figure 18. Positions of fronts, A, B, C, and D, taken from orbit 130 (44).

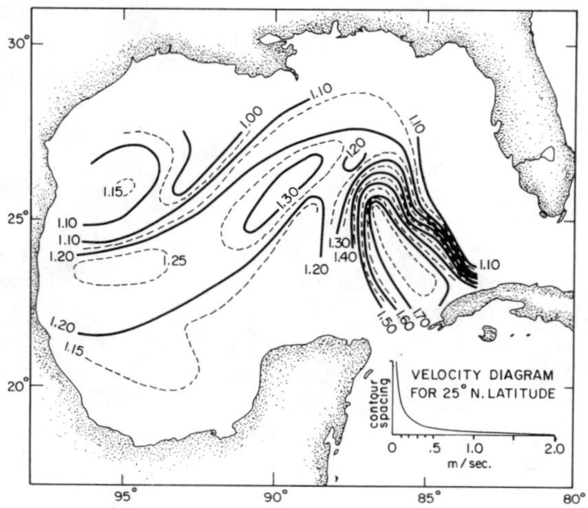

Figure 19. Dynamic topography of the region from Nowlin and McLellan (19) (44).

nals represents vertical transport of temperature as well as nutrients. From orbit 1965 on 15 March 1979, both water color and temperature were imaged over the western Gulf of Mexico (Fig. 20). At the time, the atmosphere was very clear. The upper image is uncorrected for atmospheric effects and represents the total upwelled radiance seen by CZCS channel 1 (443 nm). The lower image is the thermal presentation by CZCS channel 6.

The thermal front of the Loop Current has been observed to penetrate the Gulf of Mexico following a seasonal cycle (16). In the thermal image (Fig. 21), equatorial water exiting from the Strait of Yucatan can be seen penetrating, as a warm core, into the Gulf of Mexico as far north as latitude 27°N, which agrees with Mauls "spring position" for the Loop Current. Close inspection of this image shows that the surface thermal characteristics "fan out" through the western Gulf in a pattern similar to the frontal boundary of the central warm core. Along the eastern coast of Florida, the pattern of temperature appears to closely follow the Florida-Hatteras slope: colder water not being observed until the shelf widens north of Palm Beach (26°N) (Fig. 21). At this point, the thermal pattern closely reflects shelf dimensions.

Color and temperature patterns are clearly similar along the east coast of Florida, with an abrupt transition between waters of rich and poor phytoplankton associated with the cold

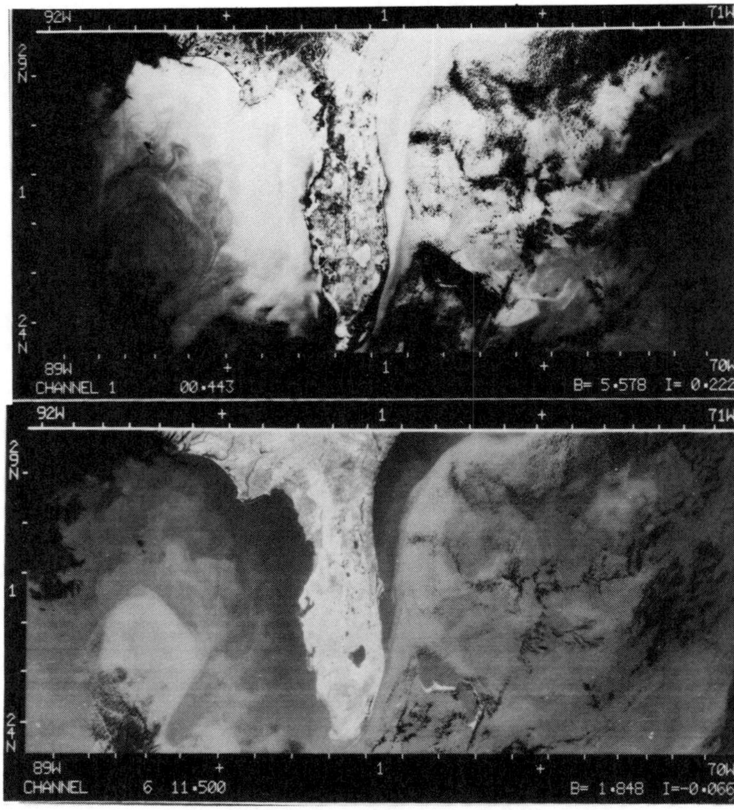

Figure 20. Orbit 1965, March 15, 1979. NIMBUS 7 CZCS channel #1 (upper). Light tone denotes high attenuation of blue (443 nm) light due to phytoplankton chlorophyll. Lower-sea surface temperature -- dark tone, cold water (44).

wall of the Gulf Stream (40). The color image for the western Gulf of Mexico indicate a large mass of chlorophyll-poor water which extends as far as latitude 30°N. However, in the region at the western Florida shelf escarpment, where the Loop Current is impinging, the water masses are rich in chlorophyll (reduced radiance at 443 nm, CZCS channel 1). The chlorophyll-rich water covers the entire shelf extending southward paralleling the escarpment. Therefore, the higher chlorophyll along this escarpment is associated with the southerly high velocity transport of the Current Loop.

Thus, this exercise demonstrates that colorimetry can easily resolve the global patterns of phytoplankton. Therefore, it is possible to study the interaction of global climate and primary production on scales for those interested in heat and carbon dioxide fluxes.

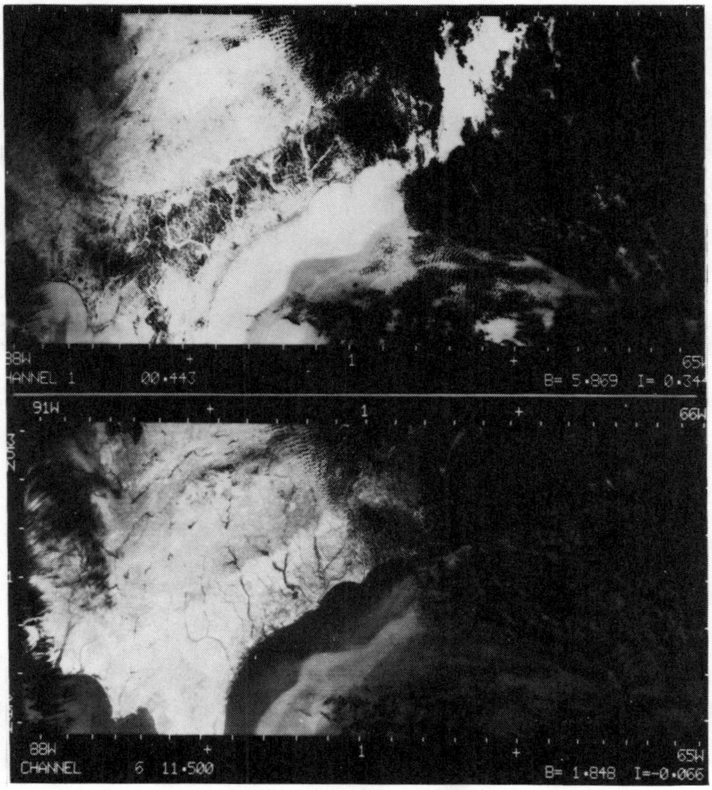

Figure 21. CZCS orbit 02425, 17 April 1979. Top: colorimetry-light, high chlorophyll; dark, low chlorophyll. Bottom: thermal-dark, cold water; light, warm water.

For anyone concerned with coastal ocean processes, the value of satellite observations cannot be overestimated. There has been a rebirth of interest in "frontal processes" observed in coastal waters primarily, I believe, because of the delineation in regional productivity produced by fronts. Whatever the cause, the efforts pioneered by workers around the British Isles have constantly used satellite thermal and now colorimetry to test their models of frontal development and movement. The reader is referred to the papers of Hunter and Pingree (20, 21, 22, 24, 25, 26).

Study of a special type of a frontal feature is now ongoing in the northeastern U.S.A. coastal waters. Periodically, warm core rings or eddies pinch off the Gulf Stream due south of Cape

Cod. The concepts of how rings are formed was put forth by Fritz Fuglister many years ago. The rings are pinched off by Gulf Stream meanders and hence are composed of two water masses- Sargasso Sea water and Slope water. These rings are not small features - the diameter of most is in excess of 100 km and hence they are ideal candidates for remote sensing (Fig. 22). Biologists and chemists are interested in these rings because of the "circular frontal" characteristics. It is believed that the rotary motion drives an upward movement of water along isopycnals thus, the rim of the ring is more rich in phytoplankton than the central core. I can testify that without satellite and LIDAR instrumentation, one would be partially blind in trying to study these features. There are those who will point out that these rings were found using ship data only. But think of the labor involved; Fuglister and his Woods Hole colleagues spent years studying the Gulf Stream before they recognized rings. It now takes the satellite a matter of seconds to map a ring structure.

COMPARATIVE ADVANTAGES - CZCS:LIDAR

Ocean color scanner and fluorescence instruments represent exciting new developments for the biological oceanographer and will provide the means for assessment of the large-scaled patchiness of the ocean in a manner only dreamed about by early oceanographers. Moreover, persons associated with the environmental management where the problems associated with blooming algae are of concern will find these devices useful.

However, both systems (CZCS and LIDAR) have limitations. In the CZCS system, it is doubtful that the effects of water color due to yellow substances and detritus can be thoroughly resolved using five wavelength sensors. The laser-induced fluorescence system presently requires massive power which eliminates its use in the satellites presently available. Hence, the areal coverage by this system is restricted to that covered by low flying aircraft.

The satellite derived information on water color will probably be in most demand by the oceanographic community. However, with the uncertainty in the interpretation of the signal, it is doubtful that the data can be used independent of sea truth. An opinion shared by many is that the best use of this research tool will be as an adjunct to seagoing operations. Simply put, this means that measurements at sea level will be given more coverage.

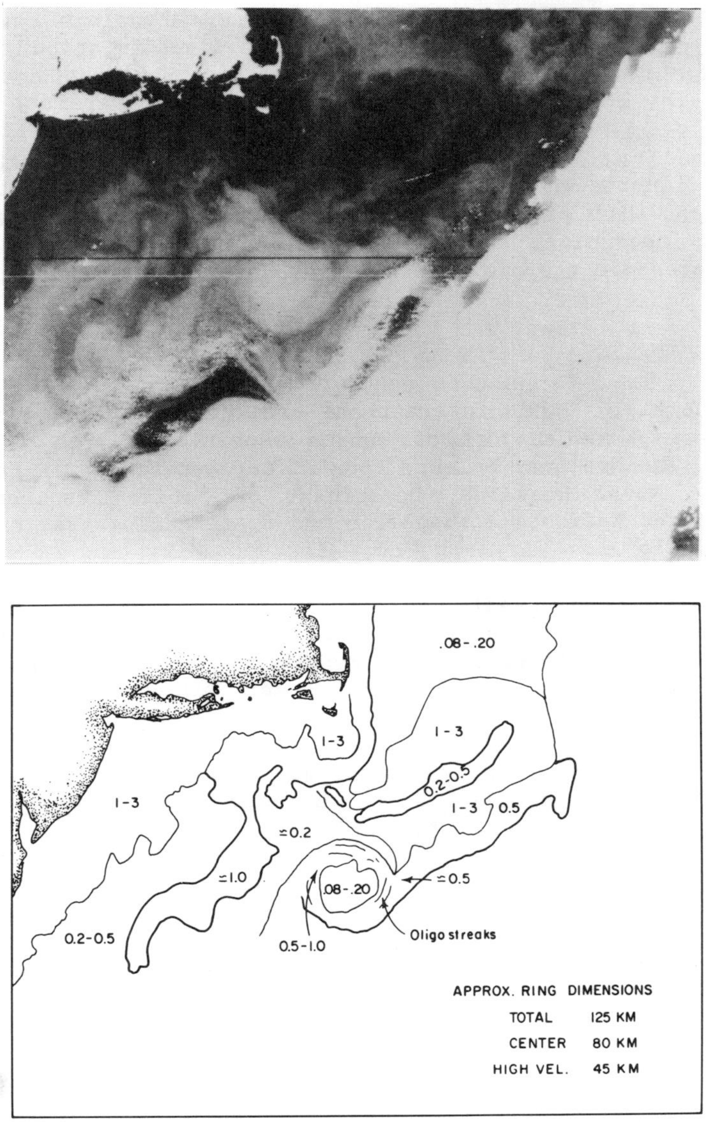

Figure 22. CZCS chlorophyll image of a warm core ring. Orbit 3309, 20 June 1979.

Even with the problem of limited coverage, the most exciting aspect of the LIDAR system is the ability to characterize the organisms present. This advantage bears directly on the problem of identification of dinoflagellate blooms. A note of caution is needed here: as previously mentioned, some distinction can be made between groups of organisms based on accessory pigmentation alone. Hence, "identification" of a bloom via remote sensing by fluorescence requires some sea truth.

Summed up, in both systems, the signal from backscattered light or fluorescence contains valuable environmental information which, with proper instrumentation can be remotely sensed. One must recognize that the present instruments and their deficiencies are the pioneers of an evolution just beginning.

ACKNOWLEDGEMENTS

Because of the interactions between water color and oceanography, the support of remote sensing research is many faceted. The National Science Foundation, Department of Energy, Office of Naval Research, Department of Marine Resources of Maine and the National Atmospheric and Space Administration have all been helpful.

Much of the credit for this product is due to the manuscript preparation by Pat Oathout. Jim Rollins prepared the figures. My wife Clarice added thoughtful comments to the text.

REFERENCES

1. Baker, K.S. and Smith, R.C.: 1982, Limnol. Oceanogr. 27(3), pp. 500-509.

2. Bannister, T.T.: 1974, Limnol. Oceanogr. 21, pp. 326-336.

3. Brooks, I.H. and Niiler, P.P.: 1977, J. Mar. Res. 35(1), pp. 163-191.

4. Duursma, E.K.: 1974, In: Jerlov, N. & Steeman-Nielsen, E. (eds.), "Optical Aspects of Oceanography," Academic Press, New York, pp. 237-255.

5. Fogg, G.E. and Boalch, G.T.: 1958, Nature, Lond. 181, pp. 789-790.

6. Garrett, C.J.R., Keeley, J.R. and Greenberg, D.A.: 1978, Atmosphere-Ocean 16(4), pp. 403-423.

7. Gordon, H.R., Clark, D.K., Mueller, J.L. and Hovis, W.A.: 1980, Science 210, pp. 63-66.

8. Gower, J.F.R., Denman, K.L. and Holyer, R.J.: 1980, Nature 288(5787), pp. 157-159.

9. Hobson, L.A., Menzel, D.W. and Barber, R.T.: 1973, Mar. Biol. 19, pp. 298-306.

10. Højerslev, N.K.: 1974, Univ. Copenhagen Inst. Phys. Oceanogr. Rep. 26, 38 p.

11. Højerslev, N.K.: 1980, Univ. Copenhagen Inst. Phys. Oceanogr. Rep. 42, 35 p.

12. Hovis, W.A., Forman, M.L. and Blaine, L.R.: 1973, NASA Ref. X-652-73-371, pp. 159-166.

13. Hovis, W.A., Clark, D.K., Anderson, F., Austin, R.W., Wilson, W.H., Baker, E.T., Ball, D., Gordon, H.R., Mueller, J.L., El-Sayed, S.Z., Storm, B., Wrigley, R.C. and Yentsch, C.S.: 1980, Science 210(4465), pp. 60-63.

14. Jerlov, N.G.: 1976, "Optical Oceanography," Elsevier, New York. 194 p.

15. Leyendekkers, J.: 1967, Koninklijk Nederlands Meteorologisch Inst. in De Bilt. 67-1 (IV-4). 40 p.

16. Maul, G.A.: 1977, J. Mar. Res. 35(1), pp. 29-47.

17. Molinari, R.L. and Yager, R.E.: 1977, J. Mar. Res. 35(1), pp. 11-20.

18. Moreth, C.M. and Yentsch, C.S.: 1970, J. exp. mar. Biol. Ecol. 4, pp. 238-239.

19. Nowlin, W.D., Jr. and McLellan, H.J.: 1967, J. Mar. Res. 25(1), pp. 29-59.

20. Pingree, R.D.: 1979, J. mar. biol. Assoc. U.K. 59, pp. 689-698.

21. Pingree, R.D.: 1975, J. mar. biol. Assoc. U.K. 55, pp. 965-975.

22. Pingree, R.D. and Mardell, G.T.: 1981, Phil. Trans. R. Soc. Lond. A 302, pp. 663-682.

23. Shimura, S. and Ichimura, S.: 1973, J. Oceanogr. Soc. Japan 29, pp. 257-266.

24. Simpson, J.H. and Hunter, J.R.: 1974, Nature 250, pp. 404-406.

25. Simpson, J.H., Hughes, D.G. and Morris, N.C.G.: 1977, Deep-Sea Res. 24, pp. 327-340.

26. Simpson, J.H., Allen, C. and Morris, N.C.G.: 1978, J. Geophys. Res. 83, pp. 4607-4614.

27. Smith, R.C. and Baker, K.S.: 1978a, Limnol. Oceanogr. 23, pp. 247-259.

28. Smith, R.C. and Baker, K.S.: 1978b, Limnol. Oceanogr. 23, pp. 260-267.

29. Smith, R.C. and Baker, K.S.: 1982, Mar. Biol. 66, pp. 269-279.

30. Smith, R.C., Eppley, R.W. and Baker, K.S.: 1982, Mar. Biol. 66, pp. 281-288.

31. Steele, J.H. and Menzel, D.W.: 1962, Deep-Sea Res. 9, pp. 39-49.

32. Tett, P.B. and Kelly, M.G.: 1973, Oceanogr. Mar. Biol. Ann. Rev. 11, pp. 9-173.

33. Traganza, E.D.: 1969, Bull. Mar. Sci. 19, pp. 897-904.

34. Tyler, J.E. and Smith, R.C.: 1970, Measurement of Spectral Irradiance Underwater," Ocean Series 1, Gordon and Breach, New York. 103 p.

35. Waterbury, J.B., Watson, S.W., Guillard, R.R.L. and Brand, L.E.: 1979, Natura, Lond. 272, pp. 293-294.

36. Yentsch, C.S.: 1962, Limnol. Oceanogr. 7, pp. 207-217.

37. Yentsch, C.S. and Reichert, C.A.: 1962, Bot. Mar. 3, pp. 67-74.

38. Yentsch, C.S.: 1963, Oceanogr. Mar. Biol. Ann. Rev. 1, pp. 157-175.

39. Yentsch, C.S.: 1971, In: Proc. Symp. Texas A&M Univ., College Station, Texas, pp. 75-97.

40. Yentsch, C.S.: 1974, Oceanogr. Mar. Biol. Ann. Rev. 12, pp. 41-75.

41. Yentsch, C.S.: 1979, In: D.L. Taylor & H.H. Seliger (eds.) "Toxic Dinoflagellate Blooms", pp. 297-302.

42. Yentsch, C.S. and Yentsch, C.M.: 1979, J. Mar. Res. 37, pp. 471-483.

43. Yentsch, C.S.: 1980, In: Morris, I. (ed.), "The Physiological Ecology of Phytoplankton," Blackwell Scientific, London, pp. 95-127.

44. Yentsch, C.S.: 1981, In: Symp. on Remote Sensing in Fisheries, NAFO Scientific Council Studies, 17 p.

45. Yentsch, C.S. and Garfield, N.: 1981, In: "Oceanography From Space", Plenum Publishing Corp., pp. 303-312.

46. Yentsch C.S. and Phinney, D.A.: 1981, In: "Bioluminescence, Current Perspectives", Burgess Publishing Co., pp. 82-88.

47. Yentsch, C.S. and Phinney, D.A.: 1982, J. Plank. Res. 4, pp. 93-102.

CHAPTER 14

OCEAN COLOR STUDIES IN ARABIAN SEA

I.V.MURALIKRISHNA

NATIONAL REMOTE SENSING AGENCY,
DEPARTMENT OF SPACE, SECUNDERABAD,INDIA

ABSTRACT

Ocean color study in Arabian sea is conducted with the obejective to establish primarily algorithm through which chlorophyll can be quantitatively determined. A method for atmospheric radiance calculation for evaluation of ocean chlorophyll concentration is developed. An algorithm relating reflectance to insitu chlorophyll pigment concentration is presented. A few chlorophyll pigment extraction models of other investigators are presented. Most of the models work well for those specific waters used to develop the algorithm.

1. INTRODUCTION

For dynamic modelling of phytoplankton and for continuous monitoring of variations in productivity and assessment of marine water quality,the remote sensing from aircraft or spacecraft of the upwelling spectral radiant energy from the ocean surface,makes it possible to determine pigment concentrations due to phytoplankton. Remote sensing of ocean chlorophyll has become possible due to the fact that its presence has got definite effect on optical properties of ocean waters and thus resulting in change of ocean color. The upwelling flux from the ocean as observed just above the ocean surface is the quantity that determines the color of the ocean which intrun is correlated to pigment concentration levels.

Because most of the algorithms for estimating marine

A. P. Cracknell (ed.), Remote Sensing Applications in Marine Science and Technology, 299–316.

pigment concentrations are emperically derived, they do not provide much information concerning the radiative processes in the sea which are essential for understanding the spectral variations of underwater irradiance. Secondly they are influenced by local conditions such as the presence of phytoplankton pigments and dissolved organic matter. The only way to improve over this problem is by varying the locations where the data were collected in order to average out the local influence. Aranuvachapun and Perry (1981) attempted to interpret some of these algorithms using theoretical model of diffuse reflectance in the sea. Bricaud and Satyendranath (1981) suggest that interpretation of ocean color appears satisfactory in oceanic waters, where phytoplankton and covarying by products are predominant. Conversely, it becomes ambiguous for coastal waters, influenced by terrigenous discharges of diverse origins. These waters therefore require special insitu studies and more sophisticated algorithms. In this context Doerffer's (1980) factor analysis is significant to find out how many independent components can be retrieved from the optical signal and which wave lengths have to be selected for the algorithm to be applied on the data.

2. CHLOROPHYLL SCANNING EXPERIMENT IN ARABIAN SEA(CSEA)

Remote sensing technology has provided us the opportunity to view the Ocean from a high altitude platform. As a result, one of the most useful applications of Remote Sensing Technology is the monitoring of chlorophyll pigments in the ocean. Chlorophyll information is related to marine productivity, thus providing the capability of detecting locations of high biological activity. Development of operational systems which provide this information should have tremendous benefits for the harvesting and maintenance of viable fisheries.

2.1 Objectives

The objectives of the experiment were:

- to establish primarily an algorithm through which chlorophyll, yellow substance and suspended sediments can be quantitatively determined.
- to experiment in waters where these three parameters would be found in steep gradients and in a variety of quantity ratios.
- to experiment with seatruth with the possibility of

optical classification of ocean waters and evaluation of bio-state.

2.2. Measurement Programme

The Ocean area selected for chlorophyll mapping is about 1500 sq.km. off Cochin in Arabian sea from 9 degrees 40 minutes to 10 degrees 15 minutes N and 76 degrees 04 minutes to 76 degrees 18 minutes E. This area is abundant in pigments and is ideal for testing the scanner responding to the entire dynamic range. From the present state of art of ocean remote sensing it is apparent that the development of reliable quantitative estimates of surface chlorophyll require that the contribution of back scattered radiation from atmosphere and sea surface had to be removed from the total upwelling radiance. This development requires support of properly collected sea truth. As recommended by the EURASEP working group on sea truth measurements, all discrete samples from the experiment should be analyzed by the same laboratory for parameters salinity, suspended matter, particulate organic carbon, chlorophyll and plankton identification.

Keeping in view, the significance and relevance of sea truth programme,Dr.E.G.Silas, Director, Central Marine Fisheries Research Institute,Cochin has been approached for boats and insitu biochemical water sample analysis required for the scanner data interpretation. All the biochemical data has been analysed by Central Marine Fisheries Research Institute. A total number of 13 stations has been covered by boat Cadalmin during December 1-4, 1980. The other boat Fishtech covered 17 stations during 3 days. Fig.1 presents the details of the boat stations where sea truth data is collected.

3. OPTICAL CHARACTERISTICS

The insitu data is collected using Techtum quanta spectrometer (QSM 2500) which scans continuously in the range 400 to 750 nm. The signals are amplified and recorded using a chart recorder on deck. Orientation of the ship was such that shadowing of the instrument did not occur. Measurements of upwelling and downwelling irradiance spectra were taken at the surface and at several depths varying from 4 to 25 meters and in water types varying from coastal waters to oceanic waters. The attenuation coefficient $c(\lambda z)$ which is function of both wave length and depth of water column, is defined here as

$$c = (1/z)\ (\ln E_o/E) \qquad \ldots(1)$$

where E_o and E are radiances at surface and depth z respectively, the depth profiles of the measurements

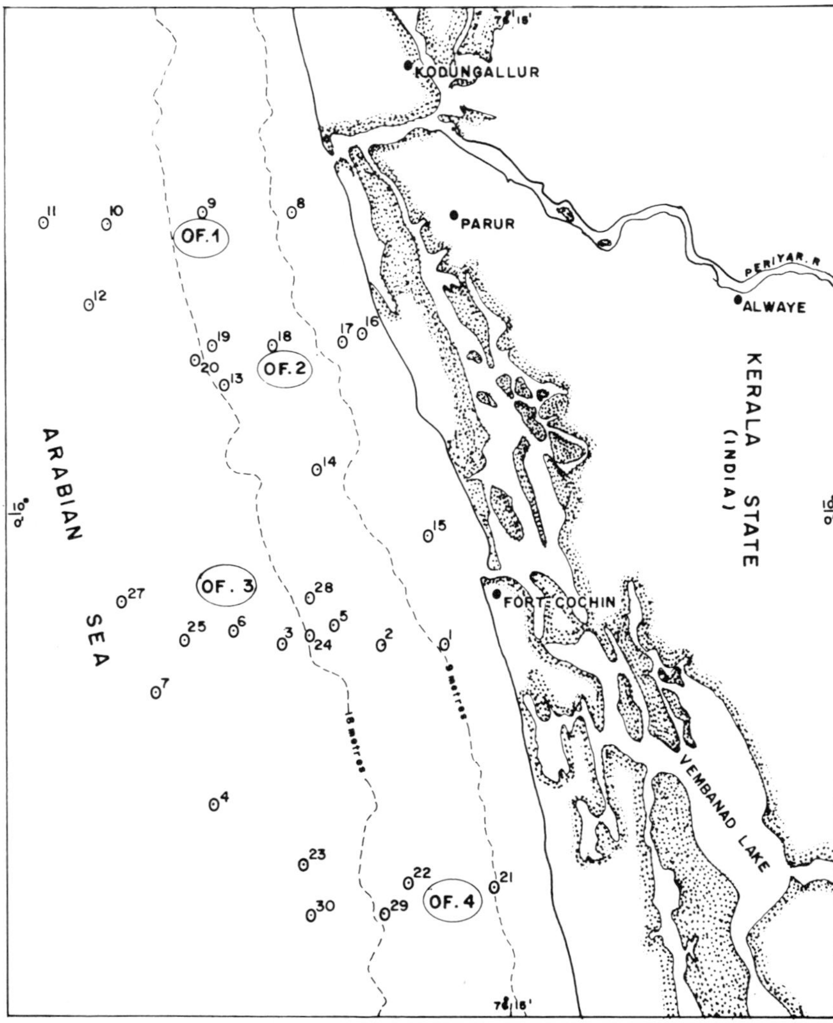

Fig.1. CSEA STATIONS WHERE SEA TRUTH DATA COLLECTED

of temperature,salinity and nutrients at the stations where spectral data were taken suggested existence of surface mixed layer. The diffuse reflectance R (λ)defines as the ratio of upwelling irradiance to downwelling irradiance at surface or any depth are computed and a typical graph of surface diffuse reflectance R (λ) for stations 9,18,27 and 29 is shown in Fig.2. In general, the curves agree qualitatively with the irradiance reflectance R (λ) presented by Smith and Baker (1978) for various values of pigment concentration.

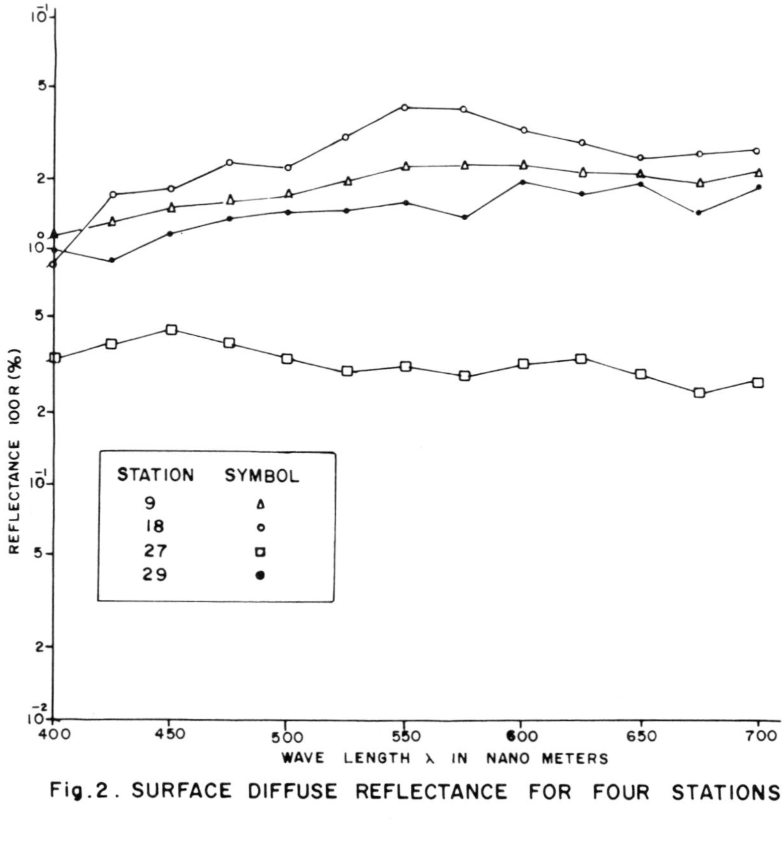

Fig.2. SURFACE DIFFUSE REFLECTANCE FOR FOUR STATIONS

The attenuation coefficient c is computed using equation 1. Defining $c=a+\beta b$ where a is absorption coefficient and βb as back scattering coefficient as obtained by Aranuvachpun and Perry(1981) and Morel and Prieur (1977)

$$R= 0.33\ (\beta b/c). \qquad ...(2)$$

From known values of $R(\lambda)$ and $c\ (\lambda)$ the values of $\beta b\ (\lambda)$ are calculated using the equation 2. These results are plotted against wavelength in Fig.3 as solid line.

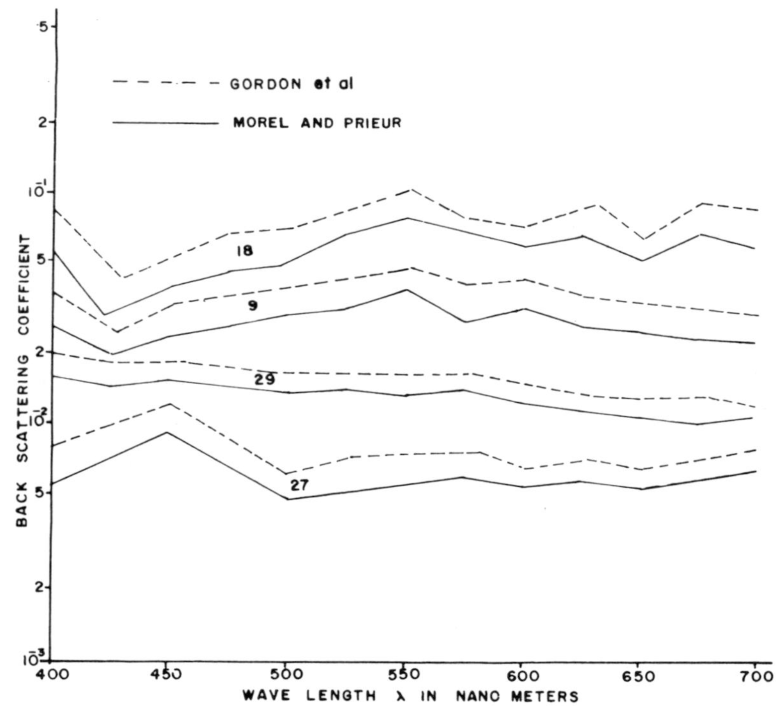

Fig.3. BACK SCATTERING COEFFICIENT SPECTRA

For the purpose of comparision the values of βb based on analysis of Gordon et al (1975) are also presented in the figure as broken line. It can be seen from the figure that βb can be taken as fairly independent of wave length as no particular inference can be drawn from the variation of βb with wave length. The low values of βb at station 27 correspond with low chlorophyll pigment concentration. Here, following the argument of Morel and Prieur(1977)it can be said that for pigment concentration greater than 2 mg/m^3 the back scattering due to particles containing pigments becomes significant and so the βb at 420 nm is smaller than βb at 560 nm as shown by curves of stations 9 and 18 in Fig.3. Here the wave lengths 420 and 560 nm correspond respectively to absorption maximum and minimum for the phytoplankton pigments(Morel and Prieur 1977). As the pigment concentration decreases the molecular backscattering of seawater dominates the particle backscattering and hence βb at 560 is less than βb at 420 nm for station 27 in Fig.3. Regarding attenuation coefficient, Aranuvachapun and Perry showed that for high chlorophyll pigments, c at 420 nm is larger than c at 560 nm. Writing ratio of diffuse reflectance R (λ_1) to R (λ_2) according to equation-2.

$$R(\lambda_1)/\ R(\lambda_2) = c(\lambda_2)\beta b\ (\lambda_1)/\ c(\lambda_1)\ \beta b(\lambda_2) \qquad ...(3)$$

the first ratio increases with decrease of pigment concentration and the second decreases with increase of pigment concentration, thus making the combined influence to give a high value of R (λ_1) to R (λ_2). Aranuvachapun and Perry suggest a linear relationship between diffuse reflectance ratio and chlorophyll pigment concentration.

4. EVALUATION OF ATMOSPHERIC EFFECTS CORRECTION FOR CHLOROPHYLL EVALUATION

The calculation of atmospheric radiance is an important task for processing of data and to arrive at meaningful conclusions regarding the bio-optical state of ocean waters. A method for atmospheric radiance calculation for evaluation of ocean chlorophyll concentration is developed. The values of ocean leaving radiance obtained after removing atmospheric radiance component agreed consistently with boat based sea truth measurements carried out by a underwater radiometer as discussed in next few sections.

4.1 Significance

Clark et al.(1969) were the first investigators to measure upwelling light from an aircraft at relatively low altitudes while simultaneously obtaining sea truth data. From these results,it became apparent that the development of reliable quantitative estimates of surface chlorophyll required that the contribution of backscattered radiation from the atmosphere and sea surface had to be removed from the total upwelling radiance. With this objective a special flying programme is planned and carried out during chlorophyll scanning experiment in Arabian sea in December 1980. The ocean upwelling radiance L_o measured at the aircraft altitude mainly consists of three components as given below.

$$L_o = L_{au} + L_{su} + L$$

Where L_{au} and L_{su} represent the radiance contribution by photons that have not penetrated the sea surface but are returned to the sensor from the atmospheric path and sea surface scattering. L is the ocean leaving radiance due to photons that penetrated the surface and its signature is associated with the water and its chlorophyll concentration. Separating the L_{au} and L_{su} from L is not possible by normal techniques. As such the proposed methodology incorporating overflights with simultaneous sea truth measurements helped to separate the path radiance component $L_p=(L_{au}+L_{su})$ from total upwelling radiance.

4.2 Correction for Nadir Level Pixel

Typical absolute spectral radiance of the open ocean should have curves which climb almost monotonically from the near infrared into the blue where the maximum radiance occurs at 430 nm. The spectral curves obtained at nadir in the absence of any noticeable sun glint at different altitudes are taken for the purpose of estimation of path radiance. Fig.1 shows the stations OF1, OF2, OF3,OF4 where overflights are carried out. The regular pattern flights are carried out in east west direction from 10 degrees 15 minutes N at invervals of 1' in latitude upto 9 degrees 45 minutes. Figures 4 and 5 (OF1 and OF3) show the contribution of path radiance in each channel. It can be seen from these figures that the amount of atmospheric contribution is minimum in infrared channels and monotonically increases with 1/wave length. The OF1,OF2, OF4 stations contain moderate to high chlorophyll concentrations whereas overflight station OF3 typically represents a region of negligible chlorophyll concentration. The influence of atmospheric column and hydrosols present in the marine environment is quite significant and about 50 to 70 per cent of the total path radiance at aircraft altitude (in the present study it is 900 meters AMSL) is coming within the first 300 meters AMSL. The contribution of path radiance increasing rapidly within the first few hundred meters altitude compared to the increase from 600 meters and higher altitudes. This is particularly applicable within the optical range of 445 to 550 nm. In the orange,

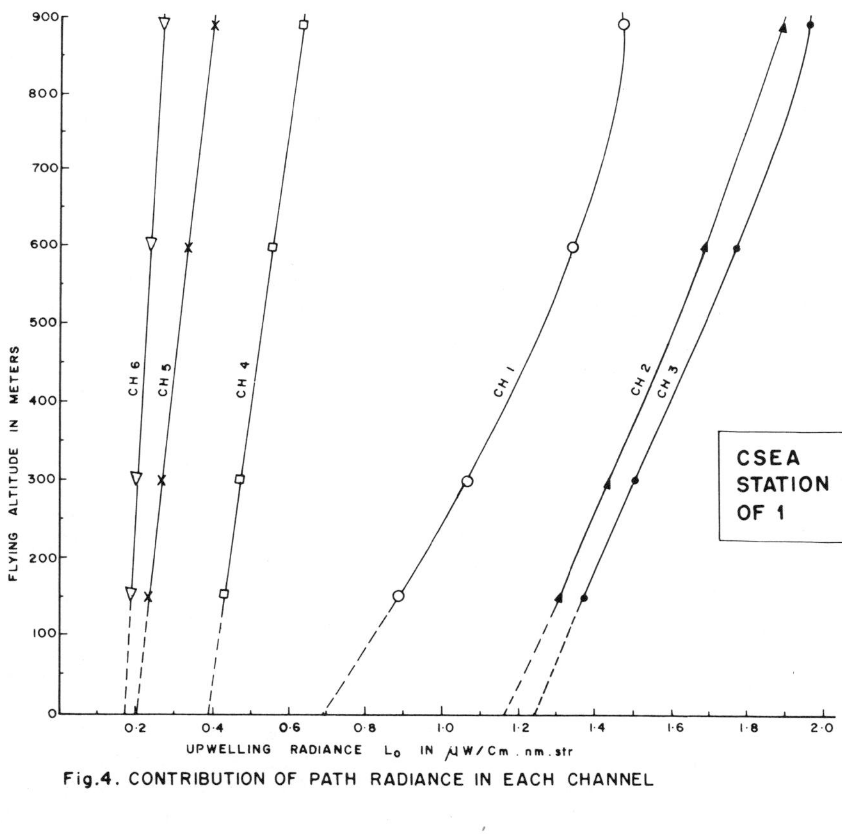

Fig.4. CONTRIBUTION OF PATH RADIANCE IN EACH CHANNEL

red and near infrared wavelength regions the increase is more or less gradual and linear. The percentage values of path radiance are evaluated for all the overflights and they are found to be consistent. The evaluated ocean leaving radiance is compared with insitu continuous measurements carried out by underwater spectrometer. The ocean leaving radiance values obtained from overflights data reduction and insitu measurement agreed well with average difference less than 9 percent. The chlorophyll concentrations at stations OF1, OF2,OF3 and OF4 are 2.67, 3.75, 0.30 and 1.15 mg/m^3 respectively. The spectral signatures for all the four cases alongwith chlorophyll concentrations are shown in Fig.6

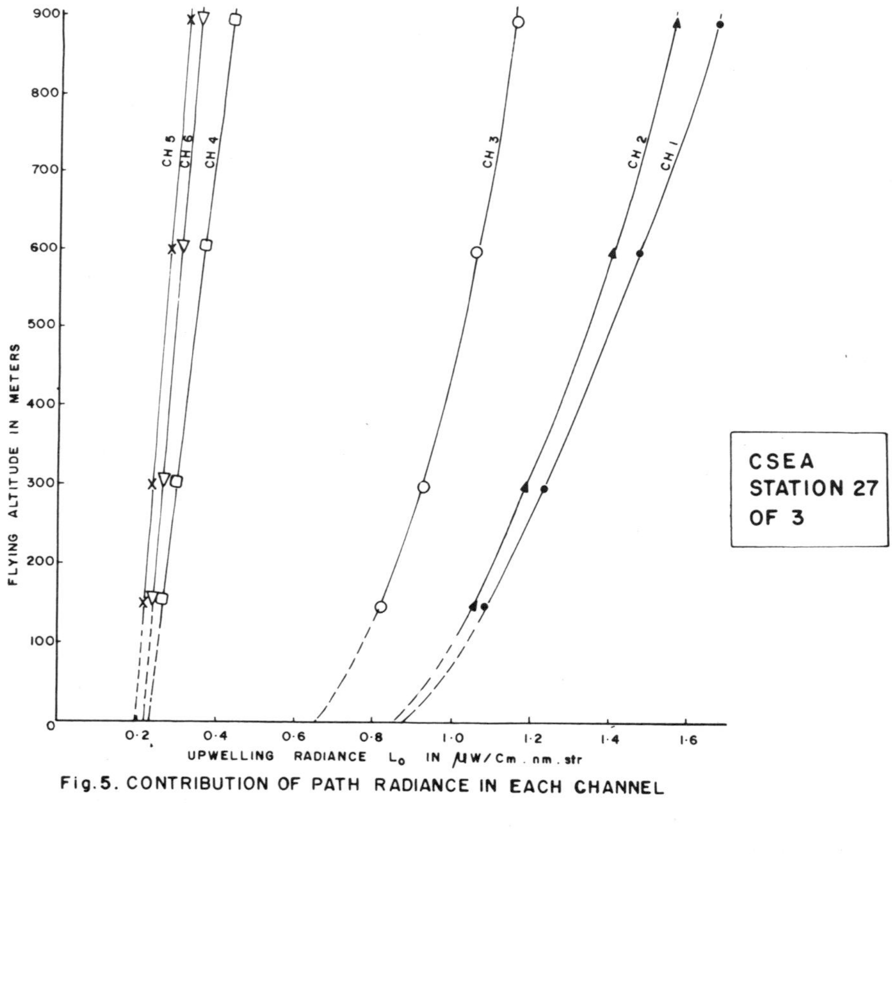

Fig.5. CONTRIBUTION OF PATH RADIANCE IN EACH CHANNEL

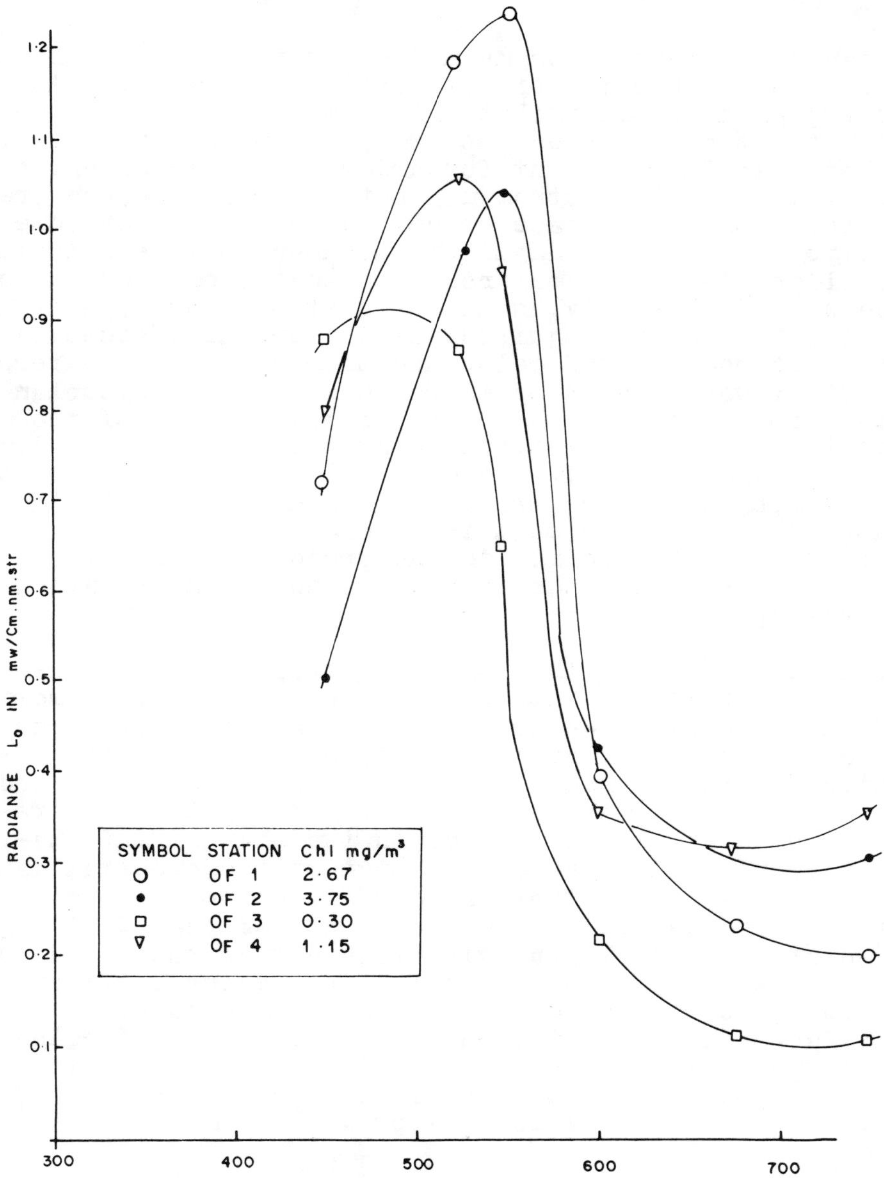

Fig.6. SPECTRAL SIGNATURE OF CHLOROPHYLL-A CONTAINING SEA WATERS (AFTER APPLYING ATMOSPHERIC CORRECTIONS)

5. OCEAN COLOR AND CHLOROPHYLL EVALUATION

5.1 Ocean Color

Ocean color represents mean distribution of radiant energy as a function of wavelength. When a phytoplankton pigment is distributed in-homogeneously over depth, its effect on the ocean color spectrum is not closely related to any straight forward measure of concentration. The color spectrum may not for instance, be predicted from the surface concentration alone for some portion of incident daylight is known to penetrate to considerable depths before being scattered back out of the water. Obviously, some depthweighted average of concentrations is required. It is generally accepted that the peak of the color spectrum of natural waters shifts toward the green in regions of high phytoplankton population and that the parametrisation of the spectrum yields a distinct relationship between chlorophyll and spectral information. Thus any attempt at monitoring the concentration of plankton, or of other suspended particles must incorporate the spectral information corresponding to the predominant species present. This incorporation leads to parametrisation of the spectral information.

It is a known fact that the oligotrophic waters are blue and eutrophic waters are green in color. So most of the attempts made by several ocean color research investigators till now are based on the evaluation of ratio of spectral data in blue and green channels and correlation of this ratio with chlorophyll concentration. However the dependence on two channel information results in under utilisation of the spectral data and gives a low statistical coefficient of determination. The ocean color radiometer at NRSA has five high resolution bands in the visible spectral range and one in the near IR. Hence a spectral parameter based on maximum spectral information would be appropriate. In this context, a statistical parametrisation of relevant spectral information is felt appropriate. With this objective, evaluation of an emperical parameter called pigment index is attempted.

5.2 Pigment Index

The NRSA Ocean color radiometer built under Indo FRG Cooperation at DFVLR has five high resolution (11-14 nm) bands in the visible spectral range and one in the near IR. Channels 1,2,3,4 and 6 are compatible with coastal zone color scanner channel alocation and an additional channel orange-channel 5 has been included in OCR. The spatial resolution at surface is 60 meters in diameter and swath width is about 2.5 km.,corresponding to aircraft altitude of 1 km. The 6 channels of OCR ocean color radiometer used has spectral bandwidth of 11 nm centered about 445,525,550,600,670 and 750 nm respectively. The first and fifth channels are pigment absorption channels, the third channel is reflectance channel. The variation of pigment concentration has no effect on spectral data in second channel. The fourth and fifth channels correspond to high reflectance due to sediment. With these features of spectral channels and specific optical characteristics of the pigment an index called pigment index (PI) is defined in 3 modes which has relevant physical explanation.

In the first mode PI is based on the ratio of channels 3 and 1. In the second mode PI is based on the ratio of sum of the pigment absorption channels and the sum of the pigment reflecting channels whereas the third mode is based on the ratio of differences of the same channels. This emperical approach gave quite interesting correlations and suggested suitable emperical equation which led to the development of reliable quantitative estimate of chlorophyll concentration. The values of PI corresponding to those modes alongwith pigment data are given in Table-1.

TABLE- 1 MODES TAKEN FOR EMPERICAL MODEL

PIGMENT INDEX			
Mode 1	Mode 2	Mode 3	Chlorophyll Concentration in mg/m^3
0.58	0.66	0.47	2.67
0.47	0.61	0.19	3.75
1.34	1.21	1.60	0.30
0.84	0.85	0.83	1.15
1.18	0.49	0.28	8.80
0.40	0.59	0.23	3.95

Mode 1 L (550)/L (445)
Mode 2 L (445)+L (670)/L(550)+L(750)
Mode 3 L (445)+L (670)/L(550)-L(750)

A statistical least square analysis has been carried out. The PI and pigment relationship is parametrised in two types with varying degrees of average relative error (A), coefficient of determination (Q) and inequality coefficient (Q_c) as given in Table-2.

TABLE- 2 STATISTICAL PARAMETERS OF EMPERICAL COEFFICIENTS

Mode	Type	Coefficients				A	Q	Q_c
		A	B	C	D			
1	1	0.66	-2.43	-	-	9.5	0.98	0.051
1	2	-	-	14.18	-2.9	3.8	0.98	0.017
2	1	0.61	-3.64	-	-	2.7	0.99	0.013
2	2	-	-	42.2	-4.13	6.8	0.98	0.040
3	1	0.73	-1.17	-	-	32.5	0.88	0.162
3	2	-	-	5.64	-1.83	6.1	0.98	0.039

The following are the two modes of emperical equations that are found to have better correlation.

Type 1 - Chl (O) = A $(PI)^B$

Type 2 - Chl (O) = C exp (D)

The statistical values given in Table 2 suggest that relation under Mode 2 and type 1 is best fit. Incidentally this mode takes the pigment index defined as the ratio of sum of the absorbing to reflecting channels. Thus the 4 channel information leads to better parametrisation required required for quantification of the chlorophyll pigment.

The significance of the parametrisation of spectral data is best understood by interpreting them from the distinct at the same time related view points. From a direct point of view the parameter PI, based on channels, 1,3,5 and 6 (Mode 2) describes all the significant modes in which the individual observed spectra differ from one another due to variations of pigment level. Another point of view is that this parametrisation proved to be effective tool as no relevant spectral information is ignored

5.3 Contour Map of Horizontal Distribution of Chlorophyll

The emperical model developed as explained in earlier section is used for quantification of chlorophyll content. First of all the area from 10 degrees to 10 degrees 15 minutes N is selected for preparation of contour map. During the period of scanning over this region light easterly winds and relatively clear skies were present. Surface temperatures were higher over the shallow mearshore areas and indicated the effect of mixing of cooler offshore sea water and warmer waters from Periyar river. The color frontal structures observed in the surface waters under lightwinds seem to separate shallow nearly homogeneous coastal waters with marked color boundary. On the seaward side, the front of greenish and offshore blue waters is also clearly found from strip chart data. The color boundaries helped in delineating the sea water into 3 distinct regions. The first region is the region effected by Periyar river. The coastal waters around the river mouth clearly show the river plume behaviour as it is entrained by surrounding water. The resultant current is pushing the plankton patch westwards in the region south off the Periyar river mouth. The second region is region of surface sea water with very high degree of pigment gradient spread over an area of about 300 sq.km. The blue green color boundary as infrared from strip-charts and emperical model corresponds to chlorophyll concentration less than 0.5 mg per cubic metre approximately running parallel to 20 meter depth contour. The third region is a region of homogeneous blue waters of oceanic type. The general distribution of the major constituents namely phytoplankton, and suspended matter indicates a high concentration just at the shoreward color boundary with chlorophyll concentrations almost constant between 3-4 mg per cubic meter along the boundary in region 10 degrees 5 minutes to 10 degrees 15 minutes N. The color coded contour map (this map will be found in the colour section, p. 453.) is shown as color plate. The highest concentration obtained in the entire region of 9 degrees 45 minutes to 10 degrees 15 minutes N is 10 mg per cubic metre just away from Cochin coastal waters. The drift of the plankton in a westward direction is clearly seen because of the interaction of the tidal current and the river plume. The plankton concentration found to be decreasing rapidly in north south direction with sharp gradients.

6. COMPARISION OF VARIOUS PIGMENT EVALUATION ALGORITHMS

The concept of using radiance ratios for various wavelengths in remote sensing applications of ocean color is well established. The upwelled radiances at 440,520 and 550 nm were grouped by Gordon and clarke (1980) into the ratios L (443)/L (550), L(440)/L(520) and L (520)/L(550) for a least squares regression analysis with the pigment data. Some selected algorithms for chlorophyll extraction from optical data are presented in Table-3 for the purpose of intercomparision. Aranuvachpun and Perry(1981) suggest despite all limitations that for low concentration (C in mg/m^3) less than (3 mg/m^3) mode 1 algorithm of Gordon et al gives more accurate estimates than mode 2

Table-3 Intercomparision between some chlorophyll pigment extraction Algorithms

Investigator	Mode of Pigment evaluation algorithm	Remarks
Gordon et al (1980)	Mode-1 $C=0.5(PI)^{-1.3}$ Mode-2 $C=0.8(PI)^{-4.0}$	PI is L(443)/ to L(550)-Mode 1 L(520)/L(550) in Mode-2
Kim et al (1979)	C = 801 exp (-20.8 PI)	PI is L(472) -L(548)/L(472) +L(548)
Deschamps et al (1977)	C=f (PI)(for low Chlorophyll)	PI based on L(525)-L(466)
Viollier (1978)	C=f (PI)(for high Chlorophyll	
Aranuvachapun and Perry(1981)	C=1.5(PI)-0.6	PI is R(560)/ R(420)

algorithm. For higher concentrations greater than (10 mg/m^3) mode 2 is suggested. A linear relationship suggesting phytoplankton pigment concentrations have linear effect on the underwater irradiance is presented by Aranuvachapun and Perry(1981). It is clear that our ability to relate pigment concentration to optical characteristics is seriously impaired by the relatively small quantity of good quality optical data on which model can be tested. Also the models proposed work well for those specific waters used to develop algorithm. Thus there is an element of uncertanity until the bio-optical state assessment model takes into consideration all possible parameters that can be applied to waters of different types under various environmental considerations.

Acknowledgement:

The author wish to thank Prof.B.L.Deekshatulu, Director, National Remote Sensing Agency for giving permission to deliver the lectures. Thanks are due to Dr.E.G.Silas, Director, Central Marine Fisheries Research Institute for providing the insitu biochemical water quality data. The NRSA ground truth groups efforts for collection of insitu optical data are verymuch appreciated. Thanks are due to P.Santhakumari for typing ably the camera ready version of the lecture notes.

REFERENCES

Aranuvachapun,S. and Perry, R.I (1981) International Journal of Remote Sensing, Vol.2, No.4, P 299-312.

Bricaud, A and Sathyendranath Shubha (1981) International Colloquim on Spectral Signatures of Objects in Remote Sensing,Avigon, France, September.

Clark, G.L., Ewing, G.C., and Lorenzen, C.J.(1969), Science, 167, P 1119-20.

Deschamps, P.Y., Lecomte, P and Viollier,M (1977) Proc. 11th International Symposium on Remote Sensing of Environment, Ann Arbor, P 1021-1033.

Doerffer R (1980) presented at IUCRM-SCOR-COSPAR Symposium Oceanography from Space, Venice.

Gordon, H.R., Broun.,O.B., and Jacobs,M.M.,(1975), Applied Optics, Vol.14, P 417-432.

Gordon, H.R., and Clark. D.K.(1980) Boundary Layer Metereology, Vol.18, P 299-313.

Gordon,H.R.,Clark., D.K.Mueller,J.L. and Hovis,W.A (1980) Science, New York Vol.210, P 63.

Kim H.H.McClaim, C.R. Blaine,L.R, Hart, W.D.Alkinson, L.P. and Yoder, J.A.(1979) NRSA Technical Memorandum-80574, August.

Morel, A and Prieur (1977), Limnology and Oceanography, Vol. 22, P 709-722.

Mueller (1974), Ph.D.Thesis, Oregon State University, USA.

Prieur, L and Sathyendranath, Shubha (1981),Limnology and Oceanography Vol.26, P 671-689.

Smith, R.C. and Baker, K.C.(1978.)March, Vol.123 (2), P 260-267.

Violler, M, Tanre, D and Deschamps, P.Y.(1981), Boundary Layer Meteorology, Vol.18 P 247-267.

CHAPTER 15

LANDSAT APPLICATION TO SUSPENDED SEDIMENTS EVALUATION

I.V.MURALIKRISHNA

NATIONAL REMOTE SENSING AGENCY.
DEPARTMENT OF SPACE, SECUNDERABAD,INDIA

ABSTRACT

Suspended sediments affect aquatic environment. In the visible range the suspended sediment concentration can be correlated with reflectance level. A ratio of Landsat MSS bands is used to develop model for evaluation of suspended sediment concentration. A six level color coded contour map of surface distribution of suspended sediments off Saurashtra coast in Arabian sea is obtained using the proposed model.

1. INTRODUCTION

Suspended sediments affect aquatic ecosystems and the productive potential of coastal water bodies. Study of suspended matter in sea water is important in understanding the genesis of sediments in seas and advection and diffusion of suspended material into a basin. Main sources of suspended matter in the ocean are river runoff, biological productivity and eolian dust. It is also necessary for assessment of fishery resources of a particular region as the particulate matter is the food for a large variety of marine fauna. The conventional techniques for measuring suspended sediments involve gravimetrical determination. With the objective of monitoring the natural resources, the National Aeronautics and Space Administration (NASA) of USA launched Landsat series of satellites. The multispectral scanner (MSS) on board the Landsat provides 80 meter resolution data in four bands each band corresponding to a fixed range of wave length

A. P. Cracknell (ed.), Remote Sensing Applications in Marine Science and Technology, 317–322.

corresponding to the electromagnetic spectrum. The motion of water masses in the vicinity of estuarine mouths is an important part of any study of the physical properties of such region. Suspended sediment in the fresh river water and the availability of synoptic coverage by Landsat satellite imagery allow determination of the circulation in this region. Landsat and aircraft based sensors have been able to respond to variations of reflected light due to varying amounts of sediments in seawater. These studies have shown the potential for use of remote sensing techniques to obtain qualitative and quantitative synoptic views of suspended sediments in the surface waters of water bodies. A review of these publications on remote sensing of various features in lacustrine and marine waters provide the following information (Sorensen et al 1976). Around 470 nm wavelength the absorption by water is minimum. An increase in particle size of suspended matter increase the upwelling irradiance and suspended matter can be quantitatively determined with reasonable accuracy from multispectral imagery by density slicing.

With the available information as given above a study is planned to evaluate suspended sediments concentration in sea water from remotely sensed Landsat satellite data. This programme of study encompasses collection of optical and biochemical data and development of a model between Landsat bands reflectance and the suspended sediments concentration. This model is applied to Landsat scene off Saurashtra coast to obtain the contours of distribution of suspended sediments in Arabian sea waters off the North Western shelf of India.

2. SEA TRUTH DATA

Any remote sensing study of earth features requires some sea truth information for evaluating the path radiance and correlation of radiometric data with insitu optical and bio-chemical data. In the present study a series of radiometric measurements are performed from boat platform with an exotech radiometer model 100 A to determine the beam transmittance and global irradiance on the ground in the four optical bands of the Landsat MSS and to determine upwelling irradiance and reflectance ratio and plot with insitu suspended particulate matter for further application to Landsat MSS to obtain the contours of suspended sediments.

The exotech radiometers recorded upwelling, downwelling and sky irradiances. The spectral reflectance and transmittance coefficient(τ) are computed from these radiometers data. For a remote sensing system looking vertically downward, τ is the atmospheric transmission of one air mass. If m is the number of air masses referred to the zenith air mass (for which m=1). The value of τ is computed using the following relationship.

$$\text{Global - Sky component} = m\, H \tau^{m}$$

where the values of H the solar constant are taken from Thekaekara et al (1971). Table 1 gives the values of these parameters in bands 4,5,6 and 7 corresponding to wave lengths 500-600,600-700,700-800 and 800-1100 nm. The measurement of sediments at these stations is done by collecting samples of water from boats. The samples were filtered using preweighed membrane filters (46 mm dia with 0.45 pores). Samples were weighed to the nearest 0.01 mg for estimating the concentration of suspended matter. These values are correlated with reflectance and possible variations of optical parameters are studied for application to Landsat MSS data.

Solar radiation reflected from sea water surfaces varied with the amount of suspended sediment and wave length. In general, reflected solar radiation between wave lengths 500-700 nm increased as the concentration of suspended sediment increased. The data indicated a change in the concentration of suspended sediment caused a change in the amount of radiation reflected from the water surface. Various ratios using band 4, band 5 and band 6 are found to give varying degrees of correlation. The correlation of spectral reflectance values in each band and the corresponding insitu suspended particulate matter values suggested that a linear regression best fit the relationship between the concentration and reflectance. This formed the basis for application of sea truth model to evaluate suspended matter contours from Landsat satellite data. The percentage organic matter, recognized as shell derbis, sponge spicules, diatom frustules, egg sacks, and lignaceous matter, varies from 25 to 35 per cent. These materials show a wide variety of differing shapes, sizes and surface textures quite different from inorganic floccules of clay or the siliciclastic particles. The variations observed in radiance and the scatter in the ratio diagram can be attributed to these different shapes, sizes and surface textures. Munday et al (1979) implemented an automated multidate Landsat CCT MSS measurement of suspended sediment concentration employing chromacity analysis. This system provided automatic pixel by pixel adjustment of atmospheric variations. In these studies Munday et al found that the effects of sediment type and size and solar angle are negligible. According to Jerlov (1976) backscatter, of which Landsat radiance is a measure, is indeed a function of grain size and shape. But one point worth mentioning here is that the grain size and shape cause changes in all bands in the same proportion thus a band ratio compensates for these effects as done in the present study.

3. LANDSAT SCENE ANALYSIS AND DISCUSSION

The Landsat scene No.160-046 is taken for the analysis. This scene is selected basically because about 4 rivers carrying terregeneous substances join the sea around Gulf of Cambay. Examination of results of suspended matter distribution obtained through conventional techniques reveal the fact that in the region of Gulf of Cambay and from Daman to Ratnagiri and a belt along the Kutch and Saurashtra coast the suspended matter was high in both surface and bottom waters. Also the area in the Gulf of Cambay and particularly off Saurashtra coast was reported to have a dispersal distance of 25 km from coast. This is an unusual case in tropical oceanic areas. With this back ground the scene corresponding to Saurashtra coastal waters is taken for obtaining the contours. The Landsat MSS band 4 band 5 values are evaluated for a sample scan line. This ratio is taken as basis and the corresponding sediment levels are evaluated making use of sea truth information. Here the main source of error is the nonlinear variation of spectral radiance value with Landsat CCT digital count in each band because the digital counts in each band correspond to different spectral reflectance levels. However in the present study the error due to this nonlinearity has been neglected. The correction for this can be incorporated using Landsat MSS calibration relationships. In this section this correction is not applied. The scene is analysed on a interactive computer and a six level delineation represented by Dark green, bright yellow, dark purple,

Cyan, bright green and blue colours is carried out. The color coded processed output is shown in color plate section. (see p. 454) The concentration of suspended matter in the surface water ranged from 0.5 to 3.5 mg/l. The concentrations greater than 3.5 mg/l represented by dark green color exist in a narrow strip bordering the coast. Next higher concentration (2-3 mg/l) represented by bright yellow is found spreading to a distance of 2 to 6 km from the coast. Moderate concentration, 1.5-2.0 mg/l represented by seep purple is extending upto a distance of about 25 to 30 km from coast. This is an unusual situation because such a high rate of dispersion is rarely found in tropical ocean waters. The concentration is found dispersing more towards Daman coast and the sea region west of the Saurashtra coast is having very low concentration with coastal waters containing as low as 0.5 mg/l concentration. This region is having comparitively lesser terregeneous supply and relatively lesser coastal abrasion. The study reported here amply demonstrates the use fulness of remotely sensed Landsat data for monitoring suspended matter distribution. When the scenes adjacent to the one analysed here are also taken and analysed a picture of sediment distribution which is synoptic in nature can be obtained. This helps also in estimating the strength of currents based on the dispersal of the suspended matter.

Acknowledgement

The author wish to thank Prof.B.L.Deekshatulu, Director, National Remote Sensing Agency for giving permission to deliver the lectures. The NRSA ground truth group efforts for collection of insitu optical data are verymuch appreciated. Thanks are due to P.Santhakumari for typing ably the camera ready version of the lecture notes.

REFERENCES

Jerlov, N.G.(1976) Elsevier Publishing House Oceanog. series 14, 231p.

Munday, J.C. Alfoldi,T.T.,and Amos,C.L. (1979) VIMS contribution No.909, Virginia, USA.

Sorensen, B.M.,Sturn B and Gatelli, E (1976), XIII

Congress of the International Society for photogrammetry, Helsinki, Finland, July 11-23.

Thekaekara et al. (1971) NASA Document SP-8005

Voillier. M and Banssart N (1979), 13th International Symposium on Remote Sensing of Environment, Ann Arbor, Michigan, April 23-27.

Table 1 RADIOMETRIC MEASUREMENTS AND VALUES OF REFLECTANCE AND TRANSMITTANCE

Station	MSS Band	Irradiance in mw/cm² Upwelling E_u $R=E_u/E_d$	Downwelling E_d (Global)	Secchi depth in meters	Transmittance
8	4	0.18	10.96	1.3 (8)*	0.76
	5	0.19	13.96		0.85
	6	0.12	9.87		0.90
	7	0.12	11.22		0.90
9	4	0.24	10.96	4 (16)	0.72
	5	0.20	14.40		0.85
	6	0.19	9.79		0.90
	7	15	11.52		0.65
10	4	0.34	11.85	4.5 (23)	0.75
	5	0.19	15.86		0.82
	6	0.05	10.98		0.90
	7	0.04	13.19		0.72
11	4	0.20	14.28	5.5 (30)	0.79
	5	0.13	18.36		0.85
	6	0.04	13.45		0.90
	7	0.02	15.76		0.95
21	4	0.32	10.54	4.4 (9)	0.61
	5	0.18	12.10		0.83
	6	0.03	9.10		0.90
	7	0.02	10.84		0.93
22	4	0.18	8.18	10.5 (18)	0.56
	5	0.08	9.83		0.78
	6	0.03	7.40		0.78
	7	0.04	8.50		0.81
23	4	0.09	5.44	16.5 (25)	0.49
	5	0.05	6.38		0.60
	6	0.02	4.54		0.65
	7	0.01	5.30		0.65

* Total depth in meters.

CHAPTER 16

LIGHT EMERGING FROM THE SEA - INTERPRETATION AND USES IN REMOTE SENSING

Shubha Sathyendranath and André Morel

Laboratoire de Physique et Chimie Marines
Université Pierre et Marie Curie
La Darse, F06230 Villefranche-sur-Mer, FRANCE.

ABSTRACT

In passive remote sensing of the oceans in the visible, spectral values of radiance upwelling from the sea are estimated, and analysed for the retrieval of phytoplankton and sediment concentrations in the surface waters. Models of radiative transfer, and the spectral characteristics of substances that determine the nature of the light emerging from the sea, are discussed here, in order to establish the link between the "colour" of the sea and concentrations of phytoplankton and suspended sediments. The inverse problem of estimating concentrations from upwelling radiance is then examined. Techniques and algorithms in use now are mostly statistical best fits. They are reviewed critically in the light of our present theoretical knowledge. Technical and practical information pertaining to satellite optical sensors available for ocean remote sensing and some airborne sensors are also given in this paper.

I. INTRODUCTION

Visible radiant energy which arrives at the surface of the ocean either directly from the sun, or after scattering by the atmosphere, is partly reflected at the air-sea interface, and the remaining transmitted into the sea. The transmitted energy is progressively absorbed or scattered by the water molecules and the other substances (particulate or dissolved) present in it. A part of the light scattered in the backward direction finds its way back into the atmosphere. The problem of passive remote sensing in the visible concerns the estimation and interpretation of this

A. P. Cracknell (ed.), Remote Sensing Applications in Marine Science and Technology, 323–357.

upwelling energy in order to retrieve all possible information on the nature and concentration of the substances present in the sea water.

The first attempts at remote sensing of the oceans in the visible, from airplanes or satellites, were carried out using photographic techniques or radiometers collecting data in a single, broad, spectral band (≃ 200 nm) in the visible. At this stage, no efforts were made to separate the ocean signal from the atmospheric signal, and the interpretations were qualitative, and mostly concerned ocean-atmosphere inter-relations. The satellite LANDSAT, launched in 1972, carried a multispectral scanner (MSS), with two channels in the visible. With this sensor came the era of, or the need for, atmospheric correction, and scientists started looking for quantitative information from the sea signal considered apart. However, the lack of spectral detail and the low radiometric precision of this instrument limit its applications. The MSS was followed in 1978 by the Coastal Zone Color Scanner (CZCS) on the satellite NIMBUS-7. This sensor collects data in four spectral bands in the visible. A number of airborne sensors are also available now, which provide even greater spectral detail. Besides, the radiometric precision of these sensors are also often better than the MSS. In this paper, the accent is placed on the CZCS and other multispectral scanners in the visible, and the algorithms, more or less precise, which have been developed for such sensors.

Let us now consider a remote optical sensor at a height z viewing the sea at an angle (θ',ϕ). This sensor receives a radiance $L_t(\theta',\phi,z)$, which can be split up as follows:

$$L_t(\theta',\phi,z) = \{L_w(\theta',\phi) + L_r(\theta',\phi)\}T_a + L^*(\theta',\phi,z)$$

where L_w is the upwelling radiance from the water body, L_r is the energy reflected at the sea surface into the same direction, T_a is an atmospheric correction factor, and L^* the path radiance due to the light scattered from the sky by molecules and aerosols. z is the height of the sensor above the sea surface. Though not explicitly mentioned in the above equation, these parameters are all wavelength dependent. Of these parameters, only L_w contains encoded information concerning the sea. The signal received by the sensor has to be processed, in order to recover L_w. This problem is treated by B. STÜRM in this volume, in the chapter devoted to atmospheric correction. Here, we deal only with the interpretation of L_w.

One of the principal aims of research in this field has been the determination of phytoplankton concentration (generally in terms of their main pigment, the chlorophyll a) from L_w. This concentration in turn serves diversely, as an indicator of biomass and productivity, a tracer of water masses or a control parameter in pollution monitoring. Another parameter of interest is the sediment load which is of importance to physical oceanographers and geologists, in studies of land-sea interactions such as wave action on coasts, river discharges, land drainage and sediment transport.

Specular reflection of sunlight from the sea surface has been used to study waves and winds at the sea surface. But this application of remote sensing in the visible is now generally considered to be of secondary importance, since microwave remote sensing techniques are also capable of carrying out such studies.

The radiometric terms and notations used in this text follow the recommendations of the IAPSO working group on optical oceanography (1). They are recalled in Appendix A.

II. "COLOUR" OF THE SEA, AND THE WATER-LEAVING RADIANCE L_w.

The water-leaving radiance just above the water surface, L_w, is related to the radiance L_u just below, by the equation

$$L_w(\theta',\phi) = L_u(\theta,\phi)\ \{1 - \rho(\theta,\phi)\}/m^2$$

where $\rho(\theta,\phi)$ is the Fresnel reflectivity for rays incident on the water-air interface at angle θ (with $m \operatorname{Sin}\theta = \operatorname{Sin}\theta'$) and m is the refractive index of sea water. This equation is valid for a flat surface. In the case of a wind ruffled surface, the reflectance changes significantly in the neighbourhood of the critical angle for emerging light ($\theta = 48°$) and remains little altered for small θ' (2).

$L_u(\theta,\phi)$ is related to the total upwelling irradiance E_u just below the surface, by the equation

$$E_u = \int_0^{2\pi}\int_0^{\pi/2} \operatorname{Cos}\theta\ L_u(\theta,\phi)\ \operatorname{Sin}\theta\ d\theta\ d\phi.$$

If the ocean is assumed to be a perfect lambertian reflector, the equation simplifies to $E_u = \pi L_u$, where L_u is the constant radiance for all upward looking angles. But this is not perfectly true. Austin (3) has indicated that, if L_u is the radiance in the nadir direction, π has to be replaced by a factor Q closer to 5.

E_u is sensitive to variations in the radiant energy incident at the sea surface. Since it is preferable that the studied parameter be independent of these variations, it is logical to work with the reflectance R, where

$$R = E_u/E_d\ .$$

This leads to:

$$L_w = \frac{R\, E_d}{Q\, m^2} \{1 - \rho(\theta,\phi)\}$$

where E_d is the downwelling irradiance, also just below the sea surface. In principle, E_d can be computed from the extra-terrestrial solar radiation, combined with the optical properties of the atmosphere. The fundamental and dimensionless quantity involved in the problem is the reflectance R; and its spectral variations are the physical expression of the "colour" of the sea.

Now, for two wavelengths λ_1 and λ_2, the ratio $L_w(\lambda_1)/L_w(\lambda_2)$ is proportional to $R(\lambda_1)/R(\lambda_2)$ only if $E_d(\lambda_1)/E_d(\lambda_2)$ and Q are constant. In fact, theoretically, Q is slightly wavelength dependent. For lack of experimental data, this variability is neglected. As for the ratio of downwelling irradiances, our *in situ* measurements show variations of a few percent under conditions favourable for remote sensing (ie, relatively clear skies and high solar altitudes). As an example, the ratio $E_d(440)/E_d(550)$ shows a variation of ± 8% around a mean value of 0.97. Practically, these variations are neglected, and the irradiance and radiance ratios are interpreted similarly.

II.1. Relationships between reflectance and the inherent optical properties.

Many models of radiation transfer in the sea have served to establish relationships between R and the absorption and back-scattering coeficients, respectively a and b_b. The Schuster method (4), which only considers a downwelling and an upwelling flux, leads to the expression $\frac{1}{2}\, b_b/(a + b_b) = R$ when b_b is much smaller than a.

Through a regression analysis performed on results acquired by a Monte Carlo simulation, Gordon et al (5) expressed R as follows:

$$R = \sum_{0}^{3} r_n x^n \text{ , where } x = b_b/(a + b_b) \qquad (1a)$$

The first order term is predominant in this polynomial fit, and r_1 was found to vary from 0.32 in the case of parallel illumination from the zenith, to 0.37 for totally diffuse illumination. Kirk (6), also using data generated through a Monte Carlo computation, proposed a linear relationship between R and b_b/a, with 0.328 and 0.391 as ocefficients for solar altitudes of 90° and 45° respectively.

The method of successive orders of scattering (7,8) gives

$$R = 0.33\ (b_b/a)(1 + \Delta) \qquad (1b)$$

for small values of (b_b/a). The second order term, Δ, which depends on the volume scattering function of the water and also on illumination conditions, remains small (typically less than $\pm 5\%$) and, in a first approximation, can be neglected. Equations 1a and 1b lead to comparable results for b_b and a values generally observed for sea water (and for $b_b/a < 0.30$).

Very few direct in situ experimental results are available to check these theoretical results. Some observations in the Baltic (9) yielded $R \simeq 0.2 b_b/a$. It was found that for turbid ($b_b/a > 0.3$) river waters, the above mentioned relationships could be unsatisfactory (10).

However, it will suffice to keep in mind that there exists a proportionality between R and b_b/a, since, in remote sensing applications, ratios of radiometric quantities (such as R or L) are usually preferred to absolute values.

The substances present in the water actually govern the inherent properties such as a and b_b , and so, R can now be linked to these substances, and to their variations in quality and quantity. This analytical approach could lead to "analytical algorithms" for the retrieval of the marine parameters of interest from $R(\lambda)$ values. Currently, it is much more common to follow a simplified and empirical approach, which consists of passing over the intermediary steps, and establishing direct relationships between R (or L) and the marine parameters, through statistical analyses of acquired data. Results from such analyses are discussed later on (§III), but now, the analytical approach is developed, in order to arrive at a better understanding of the phenomenon of ocean colour. Also, this would permit us to examine later, the available algorithms in the theoretical context.

II.2. Classification of ocean waters according to the substances present.

In order to clarify the following presentation, and more over to introduce a fundamental distinction useful in remote sensing applications and also in the "bio-optics" of the sea, a classification is presented (summarized in Fig. 1). The ocean waters are divided into two cases (11).

Case 1 waters are those for which phytoplankton and their byproducts play the dominant (and presumably exclusive) role in determining the optical properties of the water body. These by-products (2 and 3) include both detrital particles (more or less coloured) and "autochthonous" yellow substances. These waters range from oligotrophic (very low pigment concentration) to eutrophic waters (with very high concentrations).

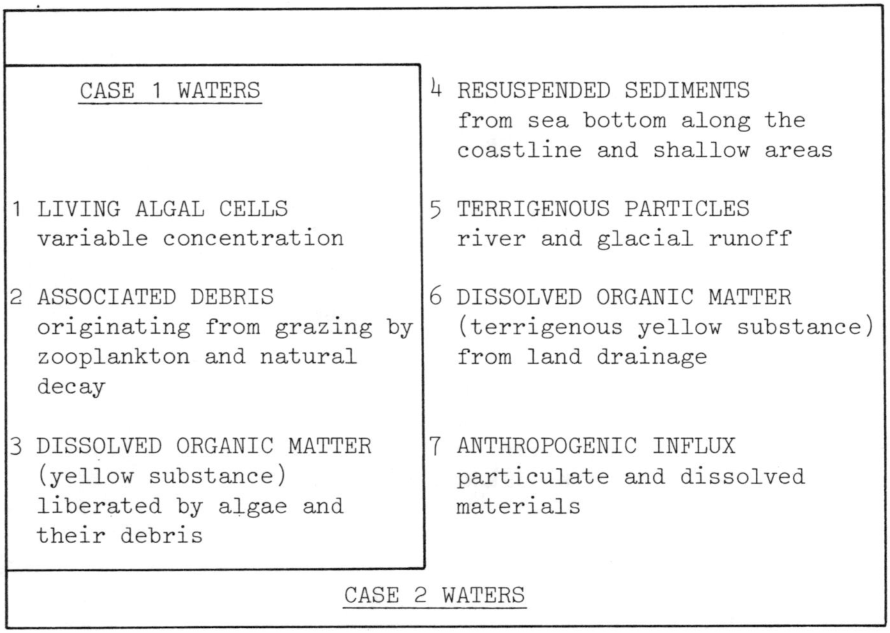

Fig. 1. Schematic presentation of Case 1 and Case 2 waters. Adapted from Gordon and Morel (83).

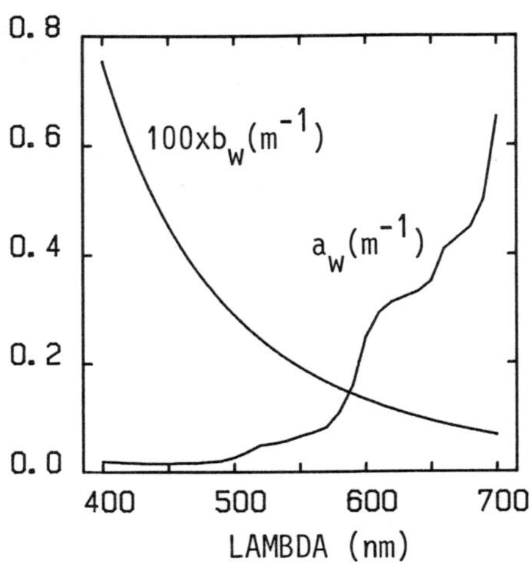

Fig. 2. Spectral absorption and scattering curves of pure sea water.

Waters depart from Case 1 and belong to Case 2 as soon as one at least of the substances numbered 4, 5, 6 or 7 is present, whereas the components 1, 2 and 3, always associated, may (or may not) be present. A high sediment load may result from land drainage or from resuspension of sediments from the sea bottom by wave action ("sediment-dominated Case 2 waters"). A high content of terrigenous yellow substances occur in some semi-enclosed zones (eg. the Baltic, some fjords and inlets). In the absence of high turbidity these waters form the "yellow-substance dominated Case 2 waters". Obviously, these Case 2 waters of diverse kinds are frequent in coastal zones and possibly offshore above banks and extended shallow shelves. Conversely, oceanic waters belong to Case 1 as a rule. In the absence of terrigenous influx and continental shelf, Case 1 waters can reach the coast line. It seems reasonable to expect that some relationships link the phytoplankton and its retinue (agents 1, 2 and 3). Consequently, the interpretation of optical properties in Case 1 waters is presumably easier than in Case 2 waters, where many unpredictable and independent influences may coexist.

II.3. Spectral "signatures" of the diverse substances which influence ocean colour.

The interest in expressing R as a function of b_b and a arises from the fact that the principle of additivity is rigorously applicable to all inherent properties (12). Thus, if n absorbing and scattering constituents are present in the medium, each with its own coefficients a_i and b_{bi} (i= 1 to n), then,

$$a = a_w + \sum_1^n a_i \quad \text{and} \quad b_{bi} = b_{bw} + \sum_1^n b_{bi}$$

where the constant terms a_w and b_{bw} represent the contributions of pure sea water to the absorption and backscattering coefficients respectively. Now, if each of these n constituents are present in concentrations c_i, and if $a_i^{\star}$ and $b_{bi}^{\star}$ are the "specific" absorption and backscattering coefficients respectively, we then have

$$a = a_w + \sum c_i a_i^{\star} \tag{2a}$$

and

$$b_b = b_{bw} + \sum c_i b_{bi}^{\star} \tag{2b}$$

(Here "specific" refers to absorption or backscattering coefficient per unit concentration of the substance). Using Eqs 1 and 2, R can now be expressed in terms of the concentrations of the different constituents of sea water, provided that all the $a_i^{\star}$ and $b_{bi}^{\star}$ values are known, and this for all wavelengths, since these parameters are all functions of wavelength. The spectral values are often referred to as "spectral signatures". They are examined in the following paragraphs.

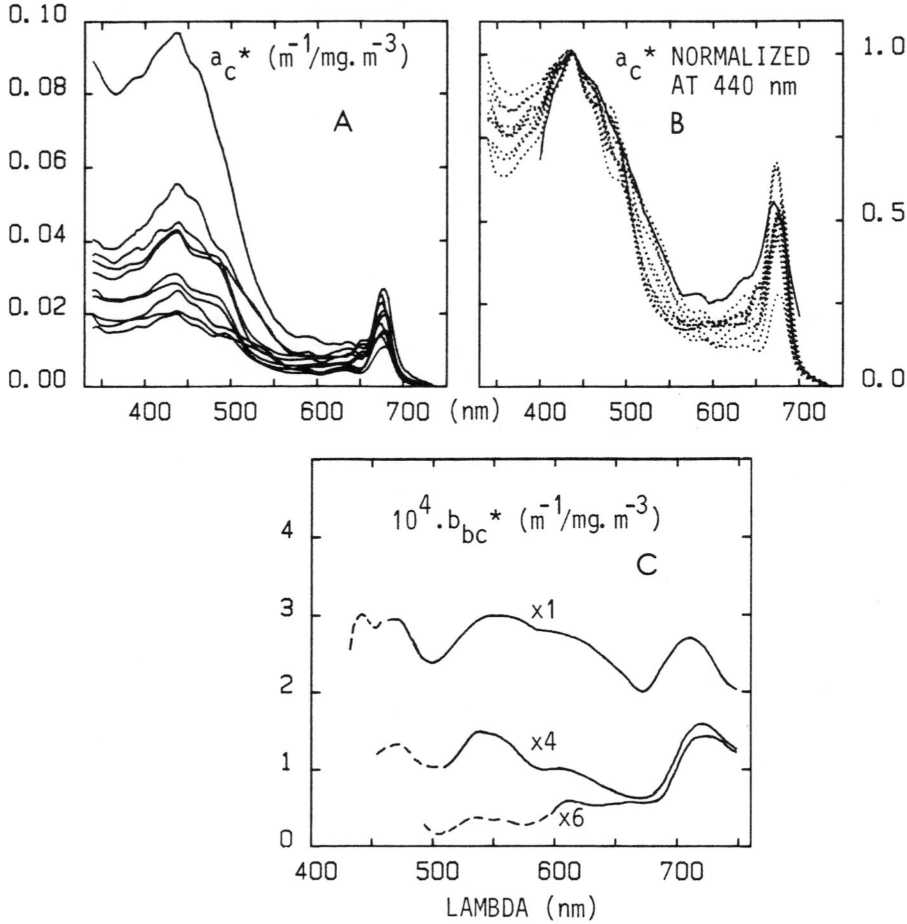

Fig. 3. Some experimental results concerning the optical properties of phytoplankton.

A- Spectral values of the specific absorption of phytoplankton (+ detritus when present), per unit concentration of (chl a + pheo a), measured in the laboratory. The 10 spectra correspond to 10 different monospecific cultures of algae belonging to diverse classes. The measurements were made by Lazzara, Bricaud and Sathyendranath of LPCM.

B- Dashed lines: the same spectra as in A, but normalized at 440 nm. Continuous line: a mean spectrum of the normalized absorption spectrum of phytoplankton, estimated from in situ absorption measurements (from Prieur and Sathyendranath, 32).

C- Spectral values of specific backscattering. Laboratory measurements on 3 monospecific cultures (adapted from ref. 27).

Pure sea water. The optical properties of pure sea water have been the subject of various experimental and theoretical studies. These studies have been recently reviewed critically by Smith and Baker (13). Their conclusions are in favour of the previous results of Morel (14) and Morel and Prieur (15). The scattering coefficient of pure sea water (b_w) has a value of 0.00288 m^{-1} at 500 nm, and its values at other λ's can be calculated from the approximate equation $b_w(\lambda) = b_w(500)(500/\lambda)^{4.3}$, and since molecular scattering is symmetric with respect to 90°, $\bar{b}_{bw}$ = 0.5. Pure sea water acts as a blue filter, with very low absorption coefficient (a_w) in the blue and blue-violet, and relatively high values in the rest of the visible spectrum. The spectral signatures of pure sea water are presented in Fig. 2.

Phytoplankton and associated detritus. The optical properties of planktonic algal cells have been measured on monospecific cultures in the laboratory. The variability of these properties from species to species has been recognized and analyzed theoretically (16-30). This information is essential when attempting to understand the optical effect of algal biomass in natural environments, but they are not sufficient, since laboratory measurements are carried out at conditions very different from those at sea. Efforts have also been made, therefore, to determine the in situ optical properties (30, 31,32). Within the sea, in fact, there is no easy way to separate the respective influences of living material and its associated detritus. This difficulty is reflected in the choice of the index used to describe the algal content in remote sensing applications: this index C is the sum of chlorophyll a and its degradation product pheophytin a. So, here, the spectral signatures of phytoplankton and their associated detritus are considered together.

Scattering by phytoplankton falls in the domain of Mie scattering (particle size bigger than the wavelength). Diverse studies (17-20,24-27) have helped to show that: 1) their specific scattering coefficient b_c^*, backscattering coefficient b_{bc}^* and backscattering ratio $\bar{b}_{bc}^*$ are all variable and wavelength dependent, 2) in the vicinity of absorption peaks, scattering appears to be depressed such that scattering spectra exhibit approximately the inverted form of the absorption spectra, 3) the volume scattering function $\beta(\theta)$ is highly dissymetric, increasing rapidly for small values of θ (forward scattering), and so $\bar{b}_{bc}$ is generally very small, of the order of 0.1%.

Photosynthetic pigments and their degradation products are the main absorbing agents in this group. Chl a is accompanied in phytoplankton by accessory pigments (mainly chlorophylls b and c and carotenoids) in proportions that vary from population to population. This in turn renders the specific absorption spectrum a variable parameter. In addition, for a given pigment composition,

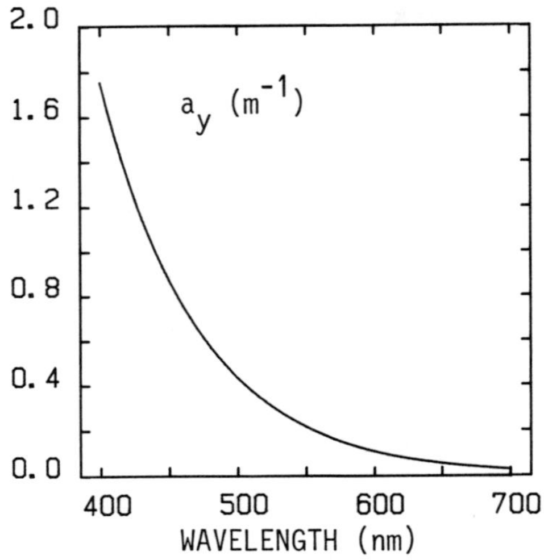

Fig. 4. Absorption spectrum of dissolved organic matter yellow substances) calculated using the exponential relationship, with p = 0.014.

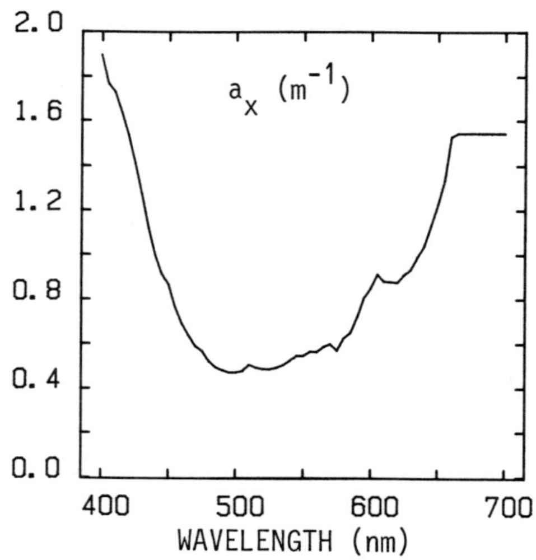

Fig. 5. The absorption spectrum of "non-chlorophyllian" particles (from Prieur and Sathyendranath, 32).

the absorption spectrum appears more or less deformed, depending on the intra-cellular concentration of the pigments and the size of the cells. The diversity of phytoplankton absorption spectra and the influencing factors have been recently discussed and reviewed (26,27,30,32). In spite of considerable variability, phytoplankton absorption spectra are generally characterized by a peak in the blue centered at 440 nm (due mainly to carotenoids and chlorophyll a) and a secondary peak in the red, at about 675 nm (due to chlorophyll a and pheophytin a). An important amount of pheo a in the medium results in a shift of the blue maximum towards 410 nm. However, generally, pheo a appears in situ in relatively small quantities compared to chl a.

Figure 3 shows some examples of the spectral signatures of phytoplankton. Taking into account all this diversity in a model of ocean colour would render the model rather unwieldly. Simplifying hypotheses are therefore needed.

Dissolved organic matter (yellow substances). Molecular diffusion by these substances is negligible compared to the contribution of sea water. Their absorption is known to have an exponential form, which is generally expressed by the equation

$$a_y(\lambda) = a_y(\lambda_0)\ \exp\{-p(\lambda - \lambda_0)\}$$

where $a_y(\lambda)$ is the absorption by yellow substances at the wavelength λ. The recent detailed study of Bricaud et al (33) has shown that the exponential coefficient p is only slightly variable, and has a mean value of 0.014 (see Fig. 4). This study has also shown that biogenous and terrigenous yellow substances do not appear very different optically.

"Non-chlorophyllian particles". This term refers to suspended matter other than phytoplankton or its immediate derivatives. This again is a highly heterogenous group, consisting of all sorts of particles brought in by rivers, land drainage and wave action. Their optical properties are known even less than those of phytoplankton.

No direct measurements of the scattering properties of non-chlorophyllian particles are available. Theoretical studies of Pak et al (34) (Mie Theory) on the scattering of clay particles with a log-normal size distribution showed hardly any wavelength dependence. Morel (35) examined theoretically the scattering of suspended particles with a size distribution that follow Junge law ($dN \propto d^{-M}$, where d is the particle diameter and dN is the number of particles with diameter d). His results showed that, for such size distributions, the variations of the scattering coefficient b with wavelengths follow a λ^{-n} law, with $n = M - 3$. Some experimental results (36-38) show that the size distribution of total suspended

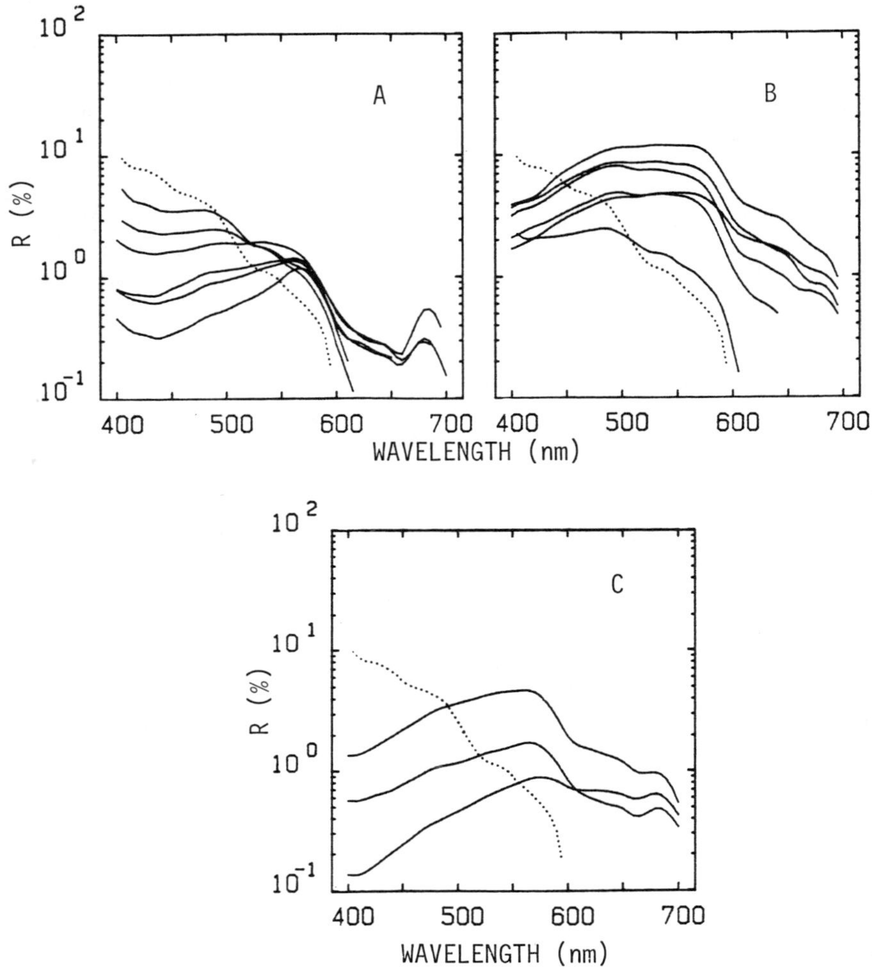

Fig. 6. Some observed spectra of reflectance. The spectra presented here have been selected to represent the three types of spectra described in the text (§II.4). In A, B and C, the spectrum of a clear deep blue sea with little suspended or dissolved organic material, is presented for comparison with the other spectra. These spectra were observed in the course of cruises DISCOVERER, CINECA 5, FOSBER and C FOX, in different marine regions (Atlantic, Mediterranean and Pacific). Different combinations of these three types also occur in nature.

A - Case 1 waters, with different quantities of phytoplankton.
B - Suspended-sediment-dominated Case 2 waters.
C - Yellow-substance-dominated Case 2 waters.
Dotted line - Clear ocean water.

particles (including phytoplankton) follow a Junge law with M varying between 3 and 5. It would then follow that n varies between 0 and 2. The experimental studies of Morel (35,39,40) on the scattering properties of total suspended particles in fact showed n to vary between 0 and 2. Burt (41) found similar attenuation variations (apparently mainly due to scattering) for suspended particles in Chesapeake Bay. The experimental results mentioned above are all influenced by phytoplankton also, when present. No direct proof exists that the conclusions are true when the non-chlorophyllian particles are considered by themselves. Like phytoplankton, the non-chlorophyllian particles are also predominantly forward scattering. But their backscattering ratio is generally higher than that of phytoplankton, being of the order of 1 or 2%.

Yentsch (42) observed a neutral absorption spectrum for suspended particles (apparently non-chlorophyllian) from Woodshole waters, and one similar to that of yellow substances for suspended sediments of deep waters (>100 m). But Prieur and Sathyendranath (32) have proposed an U shaped absorption spectrum for non-chlorophyllian particles based on observations in the upwelling regions off Mauritania (see Fig. 5).

Even if particular, and locally valid information exists concerning the optical behavior of non-chlorophyllian particles, this knowledge cannot have an ubiquitous value. Therefore, in Case 2 waters, locally valid optical properties have to be estimated.

II.4. Examination of actual data concerning $R(\lambda)$.

Figure 6(a,b and c) shows the reflectance spectra of different types of sea waters, selected to represent Case 1 waters, sediment dominated Case 2 waters and yellow substance dominated Case 2 waters.

Clear oceanic waters appear deep blue, with high reflectance in the 400 - 500 nm region, which results from the combined effect of the low absorption coefficient in this spectral domain and the backscattering coefficient which increases towards shorter wavelengths. In Case 1 waters, when the phytoplankton concentration increases, a minimum appears progressively at about 440 nm (corresponding to the maximum of pigment absorption (see Fig. 6a). A peak in the red (at about 685 nm) appears due to the in vivo fluorescence of chl a (see §II.6). Compared to those of pure (oligotrophic) waters, the $R(\lambda)$ values generally decrease with increasing pigment content.

The presence of highly refringent suspended sediments induces another kind of change (Fig. 6b): an overall increase of reflectance and a flat, slightly convex spectrum from near UV to yellow (580 nm). These waters are bright (high *R* values), discoloured, or milky green. Yellow substances (Fig. 6c) cause a continuous decrease of *R* towards violet due to increasing absorption. If the concentration is high, the spectrum is smooth in the red part. With low reflectances everywhere in the visible domain, these waters appear a dark brownish colour.

In these Case 2 waters, the presence of phytoplankton pigments is more or less masked by the other substances. Obviously, in such waters, the pigment concentration does not govern totally the shape of the *R* spectrum as clearly as in Case 1 waters. Often, sediments and yellow substances are present simultaneously, leading to intermediate situations with respect to the selected situations presented here for clarity.

II.5. Reconstruction of reflectance spectra from concentrations of the different substances.

The validity of a model of reflectance and the spectral signatures used in the model can be tested by comparing the theoretical reflectance spectra with observed spectra. Sathyendranath (30) has proposed a three component model of *R*. The three components are phytoplankton (plus their immediate derivatives), non-chlorophyllian particles and yellow substances. This model is based on Eq. 1b. The absorption coefficient is calculated according to Prieur and Sathyendranath (32). In addition, simple, but realistic hypotheses are used to calculate the spectral values of backscattering. Two degrees of freedom are required on the backscattering properties of non-chlorophyllian particles in order to get the best reconstruction. The theoretical curves compare well with *R* spectra observed in different types of sea water. The model considers spectral values of *R* between 400 and 700 nm, at 5 nm intervals.

Bukata et al (31,43,44) have established a four component model for Lake Ontario. Here, phytoplankton and their detritus are further separated into two components. The model is based on Eq. 1a, and uses spectral signatures of the four components determined for Lake Ontario waters at 4 wavelength bands centered at 443, 520, 580 and 670 nm (corresponding to CZCS wavelengths).

II.6. Fluorescence effect on reflectance spectra.

Equation 1 does not take into consideration fluorescence. If naturally fluorescing substances are present in the water, and if the magnitude of the fluorescence signal is sufficiently high, additional terms have to be taken into consideration. It is now known (15,45-48) that the peak in the red, centered at about 685 nm, which is observable sometimes in the reflectance spectra from the sea, is due to the fluorescence of chlorophyll a. Generally, this peak has a half-band width of about 30 nm, and the height of the peak gives a straight line fit with the concentration of chl a in the waters, if a restricted area is considered. However, the constant of proportionality has been found to vary (by as much as a factor of 10) with time and region. This variability is attributed to diverse factors such as nutrient stress, ambient light conditions and species composition (49-52). Besides, it has been found that the fluorescence maximum does not always occur at 685 nm but often shifts ± 8 nm, and in the case of blue-green algae, the maximum even occurs at 645 nm (51,52). Presumably, phycoerithrin which exhibits in vivo fluorescence at 578 nm and is present in the cyanophyceae group could also generate a secondary peak. It has not been detected on in situ reflectance spectra till now.

III. USE OF THE VISIBLE IMAGERY, AND ALGORITHMS.

III.1. Photographic imagery.

The first attempts at optical remote sensing of the oceans by satellites were made using colour photographs (from manned space missions such as GEMINI and APOLLO), TV cameras and IR/visible radiometers (on earth observation satellites like NIMBUS, NOAA, ESSA and ATS). The first IR/visible radiometers contained a single channel in the visible from about 500 to 700 nm. With such imagery, no efforts were made to uncouple and isolate the sea signal from the total sea-atmosphere signal, or to obtain quantitative information of the water quality. La Violette (53) and La Violette et al (54) have given reviews on the oceanographic applications of such imagery. The applications include: 1) Inference of ocean-atmosphere interactions (and related processes) from cloud patterns and their movements. 2) Monitoring of sea ice movements by judiciously chosen brightness composition to eliminate clouds, which move faster. 3) Bottom topography in shallow waters (the precision obtainable from such a method is probably not very high). 4) The delineation of watermasses and currents from associated brightness changes (which La Violette attributed to changes in sea surface roughness at times, or to the presence or absence of fogs). 5) Study of waves, slicks, internal waves, current direction and zones of convergence and divergence using reflectance patterns in the sun glitter zone. Cox (55) has given a detailed examination of the technique of wave and wind direction analysis from sun glint patterns.

II.2. LANDSAT imagery.

The satellites LANDSAT carry a Multi Spectral Scanner (MSS) which collects data at four spectral bands, of which two are in the visible (see Appendix B for further details). Thomas (56) has examined theoretically, the applicability of MSS channels 4 and 5 for the measurement of concentrations of phytoplankton, yellow substances and sediments in suspension. He showed that in MSS 4, the signal decreases with increase in phytoplankton or yellow substances, and increases with increasing sediment concentration. Consequently, he concluded that it is difficult to draw unambiguous conclusions from this signal. MSS 5, on the other hand, appeared to be sensitive only to suspended sediments (neglecting phytoplankton fluorescence) when atmospheric effects are eliminated. However, in this instrument, the saturation radiance of 2 $mW.cm^{-2}$ is split into 127 levels. This means that, at best, this instrument can only discern radiation variations of 0.7%. Thomas therefore used MSS 5 to study sediment dispersion along the French Atlantic Littoral, where sharp gradients of sediment load (and therefore, of reflectance) are present. Similarly, Mundey and Fedosh (57) have made use of MSS 5 to study the sediment dispersion of Chesapeake Bay.

Contrary to the conclusions of Thomas (56), Gower et al (58) observed that oceanic eddy patterns were noticeable in MSS 4 scenes of the oceanic area south of Iceland, which is an area of high phytoplankton concentration. They interpreted that the eddy patterns represented the distribution of phytoplankton and their associated detritus. According to them, increase in reflectance could be associated with increase in phytoplankton concentration. They found that the eddy patterns were less visible in MSS 5, and disappeared altogether in MSS 6 and 7. Similar conclusions were drawn by Viollier et al (59) who used LANDSAT images to study phytoplankton bloom in the Ushant front off Brittany. Theoretically, it seems implausible that increase in phytoplankton alone could be responsible for the increase in reflectance observed by Gower et al and Viollier et al. Another phenomenon, concomitant with the presence of algae, even if unidentified, has to be invoked to account for this increase in reflectance. One possible explanation could be the presence of highly reflecting scum created by the phytoplankton bloom.

III.3. Multispectral visible imagery from CZCS and some airborne sensors.

Launched in 1978, the CZCS is the first, and so far the only, satellite sensor in orbit, designed especially for the remote sensing of phytoplankton and sediments in the surface waters of the ocean. This scanner measures radiance in 4 channels in the visible (see appendix B). But a number of other sensors with greater spectral detail have been designed and operated from airplanes (see Appendix B for some examples). Many types of algorithms have been developed for this last generation of satellite and airborne sensors. In the following paragraphs are examined the different approaches to the quantitative estimation of oceanic

Table 1. List of algorithms of the type $M = A(r_{ij})^B$ proposed by diverse authors. r^2 is the square of the correlation coefficient, and N the number of observations used in the regression analysis.

r_{ij}	A	B	N	r^2	REFERENCES
M = Chl a + Pheo a (mg.m^{-3})					
L_{443}/L_{550}	0.776	-1.329	55	0.91	CLARK, 60
L_{443}/L_{520}	0.551	-1.806	55	0.87	
L_{520}/L_{550}	1.694	-4.449	55	0.91	
L_{520}/L_{670}	43.85	-1.372	55	0.88	
L_{440}/L_{550}	0.54	-1.13	7	0.96	CLARK et al, 61
L_{440}/L_{550}	0.505	-1.269	21	0.978	GORDON & CLARK, 62
L_{440}/L_{520}	0.415	-1.795	21	0.941	
L_{520}/L_{550}	0.843	-3.975	21	0.941	
R_{440}/R_{560}	1.92	-1.80	67	0.97	MOREL, 11
L_{443}/L_{550}	0.783	-2.12		0.94	SMITH & WILSON, 63
L_{443}/L_{520}	0.483	-3.08		0.88	
L_{520}/L_{550}	2.009	-5.93		0.95	
L_{443}/L_{550}	2.45	-3.89	6	0.61	STURM, 64
L_{443}/L_{550}	1.13	-1.71	454		GORDON et al, 65
L_{443}/L_{550}	1.216	-2.589			SMITH & BAKER, 80
M = Total suspended particles (g/m^{-3})					
L_{440}/L_{550}	0.4	-0.88	9	0.92	CLARK et al, 61
L_{440}/L_{520}	0.33	-1.09	9	0.94	
L_{520}/L_{550}	0.76	-4.38	9	0.77	
L_{443}/L_{550}	0.24	-0.98		0.86	CLARK & BAKER, per. comm.
L_{520}/L_{550}	0.45	-3.30		0.86	
L_{520}/L_{670}	5.30	-1.04		0.85	
$M = (K_{490} - 0.022)$ (m^{-1})					
L_{443}/L_{550}	0.088	-1.491	88	0.90	AUSTIN & PETZOLD, 66
$M = (K_{520} - 0.044)$ (m^{-1})					
L_{443}/L_{550}	0.066	-1.398	88	0.995	AUSTIN & PETZOLD, 66

parameters. The algorithms are applicable to data corrected for atmospheric effects.

III.4. Algorithms of the type $M = A(r_{ij})^B$

These algorithms have been obtained through the empirical method (see § II.1), i.e., a direct statistical study of the possible relationships between a marine parameter M, and a ratio r_{ij}, of reflectances (or radiances) at two wavelengths λ_i and λ_j. The marine parameters so far studied in this manner are the pigment concentration C (chl a + pheo a, $mg.m^{-3}$), the total suspended load S (dry mass, $g.m^{-3}$) and the diffuse attenuation coefficient K (m^{-1}) for a given λ.(Actually, it is K-K_w that is studied, where K_w stands for the contribution of sea water.) Because of the wide range of variations in the marine parameters, the natural trend was to perform regression analyses with log-transformed data. If a significant linear relationship is found, an algorithm of the above mentioned type is obtained. Table 1 contains a list of such algorithms proposed by diverse authors, for the retrieval of C, S or $(K-K_w)_\lambda$.

Where phytoplankton is concerned, the choice of a blue-green ratio is obvious (see Fig. 6a). The CZCS bands 1 and 3 (443 and 550 nm) are the most appropriate, being the most sensitive to phytoplankton variations. In practice, however, the signal L_w at 443 nm can become very low at high concentrations of phytoplankton. After the removal of atmospheric effect (which is high at this wavelength), the expected accuracy becomes very bad, and $L_w(443)$ becomes unreliable. $L_w(520)$ is then used despite the fact that the ratio of bands 2 and 3 is less sensitive to variations in C (eg. see Gordon and Clark, 62). At first sight, there is a considerable variation in the coefficients A and B proposed by different authors, but a detailed examination of the geographic zones where the data were acquired, and the bio-optical properties of the waters under consideration, shows that a general consensus in fact exists for Case 1 waters, which have high values for the exponent B. Lower and variable values are obtained when Case 2 waters are not excluded from the regression analysis. The existence of a representative exponent for pooled Case 1 and Case 2 waters, or for Case 2 waters only, is highly doubtful.

It is evident that, if several marine parameters (C, S or K-K_w) can be expressed through power laws with respect to the same r_{ij}, these parameters are necessarily linked univocally. In turn, this means that there is only one (independent) algorithm which gives a single "index" of the bio-optical state of the ocean which can be expressed *ad libidum* in terms of C, S or K. A set of compatible algorithms, restricted to Case 1 waters, actually leads to relationships between C, S and $K(\lambda)$, which are in good agreement with those directly observed at sea (67). To summarize this discussion, it can be said that the concomitant variations of all bio-optical properties (C, S, K) within Case 1 waters are elucidated and that the fluctuations around the average laws which link C, S and $K(\lambda)$ remain within a factor of 2.

Since several r_{ij} ratios can be formed, it can be expected that several independent parameters are attainable. In practice, with CZCS bands, only two ratios are possible, using channels 1, 2 and 3. (Channel 4 at 670 nm is used for atmospheric correction under the assumption that $L_w(670)$ is zero. In other words, water is assumed to be a black body at this wavelength. Note that this assumption is wrong for turbid waters.) The theoretical possibility of infering two parameters is, in fact, considerably reduced when we take into account the fact that $r_{443,550}$ and $r_{520,550}$ give more or less the same information. The question of how to ensure the best use of this information is, however still open, particularly for Case 2 waters.

III.5. The method of differences.

Using an airborne radiometer, Arvesen et al (68) measured the differences between the signals at 443 and 525 nm. Their measurements over diverse zones (lakes, estuaries and open ocean) showed a correlation between this parameter and the logarithmic value of pigment concentration in the surface waters. Kim et al (69) have proposed an algorithm which uses the difference $\{L(472) - L(548)\}$ normalized by the sum of the same radiances, based on 12 measurements of $L(\lambda)$ and C. Pearcy and Keene (70) found that the difference between the signals at 460 and 550 nm, as well as that between 550 and 605 nm could be used to distinguish between oceanic waters, upwelling waters and river influx along the Oregon coast. Viollier et al (71) also found a correlation between $\{R(525) - R(466)\}$ and the logarithmic concentration of chl a. But their theoretical calculations show that such a relation exists only if scattering is closely related to the pigment concentration. As mentioned earlier, this condition is approximately satisfied for Case 1 waters. In Case 2 waters, where phytoplankton and sediment concentrations vary independently, this condition is not fulfilled. For such waters, this type of algorithm predominantly describes the change in turbidity.

III.6. Three wavelength algorithms.

The algorithms proposed by Morel and Prieur (72) and Pelevin (73) make use of three wavelengths in order to be able to distinguish phytoplankton from other substances.

Morel and Prieur proposed algorithms of the form:

$$A = \frac{\alpha'(X_{12}X_{23}) + \beta'(X_{12}) + \gamma'}{\alpha(X_{12}X_{23}) + \beta(X_{12}) + \gamma}, \text{ and } B = \frac{\alpha''(X_{12}X_{23}) + \beta''(X_{12}) + \gamma''}{\alpha(X_{12}X_{23}) + \beta(X_{12}) + \gamma}$$

where A and B are the concentrations of the two substances present in the water, X_{ij} are the ratios of the absorption coefficients at wavelengths i and j, and the functions α, α', α'', β, β', β'', γ, γ' and γ'' depend on the specific absorptions. These expressions are derived from Eq. 1b and 2 according to the analytical method, taking the ratios of R at two wavelengths, and supposing that the spectral variation of b_b is known. They considered the wavelength groups (400,440,560 nm) and (440,500,560 nm). In the case of in situ observations, where direct measurements of scattering permit the estimation of the spectral variation of b_b, this method gave satisfactory results in waters influenced by both phytoplankton and non-chlorophyllian particles of a particular type (resuspended particles along the Mauritanian coast). The numerical values of α, α' etc. cannot therefore be seen to be of universal applicability, even if in this particular area they allow a good estimation of C. Only the structure of analytical algorithms for a two component system is of interest, and is valid for other situations.

Pelevin (73) used the parameter x, defined in Eq. 1a. It is known that the absorption spectrum of phytoplankton passes through a maximum at around 430 nm. Supposing that the slope of the spectral values of x due to other substances is constant between 390 and 460 nm, he calculated a parameter $x'(430)$, which is the value that $x(430)$ would take if there were no phytoplankton absorption. That is,

$$x'(430) = \frac{b_b(430)}{a(430) - Cm + b_b(430)}$$

where C is the chl a concentration, and $m = a_c^{*}(430) - \frac{1}{2}\{a_c^{*}(460) + a_c^{*}(390)\}$. m has a value of 0.0065 $m^{-1}/mg.m^{-3}$ according to his measurements. In pracice, $x'(430)$ is calculated from the measured values of x at 390 and 460 nm. Now, with $a + b_b \simeq K$, he obtained the relation:

$$C = \frac{K(430)}{m} \, \frac{x'(430) - x(430)}{x'(430)} .$$

To calculate K by remote sensing, he used the statistic relationship

$$K(430) = (0.184 \pm 0.007) \frac{R(558)}{R(430)}$$

Twelve sea truth measurements in the Baltic showed concordance with calculated values, in spite of the importance of yellow substances in these waters. This method was applied to data acquired by a low flying airplane.

III.7. Other methods.

Mueller (74,75) carried out a principal component analysis

on 36 spectra of R (55 spectral values between 422.5 and 692.5 nm). The R spectra were not corrected for atmospheric effect. The measurements were carried out along the Oregon Coast from an airplane. A multiple regression between the phytoplankton concentration and the weights of the first three components gave a significant correlation coefficient. Sathyendranath (30) has also used characteristic vector analysis to study the information contained in the reflectance spectra measured in situ. Spectral values between 400 and 700 nm at 10 nm interval were used in this study. Using a mixed sampling of Case 1 and Case 2 (suspended sediment dominated) waters, it was shown that phytoplankton concentration (chl a + pheo a) and total suspended particle concentration could be determined with good precision (r^2 = 90% on log scale) from R spectra using the weights of the first three components and their regression coefficients. Doerffer (76) also used orthogonal vectors to study chlorophyll concentration from airborne radiance measurements (at 14 wavelengths) in the North Sea.

Jain and Miller (77) and also Miller et al (78) used a two flow model of R, and optimization techniques, to estimate phytoplankton concentration. The solution accepted was that value of chl a concentration which minimized the difference between calculated and theoretical R curves. Chl a was considered to be the only variable in the model. Concordent results were obtained with sea truth measurements. The method has since then been extended for application in the four component model of Bukata et al (43) for Lake Ontario. In the iterations, yellow substance concentration is considered to be constant, and organic detritus varies proportionally with chl a.

Another approach of Bukata et al (43) has been to establish "duoisopleths" of chl a and suspended mineral concentrations for CZCS spectral bands taken two at a time. Knowing the radiance levels at two wavelengths, the concentrations are read off from the graphs. They found the 670 - 520 nm pair to be the most sensitive to variations in concentrations of the above mentioned parameters, and the 443 - 520 nm pair to be the least sensitive. (Lake Ontario waters can be considered to be Case 2. Note that the results obtained here are very different from Case 1, where the 443 - 550 nm pair is generally considered to be the most sensitive to chl a variations.) The authors tested the model on data from airborne radiometers, and pointed out that taking pairs of two wavelengths leads to a decrease in the precision of prediction, compared to optimization techniques using all four CZCS wavebands. The model of Bukata et al is designed for application to data acquired from low flying aircrafts which measure downwelling as well as upwelling radiance.

III.8. The fluorescence method.

The in vivo fluorescence of phytoplankton in the red has often been taken as an index of its concentration. Neville and Gower (45) have been the first to apply this method to remote sensing of chlorophyll from aircraft. They found a correlation between the height of the fluorescence peak at about 685 nm in the reflectance spectra, and the surface concentration of chl a. This method presents the advantage of using wavelengths where the influence of yellow substances is feeble. Atmospheric correction is relatively easier in the red than at shorter wavelengths. Besides, the form of the fluorescence peak is easily distinguished from the "noise" introduced by other substances. However, as mentioned earlier (§II.6), the fluorescence spectral form and efficiency have been found to vary a lot with time and place. Another limitation is the low fluorescence efficiency of in vivo chl a (the increment in R is of the order of 3.2×10^{-4}/mg.chl.m^{-3} according to Gower, 47). This means that it would be very difficult to measure low concentrations of chl a by this method. Gower and Borstad (48), Doerffer (76) and Spitzer et al (79) have also used fluorescence techniques for chl a determination by remote sensing.

III.9. Applications.

In section III.2, we have already seen how LANDSAT visible imagery has been used to study sediment dispersion, eddy patterns and fronts. The radiometric and spectral limitations of the MSS on this satellite series has also been discussed. Another drawback of LANDSAT is that the same flight line is repeated only once in 18 days. If, in addition, we consider that on days of cloud cover the images are hardly usable for oceanographic purposes, it follows that LANDSAT cannot be relied upon for studying any short term oceanographic phenomenon.

More widespread applications are expected from the CZCS imagery, with its greater radiometric precision and spectral detail. Besides, with CZCS, it is often possible to have images of the same scene on consecutive days, even though the exact track is only repeated once in 6 days. However, until recently, the attention of scientists working on CZCS data was focussed on the establishment of suitable algorithms for atmospheric correction and retrieval of marine parameters. Only a very small number of CZCS images have so far been used to study physical oceanographic phenomena. But already, it appears clear that, for oceanographers interested in studying patterns of eddies, fronts, meanders, water masses and circulation in the surface waters of the sea, visible imagery often brings complementary information to infra red imagery. At times, the infra red and visible imagery are found to indicate similar features on the scene studied (eg. Gordon et al, 65) have studied warm core rings of the Gulf Stream using visible and infrared imagery, and observed a good correspondance between the two). But such concordance is not always observed. Wrigley (81) observed that CZCS visible imagery could be used to identify oceanic fronts in the Pacific Ocean when the surface temperature differences (in July) were too low for obtaining sufficient detail from an IR image. Similarly, the CZCS scene for March 1979 of the NW Mediterranean shows a wealth of structural detail (see Hovis, 82; Gordon and Morel, 83), while the IR image of the same scene shows hardly any features (83), due to low temperature contrast. Mueller and La Violette (84) found a shift of 45 km between the colour and temperature fronts at the northern boundary of the Gulf Stream. They advanced horizontal cross-front diffusion, or a secondary circulation leading to weak upwelling and phytoplankton growth, as possible reasons for this shift. In this particular case, in situ measurements eliminated the possibility of a "skin effect". That is, the infra-red technique measures the temperature of an extremely thin surface layer of the ocean ("skin") of less than a mm in thickness. Under stable oceanic conditions and low winds, this skin temperature can be considerably different from that of the waters immediately below. In the visible on the other hand, the sampled depth is often of the order of a few meters, and the oceanic parameters estimated are weighted mean values over this depth (see Appendix C for further details). This difference in sampling depth also means that the information acquired from the IR or visible imagery is not necessarily the same.

Note that, in this kind of physical oceanographic applications, a great accuracy of the values of the estimated oceanic parameters is not actually necessary. All that is needed is sufficient colour contrast to identify and follow the physical phenomena of interest. But of course, when it is also possible to identify the cause of the colour contrast (phytoplankton, sediments, scum at the surface), it further helps interpret the image. The interpretation of observed features may not be exactly the same whether the parameter studied is a passive "tracer" of the water masses or not. Gower et al (58), in their study of mesoscale eddies using LANDSAT images, suggested that phytoplankton could be considered as a passive scalar, which later led to a controversy (85,86). Since phytoplankton grow and decay and are grazed upon, their concentrations evolve with time. Therefore, in physical oceanographic interpretations, it may be worthwhile to consider the speed of their evolution, compared to the speed of the physical phenomena studied. For example, upwelled waters may not be immediately identifiable by a colour difference. But if the upwelled water is allowed to stay at the surface for a few days, sufficient phytoplankton growth may occur to allow the necessary colour difference. On the other hand, in the case of infra-red images, the inverse may be true. That is, freshly upwelled waters may be identifiable by the temperature, and after a few days at the surface, this temperature contrast may have disappeared.

The only thing to emphasize is that visible and IR imageries are not redundant, but complementary. It is expected that visible imagery may turn out to be more fruitful than IR imagery, as a tool for the study of structures, and perhaps movements, of water bodies within the surface layer.

Biological applications of passive remote sensing in the visible are discussed by C.S. YENTSCH in this volume, so they are not treated here.

IV. DISCUSSION AND CONCLUSION.

A clear-cut, sure and universal method for the remote sensing of phytoplankton and suspended sediment concentrations has not yet been found, because of the following problems: 1) Phytoplankton and suspended sediments have spectral signatures that are subject to variations; this introduces a certain degree of uncertainty in the estimations. 2) Reflectance is not a linear function of concentrations, which increases the difficulty of establishing algorithms which are applicable in different ranges of concentrations of a given substance, and independently of other substances present in the water. 3) In certain wavelength ranges, the spectral signatures of different substances are similar, and if the available data are confined to that range, it becomes difficult (if not impossible) to distinguish them from each other. For example, yellow substances and phytoplankton resemble each other a lot within the spectral range 440 to 560 nm. 4) Interpretations and algorithms are also limited by the precision and spectral resolution available from the instruments used, after atmospheric correction.

However, satisfactory results have been obtained in Case 1 waters with algorithms of the type $M = A(r_{ij})^B$ (see §III.4). The main factor of uncertainty (estimated to be within a factor of 2) in this case is probably introduced by the natural variability of the absorption efficiency of phytoplankton and covarying detritus, per unit concentration of chlorophyll a. In the absence of sea truth measurements to adjust the coefficients A and B for local conditions, average values are necessarily adopted. This may lead to uncertain absolute values for estimated M. However, the estimated gradients from one point to another would still remain realistic.

Some recent studies (30, 31, 43, 44) have shown that algorithms can be obtained for the estimation of phytoplankton and non-chlorophyllian particle concentrations in some Case 2 waters, even when these substances vary independently. However, in view of the diversity of this type of waters, it is highly probable that Case 2 waters require specialized algorithms for applications in different zones. Airborne sensors, capable of measuring the reflectance with greater precision (since atmospheric effects are less important) and at greater number of wavelengths than the CZCS, have better chances of solving the Case 2 problem.

To process the CZCS imagery, it is necessary to postulate that $L_w(670)$ is zero. This is not true in sediment rich waters, as has been pointed out by Bukata et al (87). Stürm (88) and Smith and Wilson (63) have proposed methods which admit non zero values at 670 nm. The darkest pixel method of Stürm does not permit pixel by pixel correction of aerosol path radiance. The iterative method of Smith and Wilson is based on the inter-relationships between L_w of different wavelengths. These relations are however only valid for Case 1 waters and the problem remains unsolved for Case 2 waters. On other sensors, when possible, a wavelength further down in the infra red (≃780 nm) could be used instead of $L_w(670)$ for determining the atmospheric correction factors. But this leads to a different problem - to determine L_w values in the blue, it is now necessary to extrapolate the atmospheric correction over a larger spectral distance, and uncertainties on the Ångström coefficient would lead to decrease in the precision of L_w(blue). To overcome this last problem, it may be necessary to use another channel in the UV or near UV, where, in some circomstances (high concentrations of phytoplankton or yellow substances), it could be postulated that L_w be zero. Now, extrapolation from both ends towards the middle could be made with sufficient precision.

In waters rich in yellow substances, the reflectance signal from phytoplankton (when present) may be drowned in the yellow substance signal if data from 440 nm and greater wavelengths only are available. Our theoretical studies show that, including a shorter wavelength at around 400 nm increases the possibility of distinguishing between yellow substances and phytoplankton (30). Because of these reasons, in Case 2 waters, it appears worthwhile to try to measure radiances on the short wavelength side of 440 nm, even though this presents greater technical difficulties.

The full capacities of passive remote sensing in the visible remain yet to be established. The problem of the oligotrophic or phytoplankton-dominated waters (Case 1) is now reasonably well understood. With increasing understanding of the phenomenon of ocean colour, possibilities have started to open up for obtaining suitable solutions for at least some of the Case 2 waters, where, in addition, yellow substances and (or) suspended sediments are also present.

Very few scenes of analysed visible imagery of oceanic areas acquired from satellites are as yet available. But these few images have already shown an unexpected wealth of structural detail of the surface waters, and the importance of such imagery as a tool for physical oceanographers seems to be well established.

APPENDIX A

Units and symbols of radiometric quantities.

Quantity of radiant energy: Q joule.
Radiant flux: F watt. Energy per unit time. $F = Q/t$.
Radiant intensity: I watt/steradian. Flux per unit angle. $I = F/d\omega$.
Irradiance: E watt/m^2. Flux per unit surface. $E = dF/dA$. When radiance is measured with respect to a horizontal surface facing upward or downward, the quantities measured are respectively downwelling and upwelling irradiance (E_d and E_u).
Radiance: L watt/m^2/sr. Flux per unit angle per unit projected area of a surface. $L = dF/dA \cos\varepsilon \, d\omega$.
Absorption coefficient: a m^{-1}. When a monochromatic, parallel flux F passes through an infinitely small distance dr of an absorbing medium, the energy absorbed is $dF = aF dr$. Integration over distance r gives $F = F_0 \exp(-ar)$ in the case of a homogeneous medium, where F_0 is the incident flux.
Volume scattering function: $\beta(\theta)$ m^{-1}.sr^{-1}. Radiant intensity scattered in the direction θ with respect to the direction of the incident beam (parallel and monochromatic), per unit volume and unit irradiance. $\beta(\theta) = dI(\theta)/E dv$.
Scattering coefficient: b m^{-1}. Total scattered energy per unit volume and unit irradiance. $b = \int_{4\pi} \beta(\theta) d\omega = 2\pi \int_0^{\pi} \beta(\theta) \sin\theta \, d\theta$.
Backscattering coefficient: b_b m^{-1}. b_b is calculated similarly to b, but here $\beta(\theta)\sin\theta$ is integrated from $\pi/2$ to π.
Backscattering ratio: $\bar{b}_b = b_b/b$.
Attenuation coefficient: c ; $c = (a + b)$m^{-1}.
Refracive index: m. Ratio of phase velocity of radiant energy in free space to that of the same energy in the given medium. If i is the angle of incidence, and j the angle of refraction, $m = \sin i / \sin j$.
Irradiance attenuation coefficient: K m^{-1}. Defined by $- d \log E/dz$. For downwelling and upwelling irradiances, the corresponding attenuation coefficients are referred to as K_d and K_u respectively.
Reflectance: R. Ratio of upwelling to downwelling irradiance at a given depth. $R(z) = E_u(z)/E_d(z)$.

The parameters a, $\beta(\theta)$, b, b_b, $\bar{b}_b$ and c are "inherent" properties of a medium. Since they are defined for well precised incident light conditions, they are entirely determined by the nature of the medium. K and R are defined for ambient light conditions, whatever they be, and are therefore functions of both incident light field and the inherent optical properties of the medium.

Note that, in the above definitions, the volume d*v*, the distance d*r*, the surface d*A* and the angle d are all infinitely small. For further details on units and definitions of radiometric parameters, interested readers are referred to Jerlov (97) and Morel and Smith (1).

All the parameters are functions of wavelength (λ). According to convention, $dX/d\lambda$ at λ, where X is a radiometric parameter, is referred to as $X(\lambda)$. Wavelengths are indicated in nm.

APPENDIX B

Some instruments for the passive remote sensing of the oceans in the visible.

1. MultiSpectral Scanner (MSS).

Platform: LANDSAT satellites.
Orbit: Near-polar, sun synchronous.
Altitude: $\simeq$ 918 km.
Launch: LANDSAT-1 in 1972, LANDSAT-2 in 1975, LANDSAT-3 in 1978.
Distance between successive paths: 2100 km at 40° N.
Pixel size: 56X79 m.
Swath width: 185 km.

There are 14 southbound passes of the satellite per day, at about 10 am local sun time; the northbound passes are at night. Orbits repeat themselves every 18 days. MSS acquires data in 4 wavelength bands (see Table B.1). Data is split into 127 radiance levels. For further details regarding LANDSAT, see ref. 89.

Code No.	Wavelength (nm)	Saturation Radiance (mw/cm^2 μm sr)
4	500-600	24.8
5	600-700	24.4
6	700-800	20.0
7	800-1100	17.6

Table B.1. MSS spectral bands.

2. Coastal zone Color Scanner (CZCS).

Platform: Satellite NIMBUS-7.
Orbit: Near polar, sun synchronous, ascending node at local noon.
Altitude: $\simeq$ 955 km.
Launch: in 1978.

Pixel size: 825x825 m.
Swath width: 2300 km at 20° forward tilt
1600 km at no tilt
1300 km at 20° backward tilt.

The CZCS collects data in 6 channels (see Table B.2). Image registration in the 6 channels is simultaneous. The scanner of the instrument rotates fully, but the active portion is 78.7°. The instantaneous field of view is 0.05°. Scan rate is 8.0808 rev/sec. In flight calibration of the five visible channels is achieved by means of a calibration light source, which is exposed to the sensor once per every 16 revolutions of the scanner, during the non-active phase. The instrument can function at 4 gains. The signal of each channel is converted to an 8 bit word. The signal to noise level for channels 1 to 4 is estimated to be better than 150 under normal conditions, and for channel 5, it is 118. The spectral radiance from the sea is much less than 1 mw/cm^2 µm sr generally, so this high radiometric sensitivity is necessary. Channel 6 is in the infra-red.

Code No.	Wavelength (nm)	Saturation Radiance (mw/cm^2 µm sr)	
		Gain 1	Gain 4
1	433-453	11.46	5.41
2	510-530	7.64	3.50
3	540-560	6.21	2.86
4	660-680	2.88	1.34
5	700-800	23.90	23.90

Table B.2. CZCS spectral bands in the visible.

In areas of sun glint, the reflected radiance is so high that the signal from the sea gets irrecoverably lost in it. In order to avoid areas of sun glint, the CZCS has the capacity to tilt the scan plane (by tilting the scan mirror) upto 20° north or south of nadir, by 2° increments. For further details, see Hovis (90).

3. Some instruments designed for airborne platforms.

A number of airborne sensors have been designed and constructed, many of which measure radiance with much greater spectral detail than the CZCS. Some of them have additional characteristics which render them even more versatile. A few selected examples are discussed below.

The Multichannel Ocean Color Sensor (MOCS) of NASA Langley

Research Center measures radiance at 20 spectral bands within the 400-694 nm zone. The real-time ocean colour analyzer system that goes with the MOCS gives a display on an x-y oscilloscope, of the selected algorithm, and distance along the flight track, or cross plots different algorithms. During the Superflux experiments in the Chesapeake Bay, this instrument was used to guide the ships to interesting points for sea truth measurements (91).

The Langley Test Bed Aircraft Multispectral Scanner (TBAMS) is a rotating mirror scanner with the capacity to alter the spectral band locations and sensitivity according to local requirements (92).

The spectrometer of the Institute of Ocean Sciences, British Columbia (45,93) has a 256 channel silicon diode array as sensor, and measures radiance (upwelling and downwelling) in the interval 381-1065 nm. A polarizing filter is attached to the objective lens, and the instrument views the sea at Brewster angle (53° from the zenith). At this angle, the reflected light is plane polarized, and so, theoretically, this technique permits the elimination of reflected light, when the sea is calm.

APPENDIX C

Penetration depth and equivalent concentration.

Gordon and McCluney (94) used a radiation transfer model to determine the depth upto which a sensor outside the water can "see" into the sea. They defined the penetration depth as the lower limit of the water layer responsible for 90% of the diffused light which leave the sea.

Using the Quasi Single Scattering Approximation (QSSA) in the case of a homogeneous, infinite ocean illuminated by parallel light from the zenith, they formed the expression

$$L_z(\mu')/L_\infty(\mu') = 1 - \exp\{-z(a + b_b)(1 + \mu)/\mu\} \qquad (C.1)$$

where $L_z(\mu')$ is the upwelling radiance, originating from a surface layer of the ocean of thickness z, and incident at the interface at angle θ, where $\theta = \cos^{-1}\mu'$. $L_\infty(\mu')$ is the same parameter for a layer of infinite thickness. According to Snell's law, we have $\mu^2 = 1 - m^2(1 - \mu'^2)$ where m is the refractive index of water with respect to air. For QSSA, $K = (a + b_b)$, and Eq. C.1 reduces to

$$L_z(\mu')/L_\infty(\mu') = 1 - \exp\{-zK(1 + \mu)/\mu\} \qquad (C.2)$$

Figure C.1 shows L_z/L_∞ as a function of zK for different values of θ. Only angles less than 50° are considered, since light in-

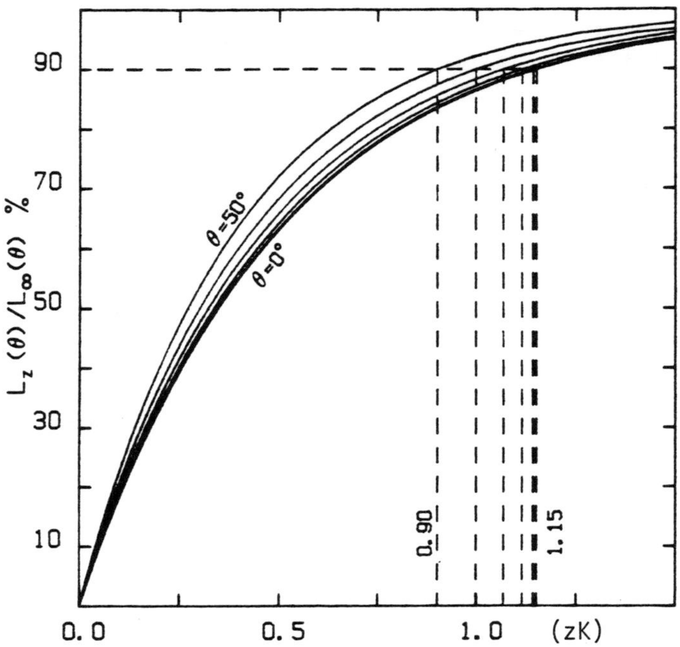

Fig. C.1. $L_z(\theta)/L_\infty(\theta)$ as a function of Kz. The 6 curves correspond to θ = 0, 10, 20, 30, 40 and 50°.

cident on the water-air interface at greater angles suffers total internal reflection. This figure shows that 60 to 70% of the light comes from the surface layer of thickness $1/2K$, and 90% of the light from the layer of thickness $1/K$. Within 15%, the function L_z/L_∞ can be considered to be independent of θ. Gordon and McCluney therefore defined the penetration depth as that depth where the downwelling irradiance is reduced to 1/e of its surface value. Therefore,

$$Z_{90} = 1/K \tag{C.3}$$

where Z_{90} is the penetration depth.

K being a function of wavelength, Z_{90} varies with wavelength (56,30,95). If the depth of the ocean is less than $1/K$, then L, as well as R, are influenced by the water column, as well as the bottom of the sea. If the depth is greater, the substances in solution and in suspension in the water can be considered as the only agents responsible for variations in L and R. This latter case only has been studied in this paper.

The intensity of downwelling and upwelling energy diminishes exponentially as a function of K_d and K_u respectively. The energy coming from depth z would be $\exp(-2zK)$ times less important than the energy from an identical layer just below the surface (with $K = K_d = K_u$; this constitutes a justifiable approximation since K_d and K_u are approximately the same for diffuse illumination). In the case of a stratified ocean, the estimated concentration of a substance would be a weighted mean of its concentration over the penetration zone (0 to Z_{90}). Gordon and Clarke (96) showed that the equivalent concentration $\bar{c}$ of a substance in a stratified ocean (which would have the same effect on R if the medium were homogeneous), can be expressed by

$$\bar{c} = \int_0^{1/K} c(z)f(z)\,dz \Big/ \int_0^{1/K} f(z)\,dz \tag{C.4}$$

where $$f(z) = \exp\{-\int_0^{1/K} 2K(z')\,dz'\}.$$

$c(z)$ is the concentration of the substance at depth z. Since $f(z)$ is wavelength dependent, $\bar{c}$ is also wavelength dependent. Equivalent concentration is sometimes referred to as effective concentration.

The concepts of penetration depth and equivalent concentration are important when comparing remotely acquired data with sea truth measurements, and when interpreting visible imagery.

ACKNOWLEDGEMENTS: The preparation of this paper received support from the European Space Agency (Contract No:ESA 4726/81/F/DD/SC). Many results of the Laboratoire de Physique et Chimie Marines (Villefranche-sur-Mer) discussed here, were acquired thanks to CNRS (ERA 278-GRECO 034), CNES (81-250) and CNEXO (81-203).

REFERENCES

(1) Morel, A., and R.C. Smith: 1982, Marine Geodesy, 5, pp. 335-349.
(2) Austin, R.W.: 1974, in "Optical Aspects of Oceanography", N.G. Jerlov and E. Steemann Nielson (Eds), pp. 317-334.
(3) Austin, R.W.: 1980, Boundary-Layer Meteorology 18, pp. 269-285.
(4) Kozlyaninov, M.V., and V.N. Pelevin: 1965, Tr. Inst. Okeanol. Akad. Nauk SSSR 77, pp. 73-79. Also 1966, U.S. Dep. Comm., Jt. Publ. Res. Ser. Rep. 36, pp. 54-63.
(5) Gordon, H.R., O.B.Brown, and M.M. Jacobs: 1975, Appl. Opt. 14, pp. 417-427
(6) Kirk, J.T.O.: 1981, Aust. J. Mar. Freshwater Res. 32, pp. 517-532.
(7) Prieur, L., and Morel, A.: 1975, Int. Union Geophys. Geod.

16 th Gen. Assembly, Grenoble, pp.278.
(8) Prieur, L.: 1976, Thesis, Doctorat d'Etat, Univ. P&M Curie, Paris. 243pp.
(9) Pelevin, V.N.: 1965, Atm. Ocean Phys. Ser. 1, pp. 539-545.
(10) Whitlock, C.H., L.R. Poole, J.W. Usry, W.M. Houghton, W.G. Witte, W.D. Morris, and E.A. Gurganus:1981, Appl.Opt. 20,pp. 517-522.
(11) Morel, A.: 1980, Boundary-Layer Meteorol. 18, pp. 177-201.
(12) Preisendorfer, R.W.:1961, In "Radiant energy in the sea", Int. Geophys. Geod. Monogr. 10, pp. 11-29.
(13) Smith, R.C., and K.S. Baker: 1981, Appl. Opt. 20, pp. 177-184.
(14) Morel, A.: 1974, In "Optical aspects of oceanography", N.G. Jerlov and E. Steemann Nielsen (Eds), Academic Press, London, pp.1-24.
(15) Morel, A., and L. Prieur: 1977, Limnol. Oceanogr. 22, pp. 709-722.
(16) Duysens, L.M.N.: 1956, Biochem. Biophys. Acta 19, pp. 1-12.
(17) Latimer, P.: 1959, Plant Physiol. 34, pp. 193-199.
(18) Latimer, P. and E. Rabinowitch: 1959, Arch. Biochem. Biophys. 84, pp. 428-441.
(19) Charney, E., and F.S. Brackett: 1961, Arch. Biochem. Biophys. 92, pp. 1-12.
(20) Duntley, S.Q., W.H. Wilson and C.F. Edgerton: 1974: Visibility Lab., Scripps Inst. Oceanogr. Ref. 7410, pp. 1.1-1.35.
(21) Kirk, J.T.O.: 1975, New Phytol. 75, pp. 11-20.
(22) Kirk, J.T.O.: 1975, New Phytol. 75, pp. 21-36.
(23) Kirk, J.T.O.: 1976, New Phytol. 77, pp. 341-358.
(24) Privoznich, K.G., K.J. Daniel and F.P. Incropera: 1978, J. Quant. Spectrosc. Radiat. Transfer 20, pp. 345-352.
(25) Kiefer, D.A., R.J. Olson and W.H. Wilson: 1979, Limnol. Oceanogr. 24, pp. 664-672.
(26) Morel, A., and A. Bricaud: 1981, In "Oceanography from space", J.F.R. Gower (Ed.), Plenum Press, New York, pp. 313-327.
(27) Bricaud, A, A. Morel and L. Prieur: 1982, Limnol. Oceanogr. To be published.
(28) Matsumoto, M.: 1981, Diplôme d' Etudes Supérieures, Univ. P&M Curie, Paris, 20 pp.
(29) Morel, A., and A. Bricaud: 1981, Deep Sea Res. 28A, pp. 1375-1393.
(30) Sathyendranath, S.: 1981, Thesis, Doctorat 3ème Cycle, Univ. P&M. Curie, 123 pp.
(31) Bukata, R.P., J.H. Jerome, J.E. Burton, S.C. Jain and H.H. Zwick: 1981, Appl. Opt. 20, pp. 1696-1703.
(32) Prieur, L. and S.Sathyendranath: 1981, Limnol. Oceanogr. 26, pp. 671-689.
(33) Bricaud, A., A. Morel and L. Prieur: 1981, Limnol. Oceanogr. 26, pp. 43-53.
(34) Pak, H., J.R.V. Zaneveld and G.F. Beardsley, Jr.: 1971, J. Geophys. Res. 76, pp. 5065-5069.

(35) Morel, A.: 1973, in "Optics of the sea", AGARD Lect. Ser. 61, pp. 3.1.1-3.1.76.
(36) Bader, H: 1970, J. Geophys. Res. 75, pp. 2822-2830.
(37) Brun-Cottan, J.C.: 1971, Cahiers Océanogr. 23, pp. 193-205.
(38) Sheldon, R.W., A. Prakash and W.H. Sutcliffe Jr.: 1972, Limnol. Oceanogr. 17, pp. 327-340.
(39) Morel, A.: 1970, AGARD Conf. Proceed. 77, Paris.
(40) Morel, A.: 1967, IAPSO/IUGG XIV Gen. Ass. Berne, Proc. Verb. 10, pp. 204-205.
(41) Burt, W.V.: 1955, J. Mar. Res. 14, pp. 47-62.
(42) Yentsch, C.S.: 1962, Limnol. Oceanogr. 7, pp. 207-217.
(43) Bukata, R.P., J.E. Bruton, J.H. Jerome, S.C. Jain and H.H. Zwick: 1981, Appl. Opt. 20, pp. 1704-1714.
(44) Bukata, R.P., J.H. Jerome and J.E. Bruton: 1981, Appl. Opt. 20, pp. 3472-3474.
(45) Neville, R.A., and J.F.R. Gower: 1977, J. Geophys. Res. 82, pp. 3487-3493.
(46) Gordon, H.R.: 1979, Appl. Opt. 18, pp. 1161-1166.
(47) Gower, J.F.R.: 1980, Boundary-Layer Meteorol. 18, pp. 235-245.
(48) Gower, J.F.R., and G. Borstad: 1981, in "Oceanography from space", J.F.R. Gower (Ed.), Plenum Press, N.Y., pp. 329-338.
(49) Kiefer, D.A.: 1973, Mar. Biol. 22, pp. 263-269.
(50) Kiefer, D.A.: 1973, Mar. Biol. 23, pp. 39-46.
(51) Loftus, M.E., and H.H. Seliger: 1975, Chesapeake Science 16, pp. 79-92.
(52) Nusch, E.A.: 1980, Arch. Hydrobiol. Beih. Ergebn. Limnol. 14, pp. 14-36.
(53) La Violette, P.E.: 1974, in "Optical aspects of oceanography", N.G. Jerlov and E. Steemann Nielsen (Eds), Academic Press, London, pp. 289-316.
(54) La Violette, P.E., P. Peteherych and J.F.R. Gower: 1980, Boundary-Layer Meteorol. 18, pp. 159-179.
(55) Cox, C.S.: 1974, in "Optical aspects of oceanography", N.G. Jerlov and E. Steemann Nielsen (Eds), Accademic Press, London, pp. 51-75.
(56) Thomas, Y.F.: 1978, Publ. Sci. Tech. CNEXO, Actes Colloq. No. 5, pp. 109-121.
(57) Mundey, Jr, J.C., and M.S. Fedosh: 1981, in "Chesapeake Bay plume study. Superflux 1980", J.W. Campbell and J.P. Thomas (Eds), NASA conference publication 2188, NOAA/NEMP III 81 ABCDFG 0042, pp. 79-42.
(58) Gower, J.F.R., K.L. Denman and R.J. Hoyler: 1980, Nature 288, pp. 157-159.
(59) Viollier, M., J.R. Grall, P. Le Corre, J. Le Fevre and C. Dupouy: 1982, To be published in Oceanographica Acta.
(60) Clark, D.K.: 1981, in "Oceanography from space", J.F.R. Gower (Ed.), Plenum Press, N.Y., pp. 227-237.
(61) Clark, D.K., E.T. Baker and A.E. Strong: 1980, Boundary-Layer Meteorol. 18, pp. 287-298.
(62) Gordon, H.R., and D.K. Clark: 1980, Boundary-Layer Meteorol.

18, pp. 299-313.
(63) Smith, R.C., and W.H. Wilson: 1981, in "Oceanography from space", J.F.R Gower (Ed.), Plenum Press, N.Y., pp. 281-294.
(64) Stürm, B.: 1980, XXVII Rassegna Internatzionale Elettronica Nucleare ed Aerospziale, Rome, March 7-16.
(65) Gordon, H.R., D.K. Clark, J.W. Brown, O.B. Brown and R.H. Evans: 1982, J. Mar. Res. To be published.
(66) Austin, R.W., and T.J. Petzold: 1981, in "Oceanography from space", J.F.R. Gower (Ed.), Plenum Press, N.Y., pp. 239-256.
(67) Morel, A.: 1981, in "Extended abstracts presented at the symposium on the radiation transfer in the oceans and remote sensing of ocean properties", IAMAP third scientific Ass., 17-28 Aug. 1981, Hamburg (F.R.G), pp. 25-26.
(68) Arvesen, J.C., J.P. Millard and E.C. Weaver: 1973, Astronautica Acta 18, pp. 229-239.
(69) Kim, H.H., C.R. McClain, L.R. Blaine, W.D. Hart, L.P. Atkinson and J.A. Yoder: 1980, J. Geophys. Res. 85, pp. 3982-3990.
(70) Pearcy, W.G., and D.F. Keene: 1974, Limnol. Oceanogr. 19, pp. 573-583.
(71) Viollier, M., D. Tanré and P.Y. Deschamps: 1980, Boundary Layer Meteorol. 18, pp. 247-267.
(72) Morel, A., and L. Prieur: 1978, Publ. Sci. Tech. CNEXO, Actes Colloq. No. 5, pp. 67-92.
(73) Pelevin, V.N.: 1978, Oceanology 18, pp. 278-282.
(74) Mueller, J.L.: 1976, Appl. Opt. 15, pp. 394-402.
(75) Mueller, J.L.: 1973, PhD thesis, School of Oceanography, Oregon State Univ., Corvallis, Oregon
(76) Doerffer, R.: 1981, in "Oceanography from space", J.F.R. Gower (Ed.), Plenum Press, N.Y., pp. 339-345.
(77) Jain, S.C., and J.R. Miller: 1976, Appl. Opt. 15, pp. 886-890.
(78) Miller, J.R., S.C. Jain, N.T. O'Neill, W.R. McNeil and K.P.B. Thomson: 1977, Remote Sensing of Environment 6, pp. 183-200.
(79) Spitzer, D., C. Rappoldt and L. Nykjaer: 1982, in "Preprint volume: First international conference on meteorology and air/sea interaction of the coastal zone", May 10-14, 1982, The Hague, Netherlands. American Meteorolog. Soc., Boston, pp. 296-297.
(80) Smith, R.C., and K.S. Baker: 1982, J. Mar. Biol. 66, to be published.
(81) Wrigley, R.C.: 1980, Trans. Amer. Geophys. Union 46, pp. 1001.
(82) Hovis, W.A.: 1981, in "Oceanography from space", J.F.R. Gower (Ed.), Plenum Press, N.Y., pp. 213-225.
(83) Gordon, H.R., and A. Morel: 1982, in "Lecture notes in coastal and estuarine studies" series, Springer-Verlag, N.Y., in press.
(84) Mueller, J.L., and P.E. La Violette: 1981, in "Oceanography from space", J.F.R. Gower, (Ed.), Plenum Press, N.Y., pp. 295-302.
(85) Lesieur, M., and R. Sadourny: 1981, Nature 294, pp. 673.
(86) Denman, K.L., and J.F.R. Gower: 1981, Nature 294, pp. 673-674.

(87) Bukata, R.P., J.H. Jerome, J.E. Bruton and S.C. Jain: 1980, Appl. Opt. 19, pp. 2487-2488.
(88) Stürm, B.: 1981, in "Remote sensing in meteorology, oceanography and hydrology", A.P. Cracknell (Ed.), Ellis Horwood, Chichester, pp. 163-197.
(89) NASA LANDSAT data users' handbook, 1976, Document No. 76SDS4258, NASA, Greenbelt.
(90) Hovis, W.: 1978, in "The NIMBUS-7 users' guide", C.R. Madrid (Ed.), Management and Technical Services Company, Beltsville, Maryland.
(91) Grew, G.W.: 1981, in "Chesapeake Bay plume study. Superflux 1980", J.W. Campbell and J.P. Thomas (Eds), NASA Conf. Publ. 2188, NOAA/NEMP III 81 ABCDFG 0042, pp. 301-322.
(92) Bowker, D.E., C.A. Hardesty, D.J. Jobson and G.S. Bahn: 1981, in "Chesapeake Bay plume study. Superflux 1980", J.W. Campbell and J.P. Thomas (Eds), NASA Conf. Publ. 2188, NOAA/NEMP III 81 ABCDFG 0042, pp. 323-337.
(94) Gordon, H.R., and W.R. McCluney: 1975, Appl. Opt. 14, pp. 413-416.
(95) Metzig, G., and E. Raschke: 1979, in "Proceedings, Workshop on the Eurasep Ocean Color Scanner Experiments", Joint Research Centre, Ispra, Italy, 30-31 Oct.
(96) Gordon, H.R., and D.K. Clark: 1980, Appl. Opt. 19, pp. 3428-3430.
(97) Jerlov, N.G.: 1976, "Marine Optics", Elsevier, Amsterdam.

CHAPTER 17

LAND INTO SEA DOES NOT GO

Richard K Bullard

Department of Land Surveying, North East London Polytechnic

ABSTRACT

The object of this paper is to outline the differences encountered between the recording of the three dimensional surface of the land and the three dimensional surface of the sea bed. The further detection of the 'fourth' dimension, or what the surface consists of will also be considered, both the land and the sea bed.

Most scientists using remotely sensed imagery are often unaware of the methods by which the data base to which they add their interpreted data is constructed. It is important to appreciate the relationship between the remotely sensed image and the three dimensional surface of the land or the sea bed. There are many advantages in collecting all four values (WXYZ) at the same time from the same platform.

1. INTRODUCTION

Methods for determining the position and depth of the sea bed have their own special problems, though, it is suggested, with remote sensing imagery the techniques for analysis can be updated so that the methods become more similar to methods of land based surveys.

The earliest recorded hydrographic sounding was in 1422 BC when the activity was recorded on the tomb of Menna an Egyptian scribe.

Methods of position fixing and sounding of those Ancient

A. P. Cracknell (ed.), Remote Sensing Applications in Marine Science and Technology, 359–372.

Egyptian days remained virtually unchanged until the early 20th Century. Despite the effort put into hydrographic surveys, only a fraction of the world's 360 million square kilometres of sea bed was surveyed, and of importance to the large ships of today, the surveys were only concerned with slow moving ships of less than ten metres draught.

On a relative basis, the early surveys, particularly those close to the shore, and considering the equipment available were surprisingly accurate. However, the absolute accuracy of these surveys and the relative accuracy of those made out of sight of land were dependent on fixing position which could at best be within several kilometres and where unknown currents existed the resultant survey could be even less accurate.

Hydrographic surveys had very much in common with land surveys until this century, when with the advent of photogrammetry, land surveys no longer relied on a point of dimensional detail from which to interpolate contours (or in hydrographic surveys between soundings), but a model of the ground could be viewed stereoscopically and this overcame the problem of spot heights. Until recently hydrographic surveying has been unable to use similar techniques as those used on land, but with the advent of the remotely sensed image of the sea bed this backlog may be made up.

2. THE PROBLEM OF UPDATING THE HYDROGRAPHIC CHART

In 1976 the International Hydrographic Bureau estimated that only 16 per cent of the oceans of the world had sufficiently accurate soundings to determine sea floor topography. Another 22 per cent of the oceans had data sufficient only for the determination of the major sea floor features, and the remaining 62 per cent of the ocean area had data that is too sparse for determining sea floor topography (1). A French expert (2) has stated that any chart based on surveys before 1970 carries with it a risk that the surveyors may have missed some dangerous wreck or other obstacle to navigation.

The present sea lanes do not always follow the most direct routes, and some of the routes are too shallow for the new deep draught vessels. These vessels must therefore seek out longer, less well surveyed, but hopefully deeper routes for their destinations.

The requirements for accurate and adequate hydrographic surveys over large areas of the world's oceans would therefore appear to be of utmost importance. The hydrographic survey resources are limited and the cost of the surveys tend to be expensive. To show how far behind the surveys are; in 1975 the

Hydrographer of the Navy (UK) stated in his report that 284 ship-years of work will be required to bring the United Kingdom's territorial waters up to modern standards and that the foreign areas of responsibility would require an additional 300 ship-years of effort.

3. REMOTE SENSING AND HYDROGRAPHIC SURVEYING

It is important to ask how remote sensing can assist the hydrographer in his task of charting the oceans? One form of assistance is to carry out hydrographic surveys by means of remotely sensed imagery to a depth of 30 metres. This figure assures that the draught of the largest ships is 30 metres and that light waves can penetrate to this depth on the shallow water zones? If this depth of water is considered suitable then a major portion of the water surface of the world would be excluded. The problem that arises is how best to cover the total water area and then how to exclude the deep water portion and concentrate on the shallow water zones?

Remote sensing of the water surface and of the ocean bed can be achieved by conventional aircraft and by satellite. Using conventional aircraft, large scale imagery can be obtained but the cost and the time taken to collect imagery for the total water area would be excessive. The current satellites being used for earth resources, Landsat 2 and 3 could theoretically cover the earth (except the polar regions) every 18 days, and if the imagery was cloud free then a complete coverage could be obtained in a short time and at a comparatively low cost. However, the resolution of the imagery and the depth at which the penetration of light waves occur and thereby record the sea bed are the limitations of Landsat imagery, and for that matter a limitation of passive systems.

In experiments carried out by the NASA/Cousteau Bathymetry Experiment (3) depths up to 22 metres in the Berry Island test area were measured from satellite data and verified by the ground data team on the Calypso. Landsat also revealed shoals that were later determined by Calypso to be at a depth of 40 metres. In the less clear waters off the Florida Coast, Landsat data were used to map shoals as deep as 10 metres. Additional research and comparison of Landsat high-grain data with aircraft acquired data and recent charts indicated that depths could be deduced without knowledge of the optical properties of the waters in the area investigated. This capability could be very important when studying remote areas that have unreliable charts.

4. CHART REVISION BY REMOTE SENSING

Examples of chart revision in the Chagos Archipelago in the Indian Ocean (72° 30' E and 5° 00' S approximately) using Landsat imagery revealed a major reef 8 kilometres long as well as a number of other variations not portrayed on the existing chart of the Archipelago. This discovery resulted in an application of the proven cartographic capabilities of the Landsat film imagery to the evaluation and the correction of the horizontal positions of features on a medium-scale nautical chart. Film images at 1:1 000 000 scale were obtained for the Chagos scenes and the chart of the area was reduced to the same scale in order to compare features (4) (5). When the imaged island features were registered to the charted features, significant variations were immediately apparent. There was a major reef or bank where the chart showed deep water and some banks appeared to be out of position by more than 15 kilometres relative to the nearest land. As a result of these findings radio warnings were sent out to mariners and a new chart was prepared. The chart used the Landsat images for the position of the islands, reefs and banks, and geodetic control was plotted and a new mosaic of the Landsat photography was prepared at the original scale. The previous survey of the area had been carried out by the Indian Navy in 1837.

In another example Landsat imagery was used to locate uncharted coastal features on the Labrador Coast. The Canadian Hydrographic Service carried out a survey of several offshore islands, rocks, and shoals from Cape White Handkerchief to Cape Chindley. The purpose of this project was to determine the positions and characteristics of these offshore points, which were required for locating the territorial boundaries. On this survey for the first time Landsat imagery was used to assist in the location of uncharted hydrographic features. As a result of this work an uncharted island and eight uncharted drying rocks were verified and positioned. To check the accuracy of the co-ordinates derived from Landsat, three islands were positioned by standard ground survey methods from previously established shore stations. The positional differences for these points were from 10 metres to 140 metres in Latitude, and from 42 metres to 91 metres in Longitude. These differences are not plottable at the scale of the existing offshore charts (6).

The above examples show that Landsat imagery has been used for three dimensional control, that is position and depth, and it is possible that the fourth dimension, the remote sensing value (W) can be determined (7). This 'W' value can be the nature of the sea bed, the size of the waves, the temperature of the water, the salinity of the water, or the amount of sediment in suspension. The 'W' value will depend on the sensor, the wave band, the filter, water surface conditions, depths of water, sea conditions, as to what 'W' represents. When 'W' is the sea bed the depth can be interpreted from the way that this feature is recorded on the imagery.

5. RECORDING THE THREE DIMENSIONAL SURFACE

In photogrammetry the aerial photograph is overlapped by 60 per cent, and this enables a stereoscopic image of the ground surface to be constructed. With the required number of ground control points the model that is produced in the photogrammetric instrument can be corrected for non-verticality of the camera axis at the time of exposure, for height variation, earth curvature, and for scale. The model can therefore become a scaled model of the land surface.

Using aerial photography over water surfaces will introduce a number of additional problems that are not encountered over a land surface.

a. There will be little or no opportunity for pre-marking ground control points except along a coastline.
b. Post-marking of ground control points will be difficult except along a coastline.
c. Height control points, other than the water surface will only be available along a coastline.
d. As the water is a more dense medium than air the depth of water penetration will be limited by the light waves.

Water penetration and reflectance from the sea bed will depend on the sun's azimuth and elevation angle to the water surface and the sea bed, the nature of the sea bed, and the direction of wave propogation, and the sea state. Figure 1 below illustrates the interaction encountered both in the atmosphere and in the water.

When remotely sensed imagery is used for determining bathymetric contours, the depth will be indicated by a density or grey value on a grey scale. This leads to a problem of acceptance of interpretation and the value given to the shade of grey. As Figure 2 shows, because of limited penetration there is a stage where the grey scale represents minimum reflectance or maximum penetration from the surface and the depth at which contours can be interpreted. It is questionable whether the interpretation is linear?

According to the Hydrographer of the Royal Australian Navy (9), he considers that his unambiguous requirement for depth determination can only be met with an active sysyem in which depths are actually measured rather than deduced. As the present methods of determining depth rely on aerial photographs and

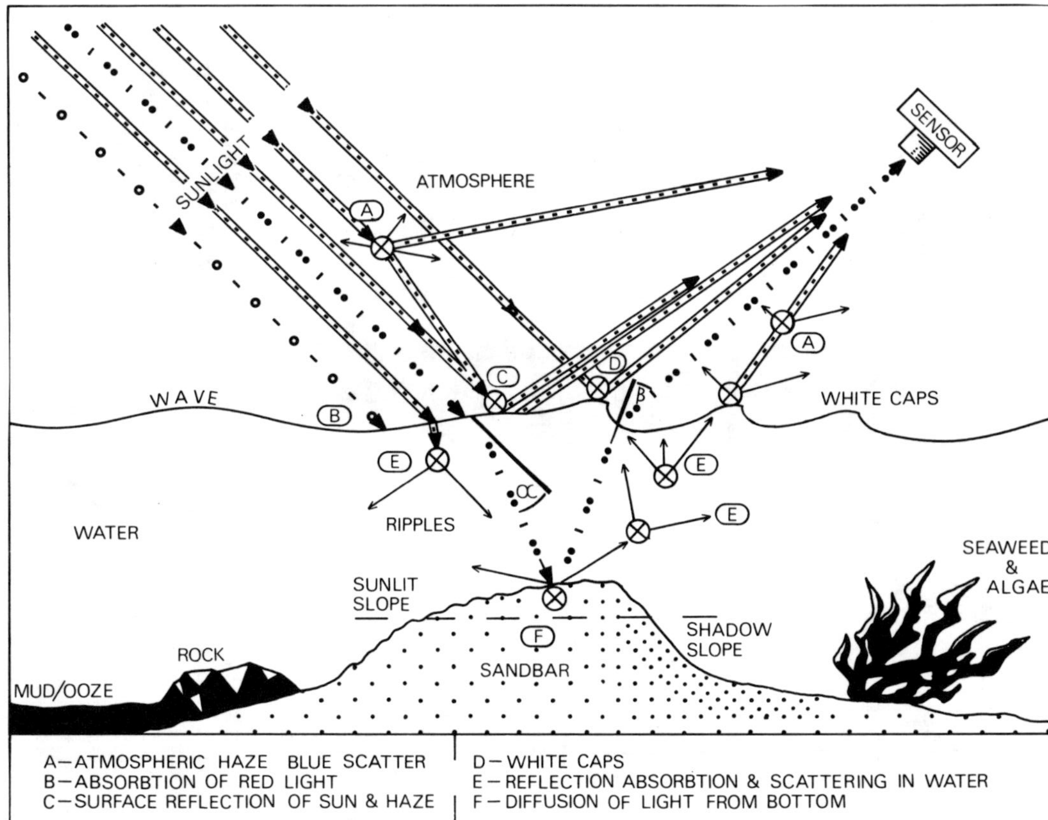

Figure 1. Scattering and Absorption of Light Waves by Atmosphere and Water

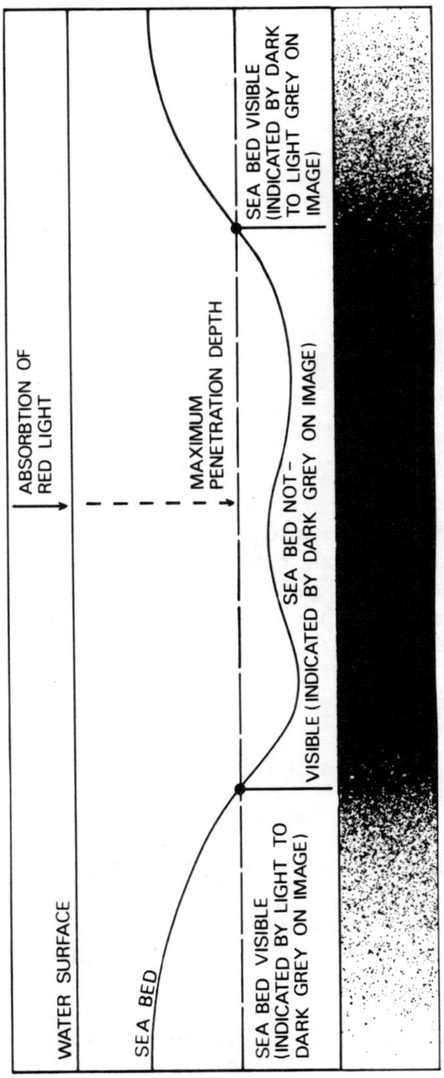

Figure 2. Depth of Water Penetration Represented by Grey Scale

satellite imagery using passive systems collecting reflected light in the visual and infrared range of the electromagnetic spectrum and depths deduced from the imagery the present systems may have to be improved. This would indicate that an active remotely sensed imaging device is required. The present position of the acceptance of passive imagery is like that of photogrammetry when it was replacing ground surveys for river boundary surveys for cadastral surveys some 30 to 40 years ago. Again the acceptance of sea survey methods are trailing behind ground surveys.

An active system for depth sounding (9) being developed in Australia is WRELADS (Weapons Research Establishment Laser Depth Sounding). Its functional specification is as follows :-

Depth measuring capability	2 - 30 metres
Accuracy of depth soundings	1 metre
Position of depth sounding	25 metres
Across track swath width	200 metres
Mean spacing of soundings	10 x 10 metre grid
Aircraft speed	70 metres per second

The principle of operation of WRELADS is shown in Figure 3 which shows the two beams, infrared and green emitted from the laser, the infrared laser beam provides the surface return signal while the green laser beam the sea bottom return signal which is subject to the depth measuring capability. The clearer the water the greater the depth measuring capability up to a maximum of 60 metres at night (9).

Figure 4 shows the comparison between the swept area covered in one second by an aircraft operating WRELADS and a surface vessel.

The geometry of the WRELADS laser is illustrated in Figure 5. The wide angle infrared beam is held vertical while the green beam has a transverse scan of 15° either side of the flight line. The infrared beam serves as the datum for depth measurements, the time of measurements will be related to a tidal gauge in the area so that corrections can be made for the state of the tide.

A system being developed in Canada is an "Aerial hydrography/laser bathymetry system", (10) which utilises a laser together with colour aerial photography. The two systems are considered complementary, colour aerial photography records shallow water while the laser records the deeper water. Trials in the Bruce Peninsula area have recorded depths of 7 and 23 metres for the photography and laser respectively.

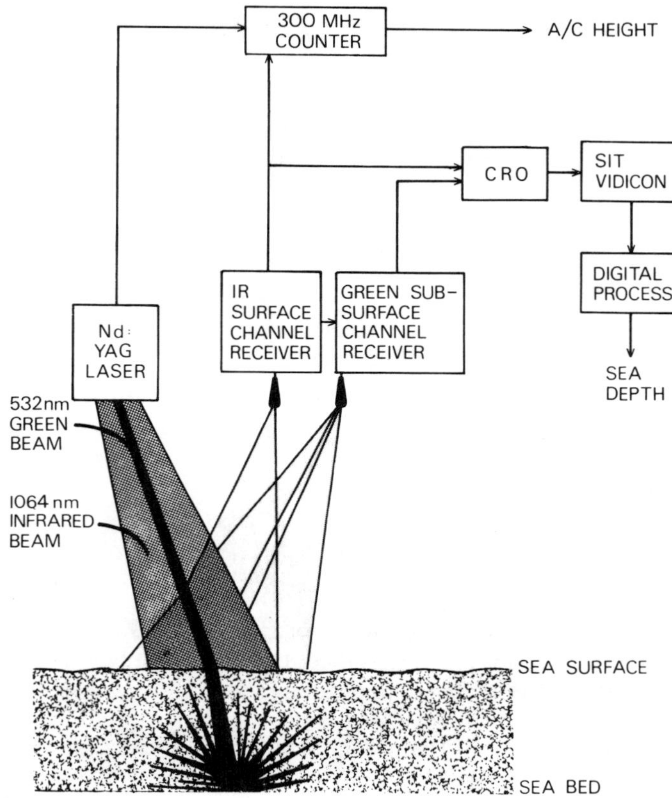

Figure 3. Principle of Operation of WRELADS

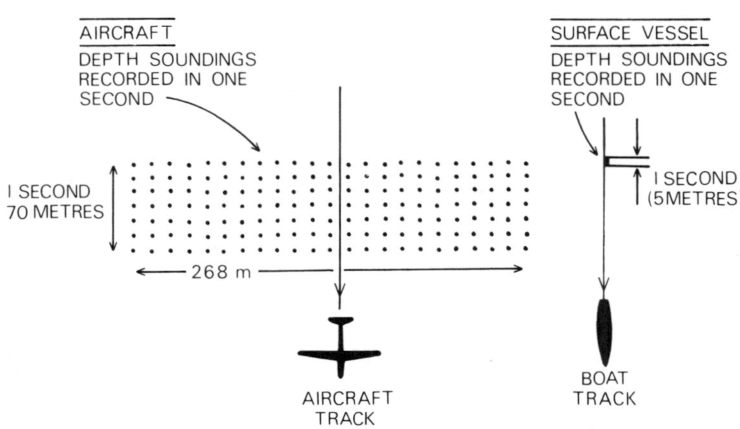

Figure 4. Comparison of Soundings by Aircraft and Boat

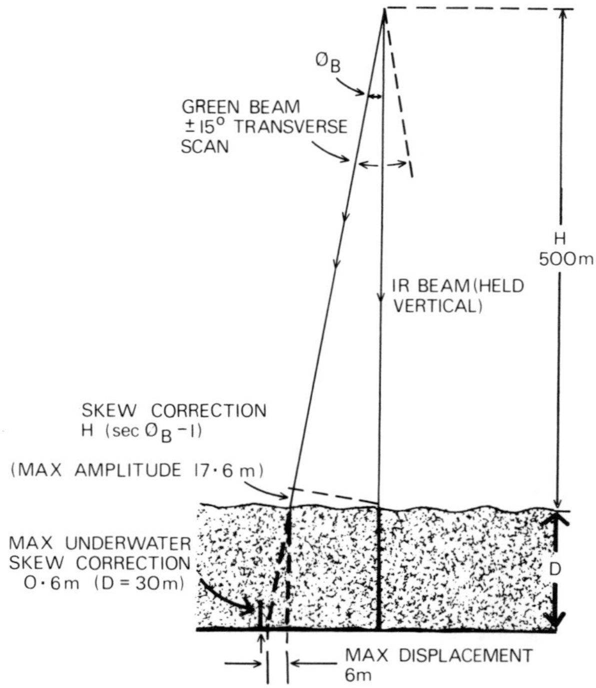

Figure 5. Scanning Geometry of WRELADS

6. AERIAL HYDROGRAPHY

By comparison with conventional hydrographic surveying, aerial hydrography is economical, speedier and more effective (10).

Three problems with aerial hydrography still remain and they have prevented the method becoming more acceptable. The problems are as follows (11) :-

a. The degree of water penetration.
b. The refraction of imaging rays.
c. The incomplete stereographic model.

When it is not always possible to carry out an aerial photographic sortie for coastal mapping so that the configuration of the land areas in a stereoscopic model provides points adequate for relative and absolute orientation, then the stereoscopic model will be unable to be completed. The orientation points required for relative and absolute orientation are difficult to select on the sea surface.

Because of the smaller scale of satellite imagery than conventional aerial photography, and therefore the larger surface coverage of these images it should have been easier to join the images to make a mosaic except that there is no stereoscopic overlap and over sea areas no opportunity to obtain ground control.

To improve the results of satellite imagery, coverage can be taken in the high-gain mode which amplifies the radiometric signals from MSS bands 4 and 5 by a factor 3 and thus provides more definitive radiometric information with respect to low-response features such as the shallow sea-bottoms (12).

The RBV (Return Beam Vidicon) imagery has an improved resolution over the MSS imagery but the amount of RBV imagery available has been very limited and has therefore not been a viable sensing device in practice.

Landsat D is expected to retain the MSS system together with a Thematic Mapper (TM) which has seven channels as listed below :-

Spectral Bands	Wavelength (μm)	Resolution (m)
Blue	0.45 to 0.52	30
Green	0.52 to 0.60	30
Red	0.63 to 0.69	30
Near Infrared	0.76 to 0.90	30
Mid Infrared	1.55 to 1.75	30
Infrared	2.10 to 2.35	30
Thermal Infrared	10.4 to 12.5	120

Grey Level quantisation : 256 levels (8 bits)

The blue spectral band 0.45 to 0.52 microns should provide a facility for water penetration that has not been available on the previous Landsat satellites where Band 4 of the MSS's green spectral band was from 0.5 to 0.6 microns.

The French satellite Spot will have a HRV (High Resolution Visible) radiometer with the four channels as listed below :-

Channel	Wavelength (μm)	Resolution (m)
1	0.48 to 0.53	20
2	0.55 to 0.60	20
3	0.62 to 0.70	20
4	0.80 to 1.05	20

Channel 1 has a similar wavelength to Band 1 of the Landsat TM; the resolution is an improvement, and of importance is the fact that stereoscopic images can be achieved by means of a tilting mirror so that coverage can be obtained from another orbit with a second scene not vertically below the satellite.

The Spot satellite HRV will have a swath width of 60 kilometres compared with Landsat D width of 185 kilometres.

With the future satellites (Landsat D and Spot) it is hoped that there will be an improvement in the penetration capabilities of the sensors and an improvement with the resolution of the surface, both land and sea bed. As Spot has a potential for stereoscopic images there will be provision for updating the methods of obtaining bathymetric contours and this will bring the hydrographic surveys to the same stage as land surveys with regard to stereoscopic models used in determining the shape (XYZ) of the land surface. It is possible that Spot images will also be used for land areas, particularly those parts of the developing world where there are no adequate topographical maps.

There is a need in developing countries to map and chart the land and the sea bed on a priority basis (13). With ship borne methods of hydrographic surveying the costs are often beyond the finances of developing countries and it is therefore even more important that an alternative technology is designed to assist these countries in mapping their priority areas.

7. CONCLUSIONS

There is a need to improve the method of detecting the sea bed and establishing the depth and position with the aid of remotely sensed imagery. The future results of Landsat D and Spot satellite imagery will be awaited with great interest to see if the improved resolution together with the blue bands will provide an increase in water penetration as suggested in this paper.

What is ideally required is an active sensor operating in the ultraviolet range as well as the blue range of the electromagnetic spectrum. The cost and time taken to obtain the information required for three dimensional control of the sea bed must be reduced. Remotely sensed imagery should reduce the cost of obtaining the sea bed data, and airborne (satellites and aeroplanes) sensors should reduce the time taken to collect this data.

The laser systems discussed in part 5 should be used for ground control and for the verification of the interpreted depths from a passive imaging system. It is suggested that the laser is far more expensive to operate over large areas than passive airborne sensors. The position of the laser (XY) and the handling and storage of data is another problem that does not occur to the same extent with the remotely sensed image.

Once an improved airborne system has been introduced and the information is being collected at a faster pace more attention should be taken of the monitoring activity which must at present take a low priority. There are certain areas where the bathymetric contours change over short periods of time and the monitoring of these changes is of importance so that charts can be quickly updated making the sea-lanes of the world safe for shipping.

The improved airborne remote sensing systems will have to be accepted by the 'Hydrographers of the Navies of the World' and it may then be said that sea based survey methods will have a common approach to land based survey methods and 'Land Into Sea will go' so that the common technique will provide four dimensional information for both surfaces.

ACKNOWLEDGEMENTS

The author wishes to thank Keith Eley of the Department of Land Surveying for assistance with the diagrams.

REFERENCES

(1) Kapoor, D.C.: 1976, "International Cooperation in Hydrography". International Hydrographic Review, LIII(2), pp. 7-15.

(2) Oudet, L.: 1973, "The Value of a Nautical Chart". International Hydrographic Review, L(1), pp. 149-161.

(3) Polcyn, F.C.: 1976, "Final Report on NASA/Cousteau Bathymetry Experiment - Remote Bathymetry Using High Gain Landsat Data". NASA-CR-ERIM-118500-1-F.

(4) Colvocoresses, A.P.: 1975, "Evaluation of the Cartographic Application of ERTS-1 Imagery", The American Cartographer, 2(1), pp. 5-18.

(5) Fleming, E.A.: 1976, "Positioning of Off-shore Features with the Aid of Landsat Imagery". USGS EROS Cartographic Memorandum No.35.

(6) Fleming, E.A., and Lelievre, D.D.: 1977, "The Use of Landsat Imagery to Locate Uncharted Coastal Features on the Labrador Coast". Proceedings 11th International Symposium on Remote Sensing of Environment, Michigan, April 1977, pp. 775-782.

(7) Bullard, R.K.: 1981, "The Land Survey Profession and Education in Remote Sensing". XVIth International Congress of Surveyors, Montreux, August 1981, Paper 202.2.

(8) Gierloff-Emden, H.G.: 1977, Orbital Remote Sensing of Coastal and Offshore Environments. Walter de Gruyter, Berlin, 1977.

(9) Calder, M.: 1979, "WRELADS - The Australian Laser Depth Sounding System". Conference of Commonwealth Surveyors, Cambridge, August 1979, Paper No. H.3.

(10) Mac Phee, S.B., Dow, A.J., Anderson, N.M., and Reid, D.B.: 1981, "Aerial Hydrography Laser Bathymetry and Air Photo Interpretation Techniques for Obtaining Inshore Hydrography". XVIth International Congress of Surveyors, Montreux, August 1981, Paper 405.3.

(11) Masry, S.E., and Mac Ritchies, S.: 1980, "Different Considerations in Coastal Mapping". Photogrammetric Engineering and Remote Sensing, 46(4),1980, pp. 521-528.

(12) Colvocoresses, A.P.: 1979, "Landsat Mapping and Charting by the United States of America". Conference of Commonwealth Surveyors, Cambridge, August 1979, Paper No. D.2.

(13) Young, F.R., and Lyons, K.J.: 1981, "Hydrographic Charting in Papua New Guinea". XVIth International Congress of Surveyors, Montreux, August 1981, Paper 405.4.

CHAPTER 18

DETECTION OF MARINE CONTOURS FROM LANDSAT FILM AND TAPE

Richard K Bullard

Department of Land Surveying, North East London Polytechnic

ABSTRACT

This paper is the product of two work sheets prepared for the summer school and written up as a contribution to these proceedings.

The examples used Landsat imagery in two forms; the first example used film, the second a print out derived from a Computer Compatible Tape (CCT) for producing the end result of the exercise. The work sheets involved two different exercises in determining water depth. The approach to be adopted together with an outline of the problems involved and a list of questions to be addressed to the end results was included on the work sheets. The only sea bed data provided with which to check the results was that derived from the existing nautical charts of the two areas.

1. INTRODUCTION

1.1 Introduction To Using Film Images

With the Landsat satellites 1 to 3 the Multispectral Scanners (MSS) cover the wavelength from 0.5 to 1.1 microns, with Band 4 covering the range 0.5 to 0.6 microns. Band 4 therefore covers the green to orange range of the visible spectrum, and while it does not cover the blue/violet end of the visible spectrum it is the band that is most often used for water penetration. The blue/violet or ultraviolet end of the spectrum may be more suitable for clear water penetration, but

A. P. Cracknell (ed.), Remote Sensing Applications in Marine Science and Technology, 373–381.

due to the atmospheric scattering effects that occur at high altitudes it may not be a practical proposition from a satellite platform. The depth to which penetration can be achieved by visible light from 0.5 to 0.6 microns will depend on a number of factors which include :-

Reflection on the water surface.
Sun angle and azimuth.
Salinity.
Amount of suspended sediment.
Nature and reflection of the sea bed.
Height of platform.
Scatter, absorption and refraction in the atmosphere.

1.2 Introduction To Using Computer Compatible Tape

The Computer Compatible Tape (CCT) has an advantage over the photographic image derived from the tape in that it is not a degraded image and the information on the tape is of the same order as that received from the satellite. Photographic images are often of the 5th order and are therefore appreciably degraded, although with the vast amount of information contained in a Landsat scene this is not always apparent even if it is appreciated. The CCT as the name implies requires a compatible computer, additional hardware, and software to be able to analyse the Landsat image.

2. MARINE CONTOURS FROM A BAND 4 LANDSAT FILM

This example was the result of a question raised at a short course in Sea Surveying, held at the author's Polytechnic. The question was,"could Landsat imagery provide bathymetric contours depicting a depth of water sufficient for supertankers to pass between Kamaran Island and the mainland of the Yemini Republic?". There are a number of articles (1,2,3) describing the determination of marine contours, most of them concluding that depth can be established under ideal conditions to a maximum of 30 metres. This depth of 30 metres is within the range required for the draught of supertankers and if this figure could be obtained the system could be considered viable for navigational purposes of the largest ships currently available.

The only Landsat image that could be readily purchased in the United Kingdom was from Overseas Development Agency at Tolworth, Surrey, where a copy of Band 4 image was available. This purchase avoided the cost and delay of purchasing an image from NASA. It is accepted that the image degredation due to the amount of copying that had taken place would limit the use to which the image could be used. However, as an example only,the film image was accepted as no other funds were available at the time to purchase imagery from NASA. This was only a first attempt to determine bathymetric contours and to answer the question posed above.

2.1 Analysis Of The Landsat Film

The Band 4 image covers part of the Red Sea and part of the Southern Yemen Republic together with Kamaran Island. The image was recorded on the 6th December 1972, with a sun elevation of 41° and azimuth of 143°. The Band 4 image has been density sliced with the aid of the IDP 3000 at the National Remote Sensing Centre, Farnborough. The image was scanned with the aid of a television camera which enabled the scene to be stored and then processed by the IDP 3000.

The first step in the processing of the image was to produce a scene which depicted the land/water interface. This scene is included with the colour plates which shows the water as red and the land as white. To enable the fullest use of the remaining eight colours to be used in the density slice the land was restricted to white in the further analysis. Figure 1 shows an enlargement of the area under investigation. The cloud over the mainland and over Kamaran Island can be interpreted because of the shadows cast on the land, this is not so obvious over the sea.

The second stage of the image analysis was to carry out a density slice of the sea area. This analysis is included with the colour plates which shows the land still as white and the sea divided into 8 colours. The colours represent 8 evenly spaced slices of the remaining part of the Landsat grey scale. (see colour section, pp. 455 and 456).

To be able to compare the results with another data base a chart of the area was obtained. The chart was surveyed between 1925 and 1930 with small corrections being applied since the original survey up to 1978.

2.2 Comparison Between Film And Chart

The problem that immediately arises is which is the most reliable, the film or the chart? As the area covered by the example is within close proximity of the coast it may be assumed that the position of points that are above the water surface are, or at least were reasonably well positioned in the original hydrographic survey. What is not clear is how well the sea bed was surveyed? Even if the survey was of a reasonable standard and the spacing of the soundings were sufficient to enable the majority of the underwater features to be recorded what changes in depth have taken place in the intervening years due to sediment, erosion, and the growth of the coral in the area.

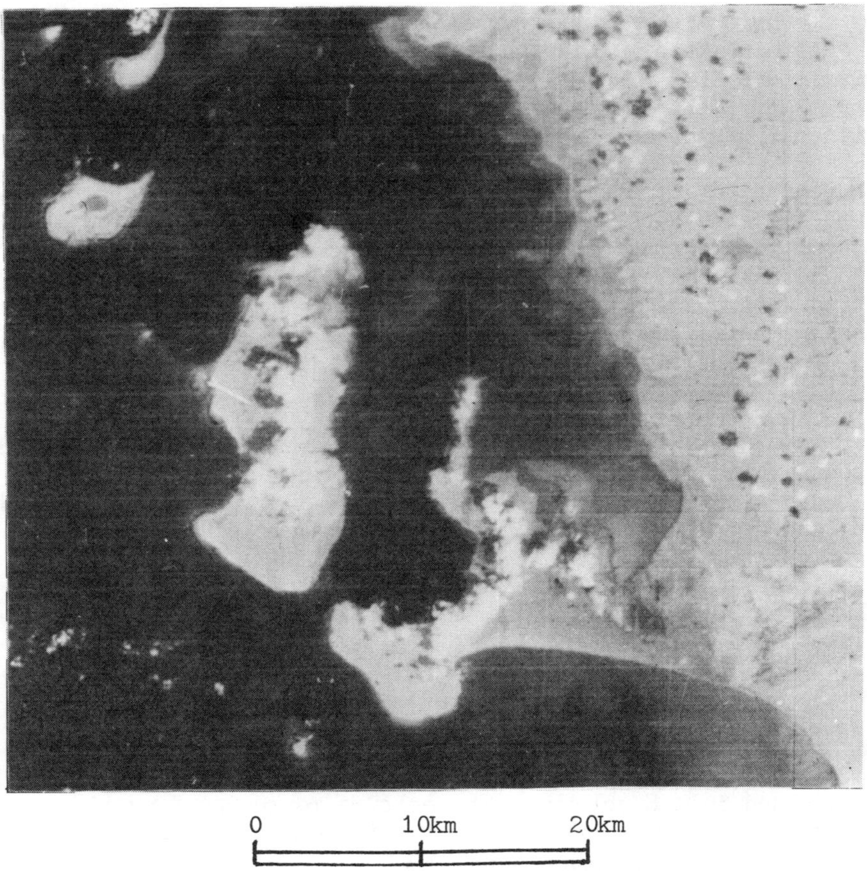

Figure 1 Enlargement of Band 4 Landsat Film

Using a slide of the island and the mainland taken from the monitor of the IDP 3000 it is possible, with the aid of a slide projector, to fuse the projected image onto the hydrographic chart mounted on a wall. There is a surprisingly good agreement between the image and the chart. The first colour plate shows the land/sea interface with the sea in red and the land white. One of the immediate problems that arises is that the height of the tide is not immediately available and as there is only one temporal image the existence of the cloud that is included in the Landsat scene will have to be detected if this is at all possible. Clouds in the Landsat scene may be incorrectly interpreted as reefs or islands and their shadows may also lead to misinterpretation, not only as a surface feature but the shadow may produce an apparent shelving below the water surface supporting the impression of an island and not a cloud.

By projecting the second slide (see Second Colour Plate) onto the aligned chart it is possible to transfer the form lines produced by the density slice onto the chart or a transparent sheet placed over the chart. A comparison between the form lines and the marine contours shows a number of ddscrepencies which need further investigation.

3. MARINE CONTOURS FROM A LANDSAT TAPE

The accompanying Band 5 and 7 Landsat images (Figures 2 and 3) cover an area of the Red Sea, part of Saudi Arabia and Port Sharmah on the coastline. The images were recorded on the 22nd August 1975 with a sun elevation of 55° and an azimuth of 110°. The two bands have been used to enable the land/sea interface to be constructed, with the aid of the Band 7 image (0.8 to 1.1 microns in the near infrared with no water penetration) and the Band 5 image (0.6 to 0.7 microns in the red part of the visible spectrum) for the determination of marine contours.

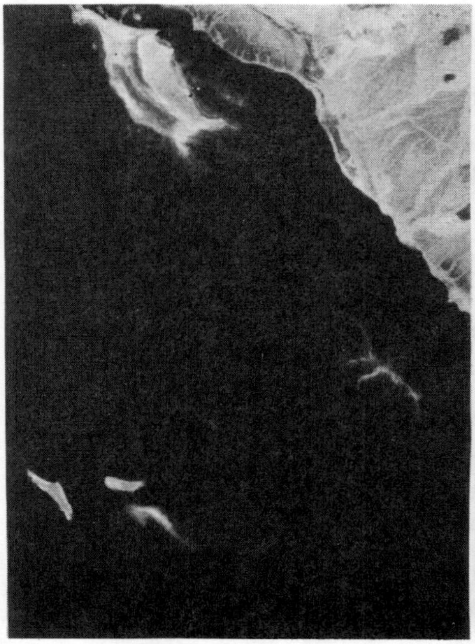

Figure 2 Band 5

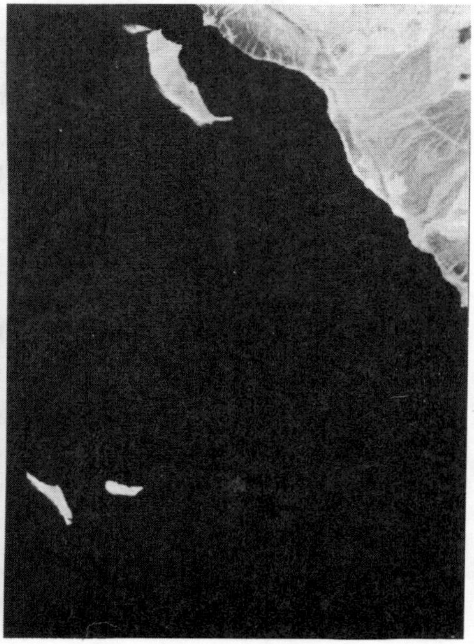

Figure 3 Band 7

An enlargement and redrawn inset contained on the hydrographic chart is provided and is shown as Figure 4. This chart enables a comparison to be made with the results of the graphic plot which is Figure 5. The graphic plot has been

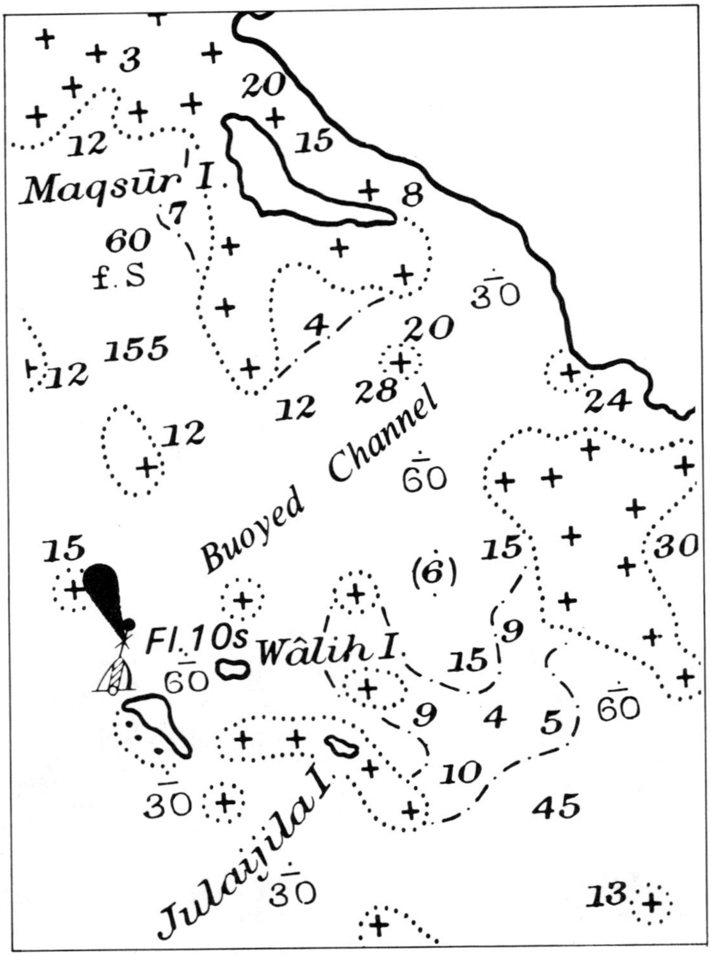

Figure 4 Redrawn Inset of Hydrographic Chart

produced from a computer analysis of the Band 5 image. The marine contours have been derived from the densities on the image. Depths of contours and their position are only an indication and are more correctly form lines at this unproven stage. A more explicit name of 'density line contours' would explain how these unproven contours are derived.

3.1 Comparison Between Graphical Plot and Chart

As in the other example the problem is to establish the reliability of the chart and the correctness of the interpretation by computer to produce the graphical plot. As the inset was at a larger scale and involved more detailed hydrographic surveying

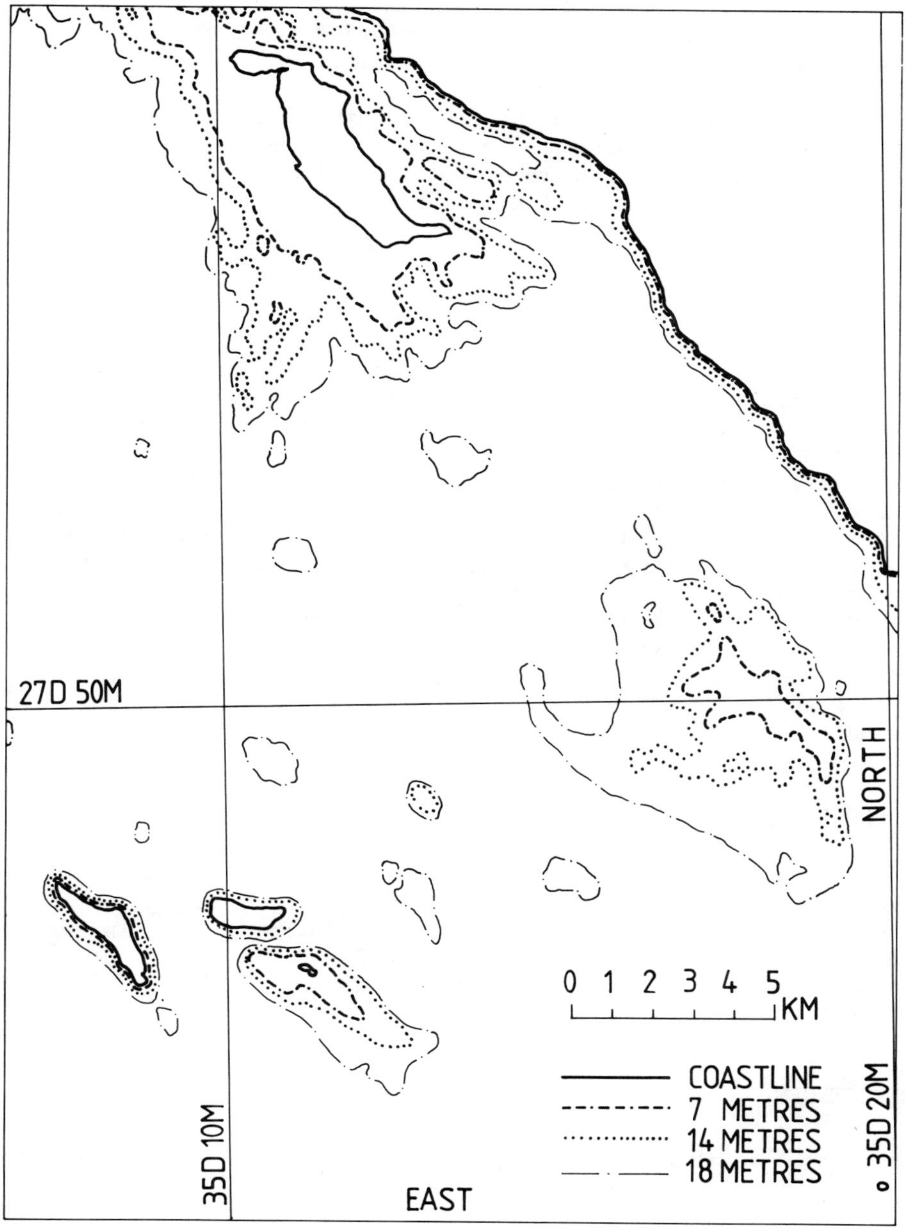

Figure 5 Redrawn Graphical Plot Showing Density Line Contours

the position of the marine contours were probably more reliably fixed at the time of the survey. The problem of the time interval between survey and the Landsat image and the state of the tide at the time the image was recorded are problems previously discussed in the other example.

To be able to compare the plot and the chart a transparency of the plot was made and was projected onto the chart. Like the other example there were a number of discrepencies which require further investigation.

4. CONCLUSIONS

Attention must be drawn to the fact that no ground (sea bed) data exists other than outdated marine contours. It would be presumptive to assume that the Landsat images in the two examples could be used as they are to correct the two charts. One or more additional images, taken with a reasonable time interval, ideally cloud free, at a known state of the tide, would improve the results and also eliminate the misinterpretation caused through cloud being present on the image. However, shifting sand banks and the deposit of sediment due to rivers and currents, together with the growth of the coral may alter the marine contours over a time period. Additional images should enable the sea bed to be monitored and changes over time to be determined. This monitoring and the updating of the hydrographic charts is an application that needs further research.

ACKNOWLEDGEMENTS

The author wishes to thank Dr Alan Benny of the National Remote Sensing Centre, Farnborough, for making the graphical plots available and for allowing these to be reproduced in this paper. To Keith Eley and David Wagstaffe of the Department of Land Surveying for assistance with the diagrams and the photographs respectively.

REFERENCES

(1) Polcyn, F.C.: 1976, "Final Report on NASA/Cousteau Bathymetry Experiment - Remote Bathymetry Using High Gain Landsat Data". NASA-CR-ERIM-118500-1-F.

(2) Colvocoresses, A.P.: 1975, "Evaluation of the Cartographic Application of ERTS-1 Imagery", The American Cartographer, 2(1), pp. 5-18.

(3) Bullard, R.K.: 1983, "Land Into Sea Does Not Go". Proceedings of a NATO Summer School - Remote Sensing Applications in Marine Science and Technology. D.Reidel Publishing Company, Holland.

CHAPTER 19

AIRBORNE REMOTE SENSING AND EXPERIMENTS WITH FLUORESCENT TRACERS

C. Valerio

Centre d'Etudes Techniques de l'Equipment, B.P. 39,
13290 Les Milles, France.

ABSTRACT

The method of using airborne remote sensing in this type of problem on coastal pollution monitoring was described in the proceedings of the 1980 Dundee summer school on remote sensing; it is briefly recalled here and merely amplified in one or two respects. An example involving the study of the design of a new sewage outfall for the town of Antibes is described in some detail and the consequential recommendations for the position of the new outfall are given. The use of two different fluorescent tracers, released simultaneously, to distinguish between the consequences of two different sources of pollution is also described very briefly.

1. INTRODUCTION

In situ measurements of oceanographic, or coastal region, parameters are usually available as observations at a rather restricted number of locations, though possibly sometimes over long periods of time. The use of remote sensing techniques enables one to obtain a synoptic view of an area and reduce, or even eliminate, the need for the spatial extrapolation. Remote sensing may also lead to a reduction in temporal extrapolation.

The general principles involved in the use of fluorescent tracers in conjunction with multispectral photography from light aircraft were described in my contribution to the 1980 Dundee summer school (1). I shall just recall a few of the main points briefly here and then illustrate the application of the technique

A. P. Cracknell (ed.), Remote Sensing Applications in Marine Science and Technology, 383–407.

to problems in coastal waters by considering two examples.

One needs to choose between aircraft-flown and satellite-flown remote sensing and several considerations are relevant. The final choice may differ according to whether one is concerned with deep-sea oceanography or with coastal waters. The nature of the information that can be obtained depends on the wavelength of the electromagnetic radiation used. For cheapness and practical convenience we shall consider photographic film as the medium for data reception and as the basis for quantitative interpretation.

2. GENERAL PROCEDURE

We consider four stages in the execution of a remote sensing project of this type. These constitute a general methodology that is capable of application to many different problems.

2.1 The study of the optical properties of the phenomenon under consideration

This involves the choice of spectral channels and some rather detailed consideration needs to be given to the factors involved. It also involves the consideration of the relationship between the intensity received at the detector and the quality or quantity of the phenomenon studied. With radar and infrared there is virtually no penetration of the surface; significant penetration is only achieved with visible radiation. Three examples of more or less operational possibilities should be mentioned: (i) charting of the sea bed, (ii) the study of transport and dispersion of suspended matter, yellow substance and chlorophyll, (iii) the study of dispersion by simulation with a fluorescent tracer such as rhodamine B.

2.2 Consideration of the absorption of visible radiation in charting the sea bed

This problem is closely connected with that of a land use inventory. The major difficulty is the intervening layer of water, which varies in quality and depth. Only wavelengths which are not excessively absorbed by water can reach the sea bed, be reflected off it and return to the sensor. It follows that only blue, green or yellow objects such as sand, water plant communities and algae can be detected. For a given quality of water, however, the resultant colour will vary according to the depth. Corrections must be made for this effect by means of bathymetric data.

The passage of light from the Sun to the sea bed and to the camera on board an aircraft is shown in Fig. 1. The light passes

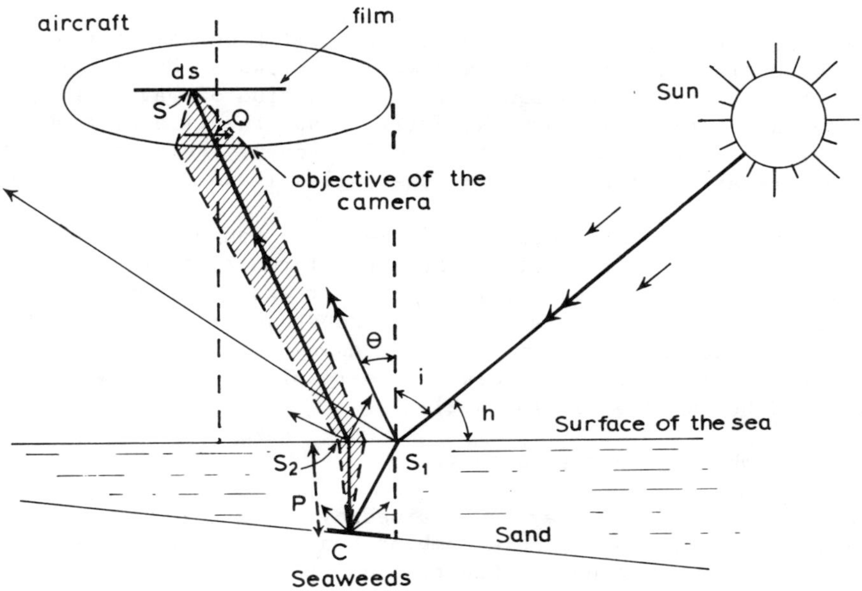

Figure 1. Diagram to illustrate the charting of the sea bed.

through the air-water interface, of which the transmission coefficient is τ, strikes the bottom at point C, after undergoing absorption and diffusion between S_1 and C where it is reemitted in every direction, especially in the direction CS_2. We call $f(\lambda)$ the coefficient giving the luminous flux reemitted in the direction CS_2 by the sea bed; $f(\lambda)$ depends on the nature of the bottom. We use $a(\lambda)$ and $b(\lambda)$ to denote the values of $f(\lambda)$ for the sand and seaweed, respectively. Along CS_2, of length p, this beam is reabsorbed and diffused by the water layer with the same coefficient of extinction $c(\lambda)$ as on the path SC_1. We use dS to denote the area illuminated both at the sea surface and on the sea bed and ds to denote the image of dS on the film in the camera. After some simplifications (2) we can find an expression for the illumination of the film. We obtain for the sea bed

$$E_f = E_o \frac{dS}{ds} f\, T^2 T^2 \tau^4 e^{-2pc} \qquad (1)$$

where f = a for sand and b for seaweed, and for the sea

$$E_m = E_o \frac{dS}{ds} T^2 \mathcal{T}^2 (r^2 + 1) \qquad (2)$$

where T and $\mathcal{T}$ are, respectively, the transmission coefficients for the optics and the filter and are functions of wavelength. E_o is the solar illumination; it is a function of wavelength, the height of the sun above the horizon, and the time of the survey. r is the coefficient of reflection of the sea. These parameters are also functions of the angles i and θ.

With the help of a microdensitometer we can measure the illumination of the film (1). From the formulae above we can see that (i) from the same value of f, for example from every surface covered by sand, we can draw isobathymetric curves and (ii) the difference between sand and seaweed can be established by selected illumination levels. The results obtained will be mentioned below (see section 3.2).

2.3 The taking of aerial surveys

A multi-spectral image may be obtained by using an array of cameras, each with a suitable combination of film and filters for its own spectral channel. One then has to make the best compromise possible between the height of the aircraft, the angle of field (on which depends the degree of solar reflection on the photograph as a function of the height of the sun and the state of the sea surface), the film-type and development used which determine the sensitivity as a function of wavelength and grain size, the aperture and speed of the exposure. It is necessary to take steps to ensure geographical registration of the photograph and accuracy of navigation.

2.4 The processing of the data

Possibilities include optical processing, automatic numerical processing including digitisation of photographs, transformation from film opacity to illumination, corrections to the data, contouring, algorithms for the retrieval of sea parameters and special processing for heavy swell. The method of presentation of processed data also needs to be selected carefully.

2.5 Interpretation

This involves converting the processed results into conclusions and recommendations that are intelligible to, and usable by, the body which commissioned the study.

This presentation illustrates the quantitative possibilities of remote sensing with a concrete example, that of the dispersion of a fluorescent tracer rhodamine B. For that it is necessary (i) to satisfy oneself that there is a good signal/noise ratio in the scene in selecting the most suitable spectral bands, (ii) to use a good digitising system, (iii) to adapt the processing to the photographs obtained, so as to make the best corrections for radiometric variations, spatial and temporal variations, independent of the phenomenon studied, and (iv) to provide oneself with a simplified mathematical representation of the interaction of the solar radiation with all the media that it encounters before its arrival on the photographic emulsion (rhodamine molecule, sea, air, instrument).

An appropriate experimental procedure enables one to study the evolution of the quantity of rhodamine on charts from photographs alone to an accuracy of the order of 95%, without it being necessary to take in situ samples. We can add to these quantitative possibilities the special qualities of instantaneous synoptic vision and say that remote sensing is thus an excellent technique for the study of aquatic phenomena which are changing rapidly in time and space. Moreover, in the spectral domain of the penetration by the light into the sea, it measures the quantity of matter integrated over the depth of penetration, giving direct access to the flux of material and not just point concentration measurements such as was the case previously with in situ sampling. An optimal utilisation of the aircraft enables one to make this method competitive from the point of view of cost by eliminating the need for expensive field work.

3. EXAMPLE: A STUDY DIRECTED AT THE DESIGN OF A NEW SEWAGE OUTFALL FOR THE TOWN OF ANTIBES

3.1 Background

As a result of the continuing increase in its population, the town of Antibes needs to improve its sewage outfall system and I shall now describe a study using remote sensing to assist in the design of a new system (3). This is a sequel to an earlier study carried out for Propriano (1,4).

At present there is an outfall which discharges at a distance 300 m offshore from Pointe Bacon at a depth of 22 m, see Figure 2. But the bacteriological analyses carried out by the C.I.P.A.M. (Cellule d'Intervention contre la Pollution dans les Alpes Maritimes) since 1976 show a level of pollution which, without being important, is sometimes on the limit of the European standards on the Plage de la Garoupe near the discharge of the outfall (see Figure 2). To improve the quality of the water along the shore, the town commissioned a study of the marine environment in the area concerned, with the object of defining

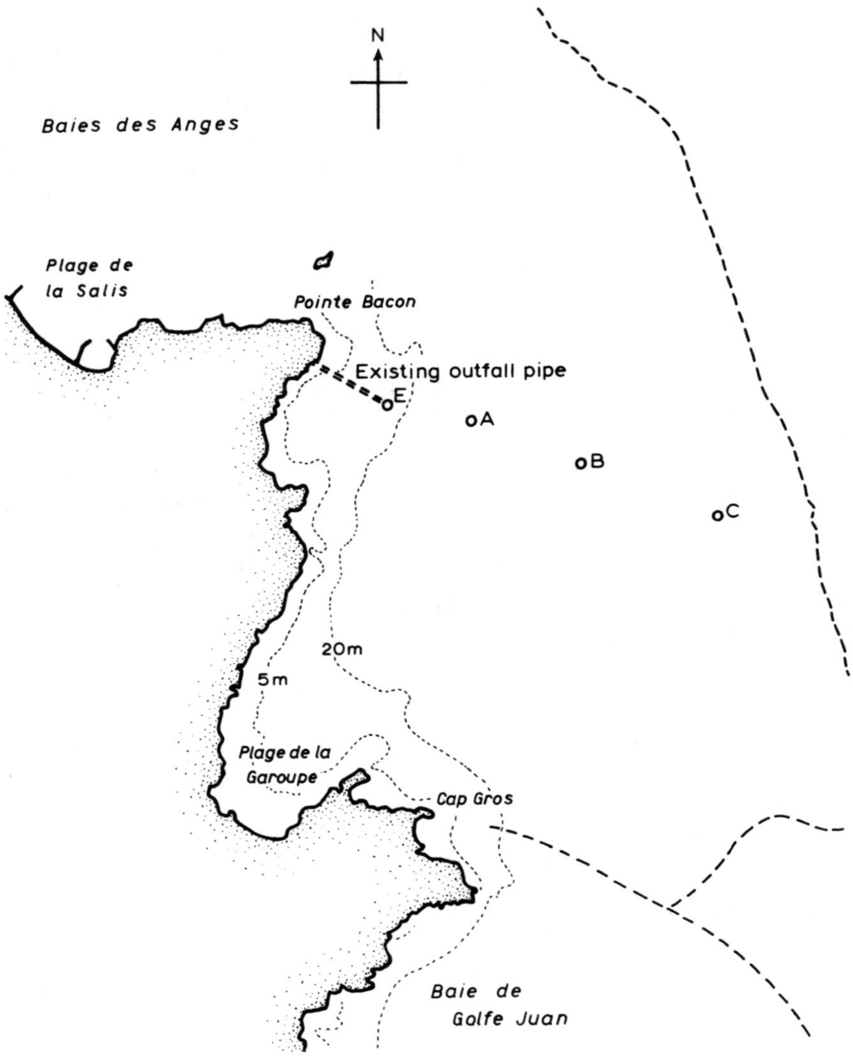

Figure 2. Sketch map of the area in question, E is the existing point of discharge and A, B and C are possible candidates for the new point of discharge. The broken lines represent natural features on photographs of the sea.

the minimum distance at which the future discharge point should be located.

With a lower density than that of the seawater, a discharge of sewage water rises rapidly to the surface and takes its place in the surface layer consisting of the first few metres. This rising can eventually be hindered by the summer thermocline (vertical stratification of the temperature) and then the discharged water becomes situated under this thermocline; thus it is necessary to know the subsurface currents at that depth.

The total dilution can be expressed as $D = D_1 \times D_2 \times D_3$. D_1 is the dilution in the upward plume, which is fairly well known for seas in which the vertical temperature stratification does not prevent the effluent from rising to the surface; D_1 depends essentially on the location and depth of the discharge point, on the speed of emission and on the existence of the thermocline. D_2 is the surface dilution due to turbulent diffusion and differential displacement of water masses; it depends essentially on the local current patterns at the surface. D_3 is bacterial dilution due to the purifying power of the sea water. D_3 is difficult to deal with; it is related to the primary production of zooplankton and phytoplankton, to the salinity and temperature of the water, to the solar illumination and to the agitation and settling of suspended materials. The meteorological conditions affect D_1 indirectly, via the existence or otherwise of the thermocline. They affect D_2 and D_3 more directly since it is fairly well established that the dynamics of the water masses at the surface are strongly correlated with the meteorological conditions. Photographic remote sensing provides information on these dynamics.

It is also necessary to establish by a study of the nature of the sea bed and its populations that the projected outfall will not destroy the existing benthic equilibrium.

The methodology used in this study involves

(a) A climatological study to identify the most frequent weather conditions and the weather conditions that are the most dangerous for the pollution of the beaches.
(b) The study of the nature and coverage of the sea bed.
(c) The study of the current at the surface.
(d) The study of dispersion at the surface.
(e) The study of the thermal stratification and of the currents at the sea bed at the point proposed for the discharge from the outfall.

Photographic remote sensing is involved in (b), (c) and (d).

With regard to (b) the nature of the sea bed in the vicinity of Antibes has been studied with the aid of side-scan sonar. However, since the results obtained were encouraging, we shall consider in section 3.2 a study of the feasibility of cartography of marine plants on the sea bed by remote sensing. It is now admitted that all sewage discharge adversely affects the plant life, and thus the end of an outfall must be situated beyond the region of the plant life.

With regard to (c) and (d) a climatological study carried out by the Centre Météorologique de Nice with data from the lighthouse of the Garoupe at Antibes showed a relatively constant division of wind conditions:

55.5% of breezes blowing from the sea towards the land (from the south) around midday and from the land towards the sea (from the North) the rest of the day

13% from the east

18% from the west

13.5% without wind.

A study of the currents has been carried out for each of these different wind conditions. An experimental study has been carried out to measure the surface dilution for the most dangerous conditions (from the point of view of a risk of return towards the coast), namely a day without wind.

3.2 Feasibility study of the cartography of marine plants

This has been carried out on a nearby marine zone, the Baie de Golfe Juan.

The coastal strip of the Mediterranean is strongly coveted for very diverse uses, going from the development of aquaculture to the large industrial and port complexes. The choice for defining these future uses depends on the actual ecological state of this littoral strip. The best indicator of this ecological state seems to be the submarine plants especially seaweeds. In effect they permit the fixing of the sea bed, the oxygenation of the water, they provide the means of support of all the coastal fauna. Their degradation would lead, in the end, to the suppression of all coastal maritime life, that is to the abandonment of aquaculture, coastal fishing and tourist activities. Conscious of the importance of these plants the C.I.P.A.M. have undertaken to prepare their cartography and to follow their evolution. For that purpose two methods have been tested up to now. The first is the use of side-scan sonar which provides such cartography between depths of 5 m and 28 m; but it does not distinguish between the state and nature of the plants and moreover the cost is very high. The second is photo-interpretation from colour photographs; this covers at most the range from 0 to 10 m depth in the water. Neither method will give satisfaction entirely (5). The present study proposes to identify the possibilities of photographic remote sensing, of which the final objective is to provide cartography of the different classes of object which one finds at the bottom of the sea, as far as possible into the deep water.

Let us consider briefly the passage of light from the Sun to the photographic film in the camera on the survey aircraft, see also section 2.2. We assume that the Sun provides plane parallel uniform illumination over the area that is being viewed. Differences in the intensity of light recorded by the film then arise from (i) differences in the depth of the water and (ii) differences in the reflectivity of the sea bed. The differences between the reflectivity of vegetation and that of sand is quite marked. From a time sequence of photographs it is possible to draw bathymetric contours and to map the regions of vegetation and of bare sand on the sea bed in shallow water. The results that have been obtained correspond well with in situ data for test areas and we are following this work through to the production of an operational method.

3.3 Study of currents

Three possible positions for the new discharge point have been tested, situated on the line of the existing outfall, at 600 m, 1,000 m and 1,500 m from the coast, see Fig. 2. In the course of a day of experimental work, we have set down at each of these three points, at several minutes separation (the time for a fast boat to go from one to another), three things:

(i) a raft of area 2 m x 2 m, fitted with a keel of 20 cm, which gives the speed of the surface currents and experiences very little direct effect from the wind.

(ii) a raft of 2 m diameter, constituted of a cylinder of 20 cm height, of which the iron framework is covered with a white fabric. With a density greater than that of sea water, it is placed at a depth of 2 m and maintained at that depth by a very small surface buoy, which is not very significantly affected by the surface current. This gives the current at a depth of 2 m.

(iii) 3 or 6 litres of a solution of rhodamine in alcohol with a density 10% less than that of sea water, which gives the mean speed of the body of water in which it diffuses ($\sim$ 3 m). Its diffusion gives information concerning the dispersion of the mass of labelled water.

This three-fold drop is carried out several times in the course of a day of experimental work.

From the paths followed by these different objects, one can deduce the directions and speeds of the relevant masses of water. Each path is extracted from a sequence of vertical aerial photographs taken at a frequency governed by the speeds of the currents and which is about once every 10 minutes. On the photographs there are differences in the colour of the surface of the sea and these help in the general interpretation. The photographs are taken with a colour filter which increases the contrast between the spot of rhodamine and the sea, and this then makes it possible to follow the rhodamine for longer.

Figure 3 presents a montage of photographs taken under conditions of a west wind. One can see spots of rhodamine dropped at different times and the discharge from the existing outfall. The coastal current gathers up this discharge and flows opposite to the general circulation offshore; one can see that this current does not penetrate into the bay of the Garoupe.

Figure 4 presents the detail of a drop under conditions of sea breezes. One can see there a rather fine time evolution of the two-horned form of the rhodamine spot. The winds on this occasion were

Time	Direction	Speed (m/s)
08.00	N	2
09.00	NE	1.5
10.00	E	0.5
11.00	SE	0
12.00	SE	0.5
13.00	S	0.5
14.00	S	1.5
15.00	S	1.5

The currents at the surface and at a depth of 2 m deduced from the successive positions of the rafts are also shown in the diagram. The emergence of the two-horned shape of the spot of rhodamine can be understood in terms of these two currents. To drop the rhodamine initially one breaks a bottle at the surface and much of the rhodamine remains very near the surface, but some of it is dragged down below the surface to a depth of the order of 2 m by the falling pieces of broken glass. Similar results were obtained with the rhodamine and rafts released at B and C.

The broken line that was marked in Fig. 2 is a natural feature visible on the photographs on this day and it represents the boundary of the large coastal current of the Côte d'Azur,

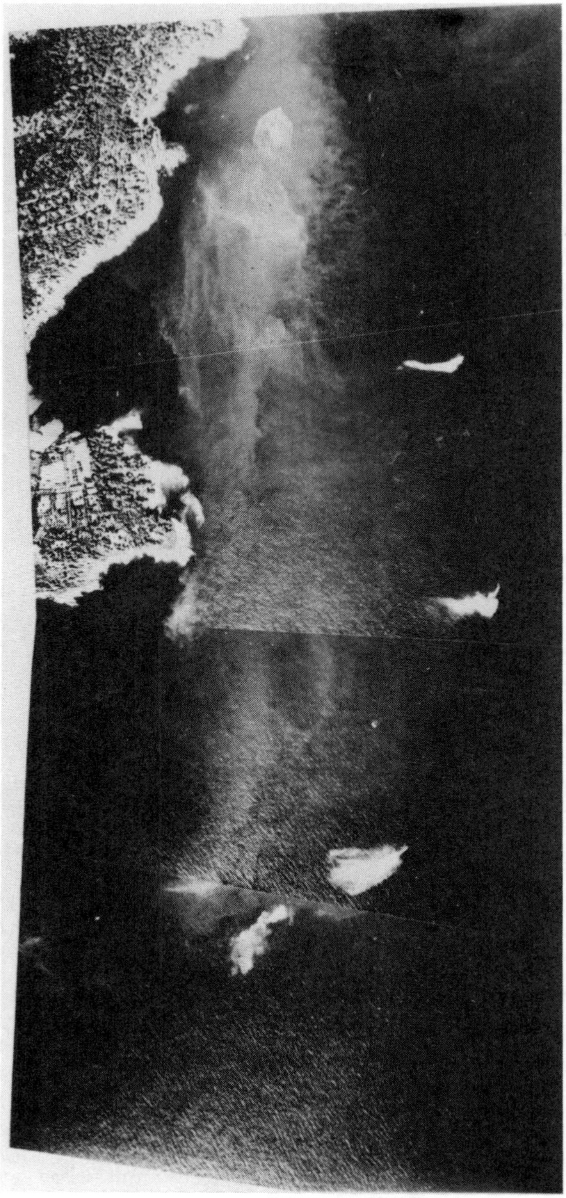

Figure 3. A montage of photographs in a condition of west wind.

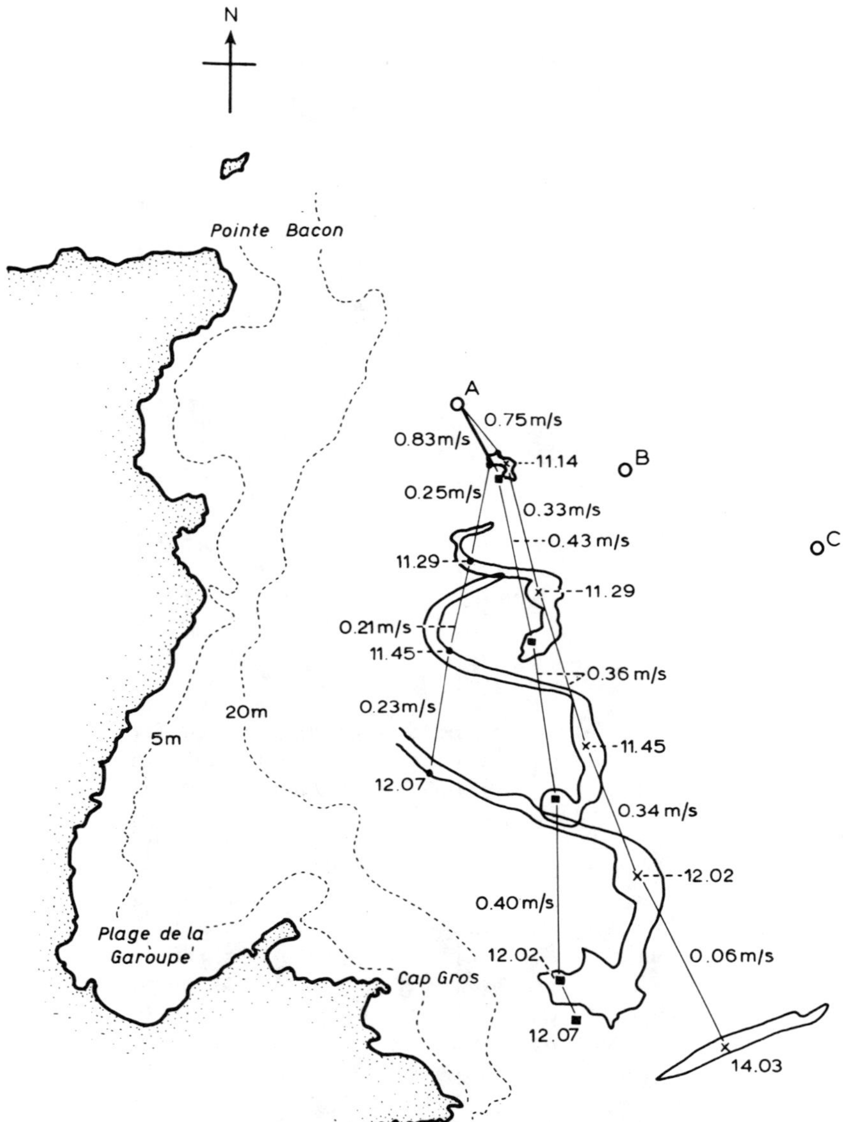

Figure 4. Movements of rafts and time evolution of a spot of rhodamine released from point A at 11.10; the positions of the point of maximum concentration of the rhodamine (X), the surface raft (■) and the raft at 2 m depth (●) are indicated, together with the calculated currents.

the Ligure, the speed of which is much greater than the speeds measured nearer the coast. Other natural features visible on the photographs (see also Fig. 2) correspond to the boundary of two water masses, one coming from the North (Baie des Anges), the other coming from the south-west (Baie de Golfe Juan) which meet off Cap Gros and flow towards the deep water and cause the boundary to the Ligure to be pushed towards the east.

Figure 5 shows a summary of the current patterns at the surface on different days corresponding to various different wind conditions. The large open arrows indicate the general circulation and the broken line represents the boundary of the Ligure current. The existing discharge and measured currents from the candidate positions A, B and C for the new discharge point are shown. One can see that the speeds of the currents at the passage of Cap Gros are relatively high for the Mediterranean; one can also see the importance of the Ligure current and of the topography in the formation of currents along the coast in the area in which we are interested.

In the light of these results, it appears that the point A at 600 m does present some risk of a return of discharged sewage at Cap Gros, but that there is no such risk for point B at 1,000 m.

3.4 Measurement of the surface dilution

To study the surface dilution of sewage discharged from a future outfall we have recourse to the use of a continuous discharge of rhodamine B at a point under consideration as a possible discharge point for the new outfall. For the experimental study we produced a continuous discharge for a period of five hours at such a point and under meteorological conditions judged to be the most unfavourable, that is from the point of view of the possibility of the return of sewage to the beaches. On the basis of the results of the studies of currents mentioned in the previous section, a point 800 m offshore (i.e. between A and B) was chosen and the experiment was carried out on a day of fine weather with a vertical anticyclone and very little wind. The rhodamine is regarded as a tracer for the masses of water in which we are interested. The hypothesis currently in vogue is that the discharged sewage follows the movement of the masses of sea water and disperses like the tracer used. This hypothesis is certainly valid for dissolved material, and for fine suspended particles carrying pathogenic germs but it is more open to doubt for larger particles of suspended matter and for oils. However, all the treatment of the sewage before discharge has the effect of considerably reducing the oils and suspended materials and the hypothesis can then be considered valid for the ensemble of the constituents.

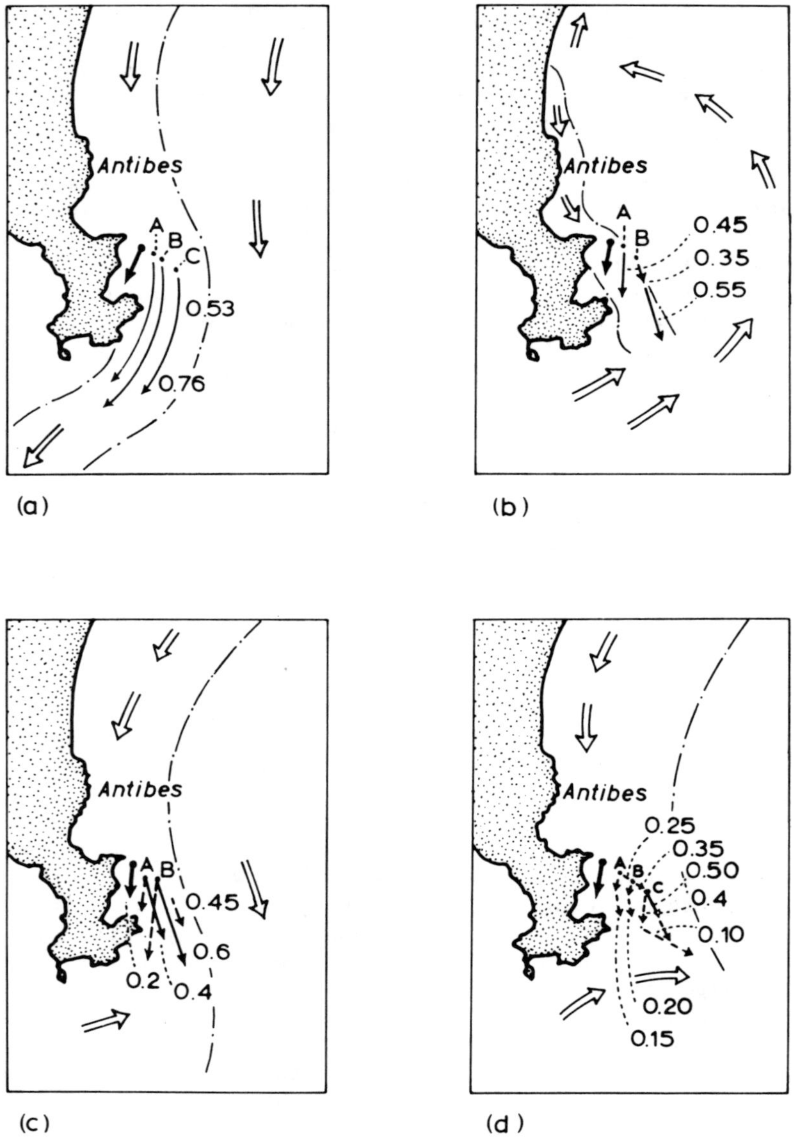

Figure 5. Surface current patterns for (a) 13 June 1980, east wind, 7 m/s, (b) 10 October 1980, west wind, 5 m/s, (c) 3 June 1980, morning breeze, NE, 1.5 m/s and (d) 3 June 1980, afternoon breeze, SE, 1.5 m/s; currents are given in m/s.

The evolution of the continuous discharge was followed all through the day by photographic remote sensing. At regular intervals survey aircraft passed vertically overhead and photographed the slick of rhodamine and part of the coast (in order to locate precisely the position of the plume with reference to the coast). 83 photographs were taken on that day; the photographs were taken and processed in the manner explained previously (1).

One thus obtains for each photograph the mapping of the spatial distribution of the rhodamine giving, at each point, the quantity of rhodamine in the whole depth of the plume. The evolution of these maps gives the evolution of the flux of rhodamine in the course of the day. The interpretation of the maps consists in seeing if there is a return to the coast and to extract from it the minimum time of return and the surface dilution between the discharge point and the coast.

We consider first the movement of water masses. Figure 6 shows some of the results from data gathered on 25 September 1980. From Figure 6(a) we see that before 13.00 the continuous discharge flows freely towards the north. Successive positions of a point discharge, released at 12.10 show that the surface current was flowing almost due N at about 0.23 m/s - 0.25 m/s. Surface rafts and rafts submerged at 2 m depth released near the discharge point also indicated currents flowing due N with slightly lower speeds ($\leqslant$ 0.22 m/s) both at the surface and below the surface. Further north the current pattern changes. A surface raft launched at 11.45 followed the curve I in a north-easterly direction from 12.13 to 12.55 and its speed decreased gradually from 0.17 m/s to 0.06 m/s. A submerged raft, also released at 11.45, followed the curve II in a more northerly direction from 12.13 to 12.55 and also suffered a similar decrease in speed from 0.17 m/s to 0.07 m/s. In the period 12.13 to 13.00 the visible line representing the boundary of the Ligure current moved in a south-westerly direction at about 0.08 m/s. Further north and east other rafts which had been dropped earlier, at 10.45 and 9.45, were observed to be moving even more slowly, with speeds as low as 0.04 m/s.

In the period following 13.00 the current flow at the point source of the continuous discharge of rhodamine is very complicated; it is illustrated schematically as follows:

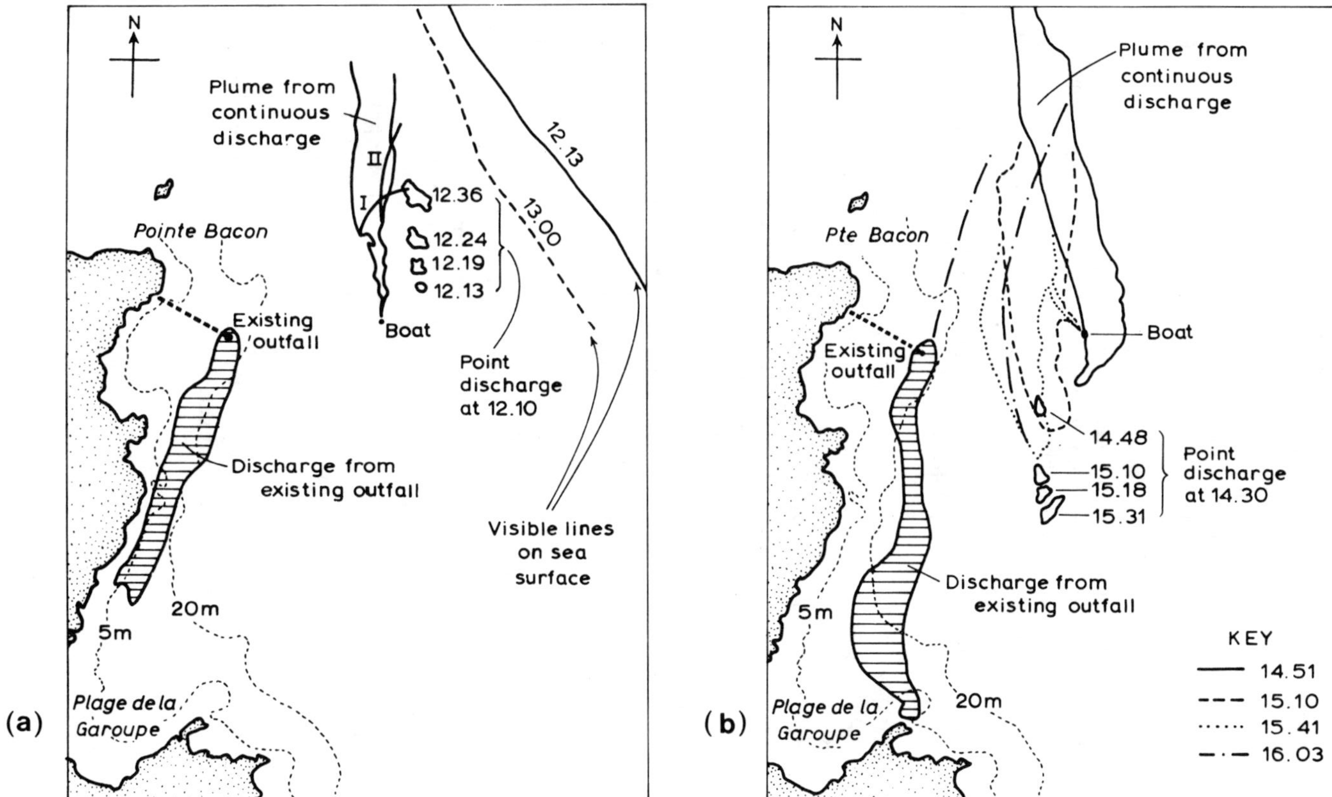

Figure 6. Evolution of the plume from a continuous discharge and a single point discharge, 25 September 1980, (a) between 12.00 and 13.00 and (b) between 14.00 and 16.00.

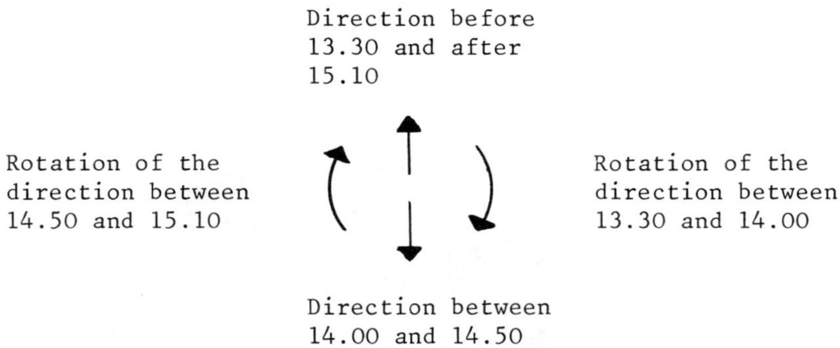

This behaviour can be interpreted in terms of the effects of the flow from the deep water approaching Pointe Bacon with a mean speed of the order of 0.10 m/s. During the period in which the current direction at the point of release of the continuous discharge is turning around, the point discharge at 14.30 between the boat and the coast at about 600 m from the coast goes from north to south with a speed which decreases gradually from 0.16 m/s to 0.09 m/s. Like the south point of the continuous discharge plume, it seems to be entangled by the coastal current which carries the water from the existing outfall. On the other hand the rafts near to the boat just moved around and meandered randomly and it did not seem interesting to us to follow their successive positions, except for two surface rafts which maintained until 14.46 a movement towards the north, but with a very low speed of 0.03 m/s.

To summarise, the vertical anticyclone on that day, produced very little wind at ground level ($\leqslant$ 1 m/s and with a variable direction) and this seemed to have little direct influence on the currents. The immobility of the air also favoured the stagnation of a layer of mist, and it was necessary to correct for the presence of this in the photographs. All during the day the plume of waters discharged from the existing outfall directed themselves towards the Plage de la Garoupe giving evidence of a coastal north-south current. On the other hand the continuous discharge from the trial point 800 m from Pointe Bacon flowed in the opposite direction (south to north) until 13.30.

We now turn to the determination of the surface dilution. To illustrate the method let us take the example of the photograph of 12.36 (see Figure 7) and consider the calculation of the surface dilution for a point M on the cut DOO at the extremity of the plume, at about 1.1 km from the discharge point. The processing involved in digitising and calibrating a photograph like this has been described previously (1). From such processing one can extract a contour map of the luminous flux due to the

Figure 7(a). Photograph taken at 12.36 on 25 September 1980.

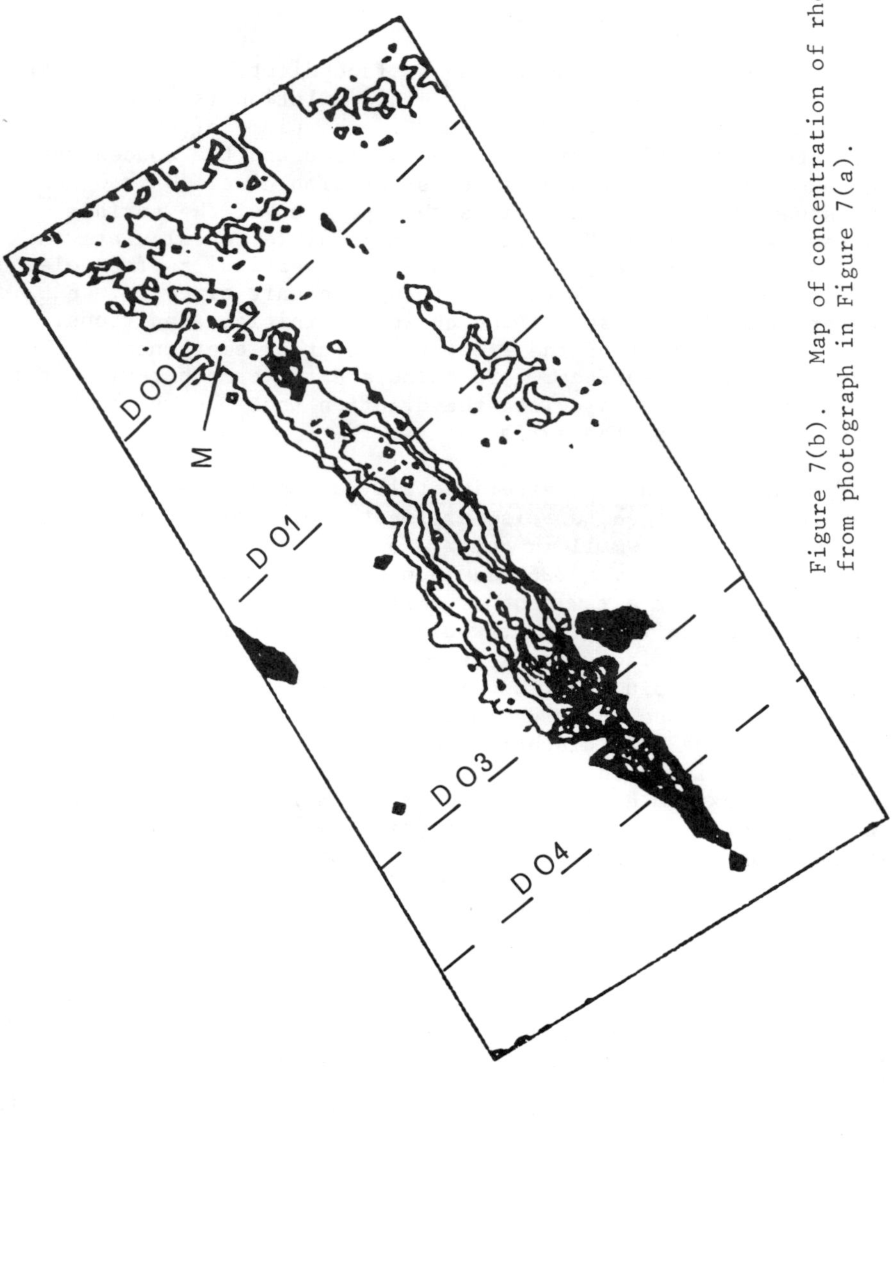

Figure 7(b). Map of concentration of rhodamine obtained from photograph in Figure 7(a).

fluorescence of the rhodamine and hence a contour map of the quantity of rhodamine (q). The quantity, q_M, thus calculated at a point M in the plume is, of course, the value integrated over the whole depth of the plume at this point:

$$q_M = \int_0^h C(z)\, dz \text{ in gm/cm}^2,$$

where C(z) is the concentration.

What we want to find is the surface dilution D_2, not just the concentration of rhodamine. This dilution is the ratio of the maximum concentration C_o found in the plume of the rising of the waters at their arrival at the surface and the concentration C_M found at the point M under consideration, i.e. $D_2 = C_o/C_M$. The study of the currents gives the speed u_M of the surface current at the point M. It is then possible to calculate the flux of rhodamine, F_M, which crosses the vertical surface element of width $d\ell$ and depth h (normal to u_M) in unit time, $F_M = q_M \underset{\sim}{u}_M . d\underset{\sim}{\ell}$. The problem then arises of the choice of initial conditions, that is to say that of the F_o at the point of emergence of the discharge on the surface. To calculate the concentration at this point one needs to calculate the dilution that occurs as the discharge rises to the surface.

When one simulates experimentally a continuous discharge the diameter of the tube for discharging the alcoholic solution of rhodamine is very small in comparison with the diameter of the future outfall pipe. But the rhodamine becomes diluted very rapidly and after a few tens of metres it occupies a volume comparable with the volume that would be occupied by the discharge rising to the surface from a future outfall. We choose a position in the plume of rhodamine at which the width of the plume is equal to the calculated width of the plume from the future outfall just as that discharge would reach the surface. In this example this is at the location D04 in Figure 7. We then take as q_o the maximum quantity of rhodamine across D04 and with u_o, the speed of the current at this point, we obtain $F_o = q_o \underset{\sim}{u}_o . d\underset{\sim}{\ell}$. In consequence we now take the surface dilution D_2 to be

$$D_2 = \frac{F_o}{F_m} = \frac{q_o \underset{\sim}{u}_o d\underset{\sim}{\ell}}{q_M \underset{\sim}{u}_M d\underset{\sim}{\ell}} = \frac{q_o u_o}{q_M u_M} .$$

Knowing all the quantities q_o, q_M, u_o and u_M we can calculate D_2. In the example chosen here we found at a point M on D00, about 1.1 km from the discharge point, $q_M = 2 \times 10^{-6}$ gm cm^{-2} and at D04, $q_o = 4.6 \times 10^{-5}$ gm cm^{-2}. If one assumes $u_o = u_M$, which is at least a reasonable first approximation, then the surface dilution is given by $D_2 = (4.6 \times 10^{-5})/(2 \times 10^{-6}) = 23$ at the point M.

The above calculation illustrates the calculation of the surface dilution but it is not actually of very great interest in practice because this plume is not directed towards the coast anyway. Only the very short period from 15.00 to 16.00 sees the plume approaching the coast, but even then there is little risk of it reaching there directly. The southern part of the plume, at the limit of the coastal current, travels so as to pass well clear of Cap Gros, but with a rapidly decreasing speed. The northern part is again travelling towards the north without any risk of returning to the coast. The central part expands towards the position of the discharge from the existing outfall. The photograph of 15.41 (Figure 8) shows that this part of the

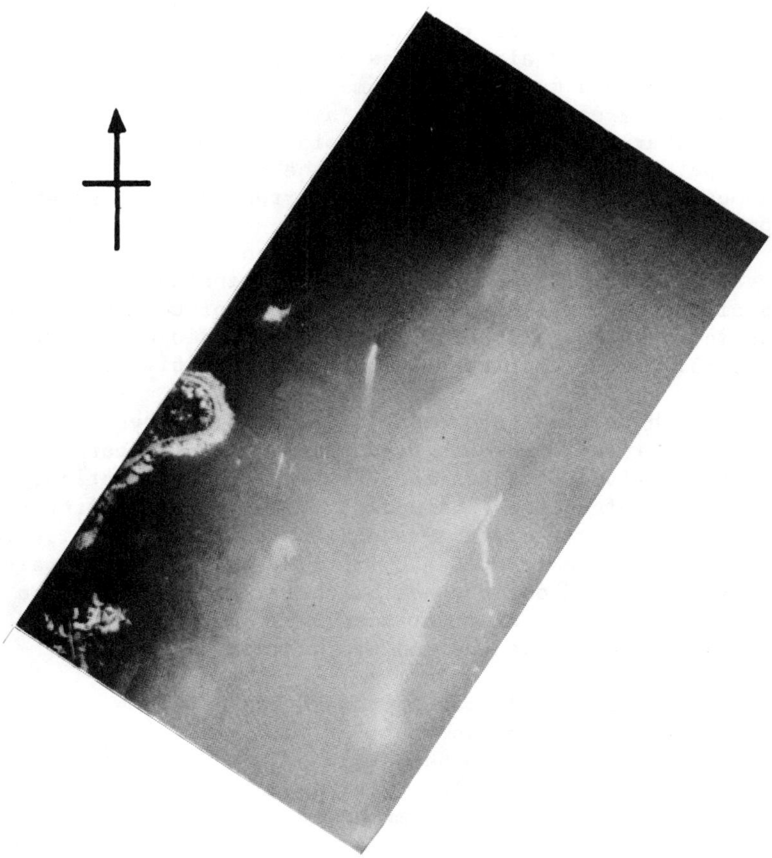

Figure 8. Photograph taken at 15.41 on 25 September 1980.

plume of rhodamine is about 400 m from Pointe Bacon. It is heading towards Pointe Bacon but with a surface dilution of the order of 10, the major part of the rhodamine remaining along a north-south line at about 200 m west of the boat. Towards 16.00, this line of dilution 10 reaches the extremity of the existing discharge and will be drawn by the coastal current towards the Plage de la Garoupe.

3.5 Conclusion of Study

To summarise, one must establish from this experimental work using airborne remote sensing that in the period during which photographs were taken from 10.00 to 16.00, the rhodamine which was released in a continuous discharge from 9.45 to 15.00 does not reach the coast.

In the morning the plume is directed towards the north, where the coast is very far away, and it possesses a dilution of 23 at a distance 1100 m along its axis from the discharge point. Nevertheless the rotation of the plume towards the west provides a weak possibility of return, of only a small part of the rhodamine discharged between 14.00 and 15.00, towards the Plage de la Garoupe, but with a dilution, at the existing discharge point, about ten times greater than is currently the case and corresponding only to one hour of risk during the day. Moreover, it must be remembered that this type of meteorological condition only prevails 13.5% of the time. Finally the studies of current patterns, mentioned earlier, have shown that for the rest of the time with a discharge point situated more than 800 m away from the coast, the discharged water will travel directly out to sea.

From all these results one can conclude that beyond 800 m offshore one is in a zone which presents a risk of return to the coast, on the surface, that is negligibly small in practice. The Town Council of Antibes has opted for a discharge point at 1000 m, taking thus a reasonable additional safety margin. The checks carried out weekly by the C.I.P.A.M. on the chosen part of the coast concerned ought to allow the verification of the sound basis of this choice, after the construction of the new outfall, that is to say at the end of this year.

It should be recalled that, although these remote sensing studies using aerial photography are of very great importance, they are not the only ones to be involved in the choice of the discharge point. In particular what happens between the sea bed and the surface is also very much involved (possible blockage by the thermocline, value of D, etc.).

4. THE USE OF TWO DIFFERENT FLUORESCENT TRACERS

I should like to mention that very recently we have started some work in which we make use of the simultaneous release of two fluorescent tracers, rhodamine B and fluoresceine. This is particularly valuable when two potential sources of pollution are present and the example being studied just now involves the discharges into the sea from the small river la Cagne and an existing outfall with a discharge point at a depth of 8 m at a point 200 m offshore. A simple measure of the pollution on the nearby beaches was insufficient to discriminate between the two possible sources of the pollution. The experiments under way just now involve releasing fluoresceine into the river and rhodamine into the sewage outfall and studying the evolution of these discharges from a quantitative study of multispectral aerial photographs. In this way it is possible to distinguish nicely between those parts of the beach that are polluted by the river water and those parts that are polluted by the sewage from the outfall, see Figure 9.

5. CONCLUSION

I would remark that from the point of view of gaining knowledge of the state of the surface of the sea, also of the sea bed in shallow water, and of the physico-chemical properties of the layer of water above the sea bed, the use of visible radiation is particularly well suited to coastal oceanography. Moreover, the possibility of synoptic and repetitive cover by airborne remote sensing, as required, provides a technique that is particularly adapted to the study of this marine environment which is capable of such rapid variation. Finally the possibilities of multispectral aerial photography make it an incomparable instrument for the collection and quantitative evaluation of data. We have discussed the use of this technique in connection with the monitoring of pollution and design of new outfall systems, but it can be put to work on other subjects of study in coastal oceanography, such as the cartography of shallow sea beds, the quantitative inventory of existing pollution, the transport of sediment etc. Finally, the system in question is an excellent tool for simulation and an aid to understanding for future satellite programmes.

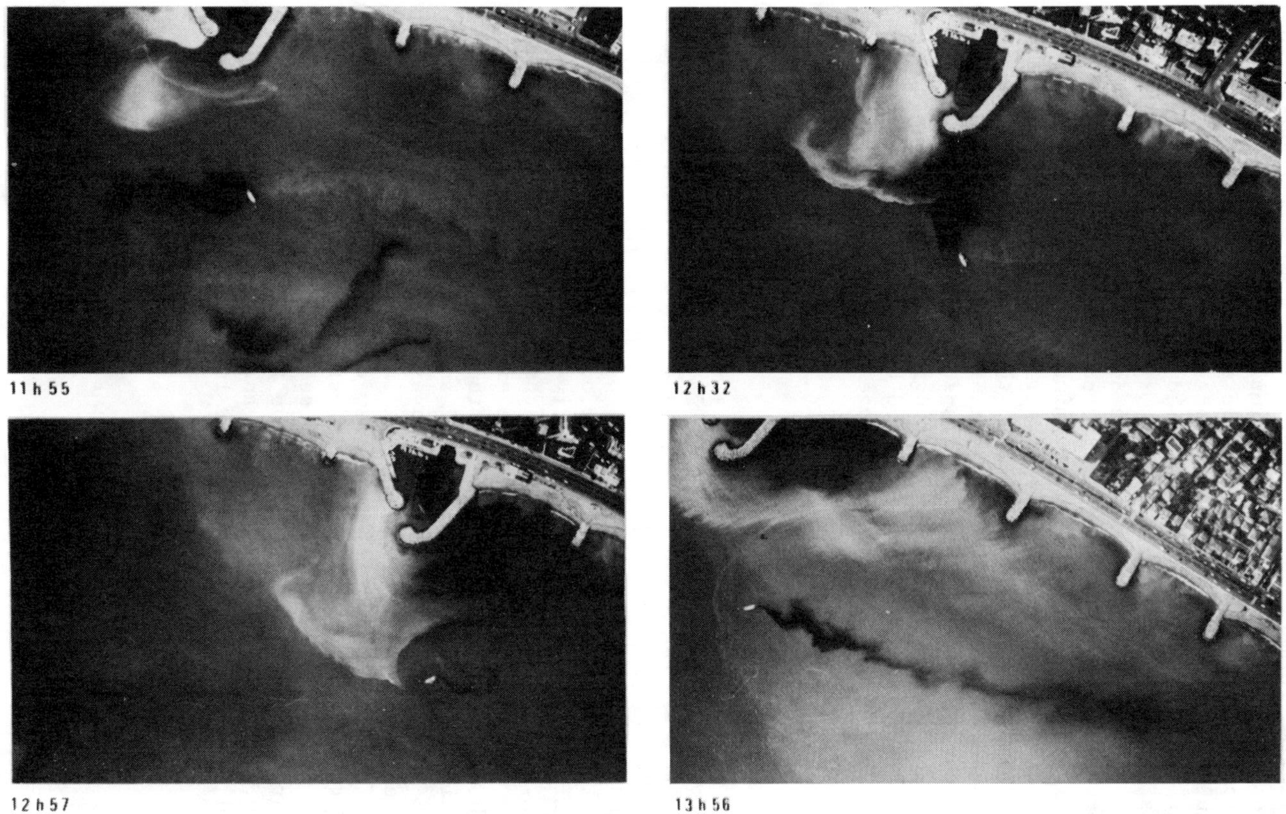

Figure 9. Cagnes sur Mer, 29 October 1981, showing the evolution of the plume from continuous discharges of rhodamine from the boat (black plume) and fluoresceine at the mouth of the river (white plume).

REFERENCES

1. Valerio, C.: in "Remote Sensing in Meteorology, Oceanography and Hydrology", ed. A.P. Cracknell (Ellis Horwood, Chichester, 1981), chapter 14.
2. Lefevre, T.R. and Valerio, C.: December 1981 "Etude de faisabilité de la cartographie des herbiers marins par télédétection photographique" pub. C.E.T.E. d'Aix pour le compte de la MIAPENM.
3. Valerio, C.: "Traitement et rejet des eaux usées de la Ville d'Antibes; étude d'implantation de l'émissaire en mer", pub. C.E.T.E. pour le compte de la ville d'Antibes.
4. Valerio, C. and Llebaria, A.: "A quantitative multispectral analysis system for aerial photographs applied to coastal planning", Int. J. Remote Sensing, 3, (1982), 181-97.
5. Meinesz, A. et al: "Methodes récentes de cartographie et de surveillance des herbiers de phanérogames marines", Vie et Milieu, 31, (1981), 27-34.

CHAPTER 20

OCEANOGRAPHY FROM SPACE

Dr. T.D. Allan

Institute of Oceanographic Sciences,
Brook Road, Wormley, Godalming, Surrey, U.K.

PART I PAST ACHIEVEMENTS

1. Narrative

In 1964 a symposium was held at the Woods Hole Oceanographic Institute entitled "Oceanography from Space". At the time, the title had an incongruous ring to it. It was but a few years since the launch of Sputnik, and a community reared in the traditional methods of studying the sea from research vessels was ready to display a healthy scepticism of what could be achieved from a satellite travelling at speeds of several kilometres per second, hundreds of kilometres above the sea. True, remote sensing as a technique for studying the sea was hardly new and could probably be traced back to the look-outs of earliest sailing vessels. In more modern times acoustic remote sensing had been used with conspicuous success and the major ocean ridges and marginal deep sea trenches had already been identified and charted by means of echo sounders. In the more recent past, other remote sensing equipment such as side-scan sonar, magnetometers and ship-borne gravity meters had been developed for geophysical investigations of the sea floor which within a few years were to reveal an ordered pattern of anomalies leading to revolutionary theories of sea floor spreading and plate tectonics.

It was also recognised that although satellite-borne sensors could not match the overall accuracy of 'in situ' instruments, their coverage of the oceans afforded a synoptic view denied the surface platform. The polar oceans could be sampled more than a dozen times per day and, for good measure, information on

A. P. Cracknell (ed.), Remote Sensing Applications in Marine Science and Technology, 409–433.

all of the other oceans could also be gathered many times in the same day - recording the data on board and transmitting to a convenient ground station. By contrast, a research vessel on a cruise to the Antartic could plan to be away for several months.

There was thus sufficient interest in the future potential of satellites to persuade 150 oceanographers to gather at Woods Hole and discuss how this emerging technology might be most profitably used.

Then, as now, satellite sensors were categorised according to that part of the electromagnetic spectrum in which they operated. Sensors using the visible and infra-red part of the spectrum - such as the multi-spectral scanners ultimately deployed on Landsat - had reasonably good resolution at the surface; atmospheric effects could be reduced by combining the different bands so as to enhance the signal and suppress the unwanted contributions to it. On the other hand optical sensors were unable to penetrate cloud so that the repetitive sampling often required in scientific studies was rarely possible - especially over temperate latitudes. By contrast, microwave sensors could operate day and night in all weathers. At the time of the Woods Hole meeting, however, they appeared to lack the resolution and precision required to make a significant contribution to scientific studies.

The radar altimeter is a case in point. The altimeter monitors changes in the height of the spacecraft above the sea surface by transmitting a very short pulse and measuring the time taken for it to be reflected back from the sea surface. Variations in the difference between local mean sea level and a true level surface (the geoid) determine the barotropic, time-dependent motion which can be measured only indirectly, and over limited areas, from surface vessels if certain assumptions are made on the 'level of no motion' in the sea. It was noted at the Woods Hole meeting that for studies of the ocean's general circulation precisions of the order of 10 cm were required from radar altimeters and this was about an order of magnitude better than could then be achieved. By the time of the second 'Oceanography from Space' symposium held at Venice in 1980 the radar altimeter flown on Seasat had achieved an estimated precision of 7 cm.

2. Oceanographic Measurements from Satellites

2.1 Geodesy. One of the earliest scientific contributions made by satellites was in the field of geodesy. It quickly became clear that an imperfect knowledge of variations in the earth's gravity field caused satellite orbits to deviate from those predicted, and by tracking a number of satellites over a long period through a global network of stations, the orbit deviations could be analysed and gravity models could be constructed with an ever-increasing number of harmonics. The 'level' surface of the earth of constant gravitational potential is called the geoid. Prior to 1958 very little was known about the shape of the geoid and even maps published in 1963 show discrepancies of over 50 m compared to present day maps.

As we shall see when we examine the potential applications of precise satellite altimetry, a good knowledge of the geoid is important if we are to detect ocean currents since, of course, the geoid corresponds to the surface of the oceans after removing the effect of all external focus acting on it such as those due to tides, currents, storm surges, wind and atmospheric disturbances. Thus, if the oceans and atmosphere were motionless the geoid would coincide the mean sea level and its extension beneath the continents.

The shape of the geoid is irregular with respect to the reference ellipsoid due to gravity variations produced by an uneven distribution of mass within the earth. The separation of the geoid from the ellipsoid varies by about $\pm$ 100 m over the the surface of the earth which is two orders of magnitude less than surface relief and about two to three more than the undulations of the sea surface.

2.2 Buoy tracking. Another application of satellites to ocean studies, which is strictly not in the field of remote sensing but which nevertheless is closely associated with it, is that of tracking drifting buoys, receiving their recorded data and re-transmitting to a convenient ground station. During FGGE - a large, multi-national experiment organised under the aegis of the UN Global Atmospheric Research Programme (GARP), the French designed system ARGOS, deployed on TIROS-N, succeeded in tracking over 300 buoys in the southern oceans between 20° and 65°S during a 9-month period - thus providing considerably more information on currents in the area than had been provided from all other sources in the period up to then.

A Japanese study of the Kuroshio current using five satellite-tracked drifting buoys was reported recently. The buoy trajectories, which had been continuously monitored for periods up to 16 months, revealed a variety of mesoscale eddies and meanders. One large and unexpected cold eddy was tracked for 190 days and later confirmed by hydrographic observations. Similar studies have been carried out in the Indian Ocean by the Institute of Oceanographic Sciences.

2.3 Multi-Spectral Scanners. Nothwithstanding their inability to penetrate cloud the majority of earth surveillance satellites have carried multi-spectral scanners that operate in the visible and infra-red part of the spectrum. The NOAA series for meteorology and the Landsat series for land-use are the best known of the operational systems. Infra-red maps of sea surface temperature (of the type received in Dundee from NOAA 6 & 7) have been produced for many years now and have revealed a wealth of detail of ocean eddies, fronts, current systems and other complex patterns which could not have been observed from surface vessels. The Coastal Zone Colour Scanner (CZCS) launched in 1978 on Nimbus-7 was designed specifically to provide useful information to marine biologists and pollution detection agencies.

Biologists share a requirement of other oceanographers for a range of surface data including sea surface temperature and circulation patterns, but their special requirement is the rate of photosynthesis in the upper sunlit layers of the sea. Considerable information can be inferred from maps of the chlorophyll concentration of the water; this chemical has a strong spectrographic signature of absorption bands making it a suitable study for photometric remote sensing. Much of the sophisticated image processing techniques that have evolved in this field have been devoted not only to identifying different constituents in the surface layer of the ocean but also to separating that part of the signal contributed from contrasts in the open ocean from that part originating in the atmosphere intervening between the spacecraft and the sea surface.

From the CZCS imagery processed to date it is clear that there is considerable horizontal structure in the colour of the ocean which CZCS is capable of delineating even in the presence of very turbid atmosphere. Algorithms developed to date are found to be more accurate if designed for specific areas. Use of 'universal' open ocean algorithms appears to provide figures for pigment concentration to within a factor of two in accuracy.

At present the CZCS has been operating for almost 4 years and presumably, its days of useful life are numbered. NASA is said to have commissioned a MARK II model so, for the marine biological community, there is hope that satellites will continue to chart colour changes in the ocean. It was originally planned that the first European remote sensing satellite, ERS-1, would carry an Ocean Colour Monitor employing several more bands than the CZCS but the idea was later abandoned in favour of a microwave payload. An instrument of this type is now proposed for a follow-up to ERS-1.

3. Microwave Sensors & Seasat

Seasat was the first satellite to carry a suite of microwave sensors dedicated to the study of the sea (Fig. 1). The primary mission of the spacecraft was to prove the concept that surface winds, waves and currents could be measured to useful accuracies and in this it was largely successful, despite a fatal power failure after no more than 100 days of operation.

One fortunate aspect of Seasat's operation was the extent of controlled surface observations made to evaluate the sensor's performance. Apart from a specially-designed 'surface truth' experiment which took place in the Gulf of Alaska, a multi-ship air-sea interaction experiment (JASIN) happened to be carried out coincidentally over a 2-month period in an area between Scotland and Iceland.

Seasat's suite of microwave sensors comprised:

- radar altimeter (ALT)
- synthetic aperture radar (SAR)
- scatterometer (SASS)
- scanning multifrequency microwave radiometer (SMMR)

The main objective of the mission was to assess the concept of making useful measurements of sea surface waves and winds from an orbiting satellite. The accuracy goals placed on the Seasat sensors were as follows:

- Wind speed $\pm$ 2 m/s
- Wind direction $\pm$ 020°
- Significant wave height $\pm$ 50 cm
- SAR resolution 25 m over 100 km swath

Studies carried out both before and after Seasat's launch have shown that if these accuracies could be routinely achieved by operational marine satellites then substantial benefits should accrue to a variety of commercial interests summarised in Table 1.

3.1 Radar altimeter. Wave height was measured by the altimeter. On 8 satellite passes over the JASIN area a pitch/roll buoy was deployed from RRS Discovery and a comparison made by Webb (1981) between buoy and satellite estimates showed agreement better than $\pm$ 20 cm for significant wave height (H_s) values up to 2.5 m.

The altimeter averaged 1000 values of H_s to produce a value every second (corresponding to a spacing of about 7 km) along the satellite's sub-orbital track. No information on wavenumber or direction was given by the altimeter but the Seasat instrument did provide estimates of surface wind speed along the sub-satellite track to complement the scatterometer measurements over wide swaths on either side of the spacecraft.

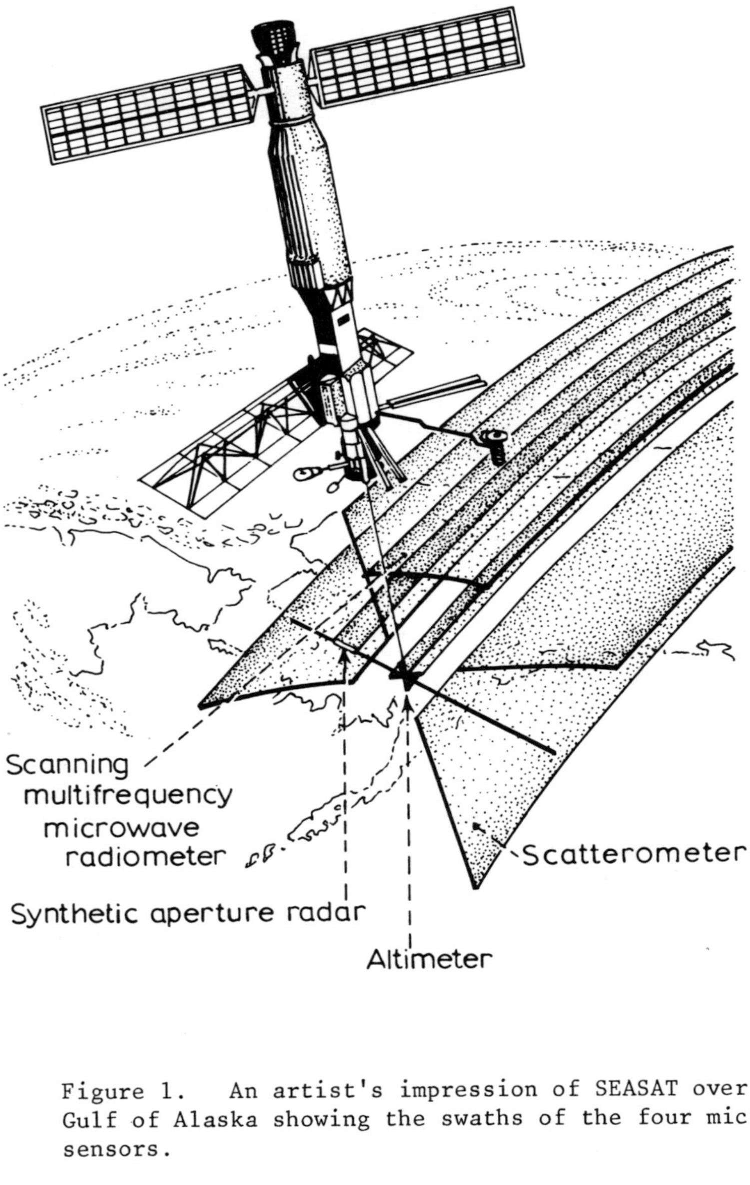

Figure 1. An artist's impression of SEASAT over the Gulf of Alaska showing the swaths of the four microwave sensors.

Table 1: ESTIMATED IMPACT OF REMOTE OCEAN SENSING ON COMMERCIAL OPERATIONS

Measurement	Primary Beneficiaries	Application
Ocean surface winds	National environmental centres	Short-range prediction
	Shipping	Ship routine/scheduling
	Deep-ocean mining	Operations/scheduling
	Off-shore oil and gas	Operations/scheduling
	Commercial fishing	Tactics/scheduling
Waves	Shipping	Ship routing
	Offshore oil and gas	Operations/scheduling
	Marine construction	Design/construction
	Deep-ocean mining	Operations/scheduling
	Commercial fishing	Tactics/safety/scheduling
Sea-surface temperature	Climate research	Monitoring/prediction
	Commercial fishing	Tactics/area selection
Sea ice	Arctic oil and gas	Operations/scheduling
	Ice patrol	Aircraft replacement

The usefulness of global monitoring of wave height was illustrated recently on the cover of Nature where a chart of 3-month average waveheight over the world's oceans was presented by Chelton et al.(1981)(Fig. 2). For all of the southern hemisphere and for the majority of oceans in the northern hemisphere it would be impossible to contour average wave heights because of the paucity of wave observations at the surface. Thus Seasat provided a brief glimpse of what future satellites will provide routinely. It is worth noting that the Seasat wave height chart shows that in the 3 months, July-October, the equatorial basin had the narrowest range of wave heights where 70% of observations were less than 2 m and none greater than 4 m; whereas for high southern latitudes only 2% were less than 2 m while 40% were greater than 5 m.

3.2 Scatterometer and SMMR. Surface wind speed was derived from the 5-frequency, dual polarisation, passive radiometer (SMMR) over a swath extending 600 km to the right of the satellite's track. A comparison with the JASIN wind data showed the SMMR winds to be biased high by about 2 ½ m/s up to the end of August at which time the bias appears to have dropped to no more than 1/2 m/s. This was borne out both in a comparison with SASS values and with 'in situ' surface observations.

By employing 2 pairs of orthogonal antennae - 1 pair pointing forward, the other backward - wind direction as well as wind speed could be estimated by SASS over resolution cells measuring approximately 50 x 20 km^2. Examples of SASS-derived, contoured wind fields, together with wind direction vectors for each resolution cell, are shown in Fig. 3. There was generally good agreement with ships' observations.

3.3 Synthetic Aperture Radar (SAR). The range of ocean surface features detected in the SAR record - many quite unexpected such as the detailed bottom topography revealed in an image of the Channel (Allan 1979) - has caused considerable speculation not only on the mechanism of the interaction between the radar emissions and the moving sea surface but also on the physical meaning of the features revealed. An analysis of the JASIN record, and others around North America, indicates that swells must exceed certain limiting values in dominant wavelength and wave height before they are imaged by SAR.

A degree of controversy surrounds the issue of the level of distortion that may be introduced into the image because of the movement of the water particles during the comparatively long time a wave remains in the SAR beam. The effect of any distortion can be shown to be more serious for waves travelling in the same direction as the satellite than for waves travelling across the satellite's track. It has been argued that under certain conditions

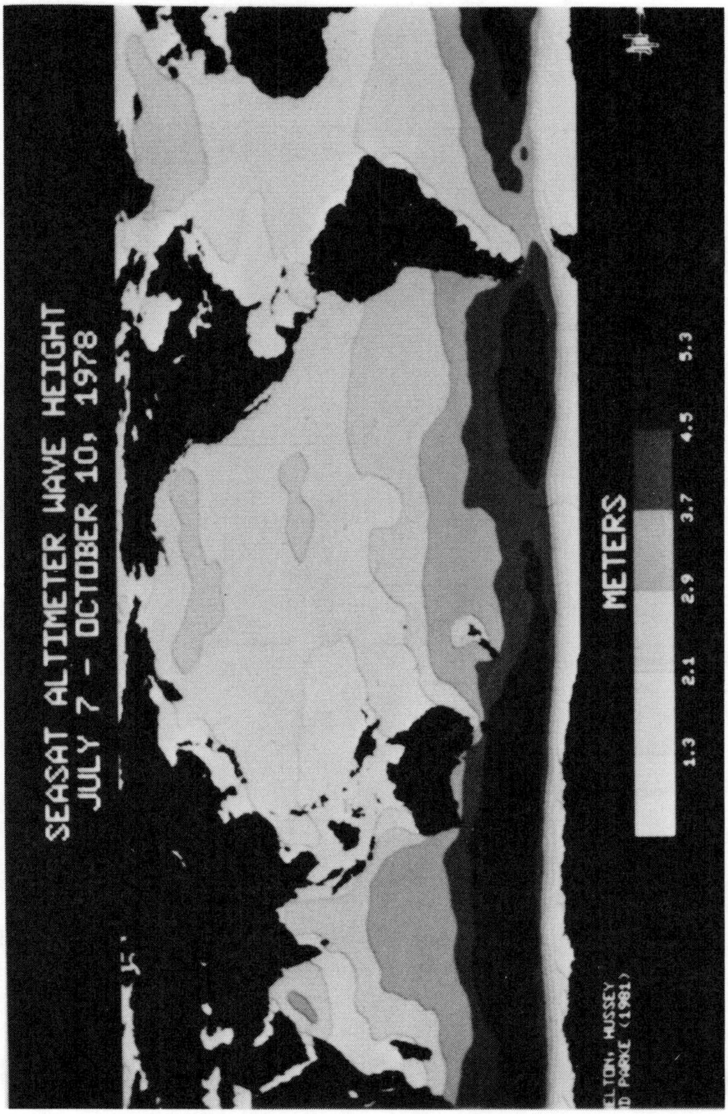

Figure 2. Global wave climate from SEASAT's altimeter (Chelton et al 1981).

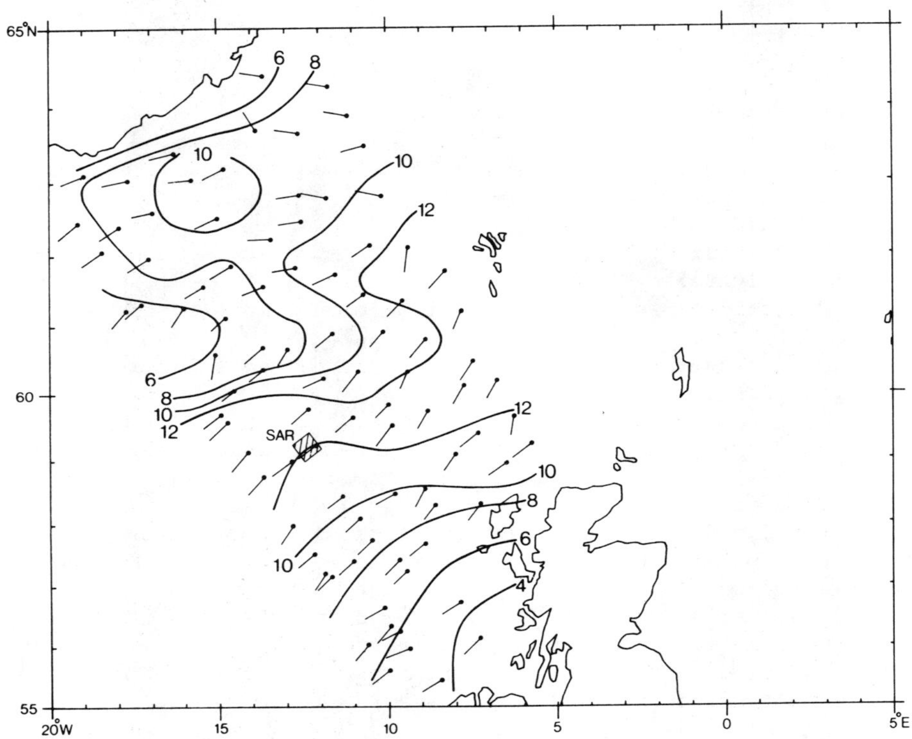

Figure 3. Contours of wind speed plotted from the vectors derived from SEASAT's scatterometer over the JASIN area.

the effect may be severe enough to smear out a significant part of the total wave spectrum. At the time of writing, the analysis carried out on the JASIN imagery has failed to resolve this question. (Allan and Guymer, 1983). Digitally-processed SAR data and digital Fourier Transforms of the record are required. These techniques are comparatively new and time-consuming; much of the SAR imagery studied to date has been optically processed and this technique, though much faster, introduces its own distortions. An example of a digitally-processed image of surface swell together with its Fourier transform is shown in Fig. 4.

What is clear from the examples of digitally-processed imagery over the JASIN area is that surface swell can be clearly distinguished. In some cases the dominant wavelengths revealed in the imagery correspond well to the period of spectral maximum measured by surface buoys, but the agreement between ships' visual estimates of wave period and SAR wavelengths is less good. This probably reflects the quality of the visual observations since the ships' records demonstrate a significant scatter.

Table 2 summarises the overall performance of Seasat's sensors over the JASIN area. It appears that, by and large, surface wind speed and direction plus significant wave height, wave period and swell direction can be measured as accurately from a satellite as from a surface platform though the extent of SAR distortion of waves is not yet resolved. Notwithstanding these few remaining doubts the extent of global coverage achieved daily by a polar-orbiting spacecraft should cause major changes to be introduced in present day forecasting techniques. Table 3 shows a comparison of the number of wind and wave observations made by Seasat compared to the number of surface reports available to the US Fleet Numerical Oceanographic Centre at Monterey during the time of Seasat's operation.

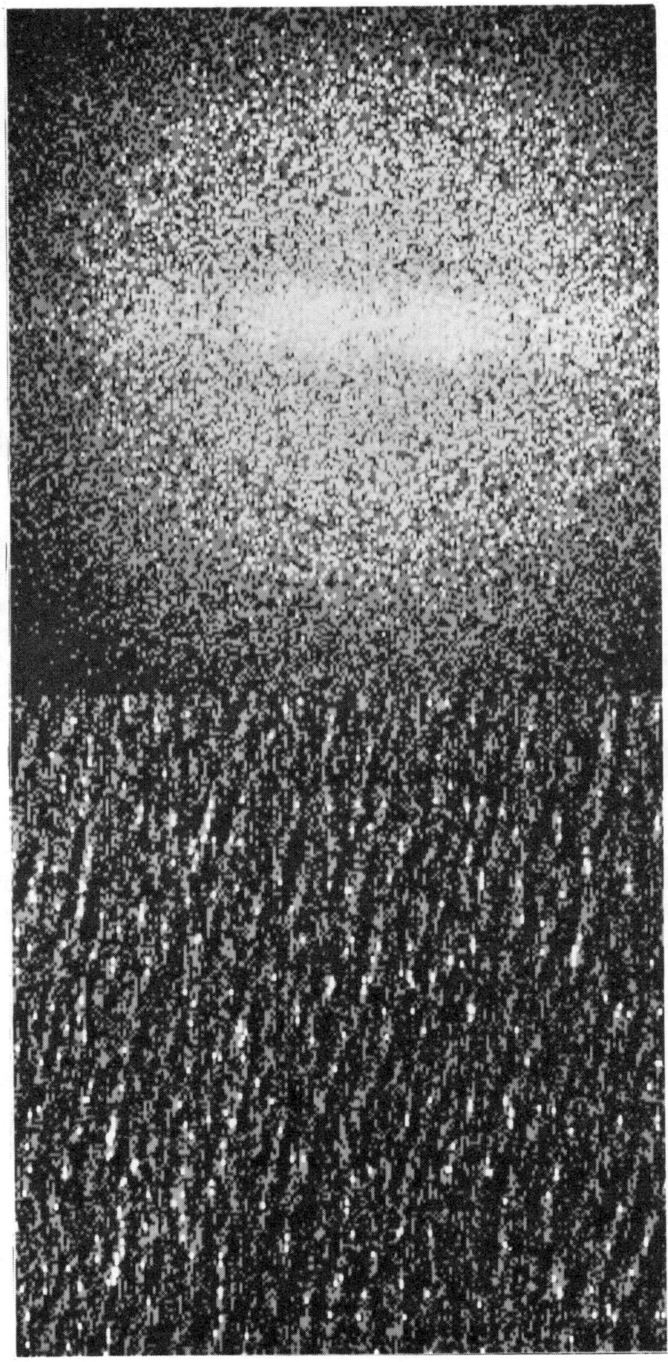

Figure 4. Atlantic swell waves imaged by SEASAT's SAR over JASIN and (on the right) a 2-D Fourier transform of the same scene showing the spectral spread of wavenumbers and wave direction (rev 1049).

Table 2:

ANALYSIS OF SEASAT SENSOR PERFORMANCE OVER JASIN

Sensor	Objective	Jasin analysis
1) SASS	± 2 m/s, 020° for $0<V<25$ m s	0 ± 1.6 m/s, 18° for $3<V<16$ m s
2) SMMR (wind speed only)	± 2 m/s	Before rev 900 SMMR surface 2.5± 1.8 m/s
	for $7<V<50$ m/s	After rev 900 SMMR surface 0.6± 0.9 m/s for $3<V<16$ m/s
3) Altimeter		
H_s	0.5 m or 10%	± 10 cm $0<H_s<4$m
Wind speed	± 3 m s	± 2 m/s $0<V<10$ m/s
		Begins to saturate above 10 m/s
4) SAR		
a) Directional Wave Spectra		Dominant wavelength to ± 15%
		Dominant direction to ± 10°
		(when waves clearly visible)
b) Surface features		
: Surface streaks		λ 2-3 BL depth. When observed wind speed > 15 m/s
: Internal waves		In vicinity of seamounts.

Table 3:

Ocean Basin	Surface Winds		Wave Heights	
	FNOC	SEASAT	FNOC	SEASAT*
North Atlantic	982	3365	718	5960
South Atlantic	60	3890	52	6655
South Pacific	96	8360	85	14768

*Concentrated along satellite track

PART II FUTURE PROGRAMMES

1. Narrative

In part I we took a broad view of the marine applications to which satellites have already made a significant contribution. We saw that geodesy and the drifting buoy programmes had advanced remarkably in the last decade as a direct result of satellite programmes and that more recently, the experimental satellites Nimbus-7 and Seasat had demonstrated respectively

i) the potential capabilities of monitoring the colour changes in the ocean (the CZCS) and

ii) the accurate monitoring of surface winds and waves by means of all-weather radar sensors. We looked at examples of the outputs of some of those sensors and illustrated the type of marine commercial activity which could exploit these measurements.

In part II I shall highlight some of the scientific programmes which are most likely to benefit from future satellite missions and I shall attempt to summarise the scope of these missions at the present stage of planning.

The physical processes that occur within the oceans are observed across a wide spectrum of space and time scales from small ripples on the sea surface to ocean-basin fluctuations at an appropriate range of frequencies from seconds to years. Sometimes the results of local investigations are sufficient to provide insight into the physics of a global phenomenon. Air-sea interaction processes such as wave generation may be modelled and tested from a single platform at sea. Again, if processes are relatively time-invariant then ocean-wide surveys carried out by a few research vessels may reveal patterns of a kind which, as mentioned in part I, led to the formulation of theories of sea-floor evolution and plate tectonics almost 20 years ago.

But oceanographers wishing to study time-varying processes at ocean-basin scales are faced with virtually insurmountable sampling problems. The oceans cover some 360 million square kilometres and although more buoys are deployed now than ever before and individual research vessels participate in more international, multiship investigations, the inadequacy of the sampling from surface units remains a serious obstacle to understanding the complexities of ocean systems.

Oceanographers have no equivalent to the global network that reports twice daily around the globe on atmospheric conditions. There are little more than a handful of permanent oceanographic observing stations. The reasons for this have been summarised by Wunsch (1981).

i) Radio waves do not propagate through the ocean and neither, of course, does light.

ii) Ocean investigations must therefore rely heavily on research vessels which are slow, comparatively expensive and sometimes uncomfortable. It has been said that, in some respects, the worst place from which to study the sea is the sea surface.

iii) The pressures to produce reliable forecasts is less for the ocean than for the weather though the increase in marine exploration, its advance into deeper water plus the increasing requirement for large freighters and super-tankers to save passage time is beginning to bring about a change in this state-of-affairs.

But if we are ever to understand the physical processes which control our climate then it is imperative to know much more than we can guess at present of the transport of heat around the globe by ocean circulation. Satellite remote sensing appears to offer an important contribution; in fact it may be more true to say that it offers the only effective way of obtaining the necessary repetitive observations at global scales.

Let us examine more closely what role a precise altimeter, deployed from an earth-orbiting satellite, might play in monitoring ocean circulation.

2. The Promise of Altimetry

If the synthetic aperture radar is an example of a sophisticated space instrument whose time is yet to come the radar altimeter is an instrument which has been progressivly improved from the first model flown on Skylab to the remarkable $\pm$ 7 cm accuracy model deployed on Seasat. The question now posed by the post-Seasat analysis is how well can such an instrumental precision be used for studies of the ocean's circulation in the face of other uncertainties which enter into satellite altimetry?

Ocean currents are associated with pressure gradients produced to balance the Coriolis force of the earth's rotation. In the case of the open ocean this slope is very small. It can be calculated from the simple Coriolis equation

$$2\,\omega v \sin\lambda = g\tan\theta$$

where ω = angular velocity of the earth

v = current speed
λ = latitude
g = acceleration due to gravity
θ = slope of the sea surface

Thus, at the equator, the slope is zero. At a latitude of 30° writing

$$\omega = 7 \times 10^{-5} \ s^{-1}, \ g = 10 \ ms^{-1}$$

then

$$\tan \theta = \theta = 7 \times 10^{-6} \ v$$

Thus for $v = 0.1$ m/s (~ 1/5 knot)

$$\theta = 7 \text{ cm}/100 \text{ km}.$$

Across the Gulf Stream where currents can reach 1.5 m/s then slopes greater than 100 cm across 100 km are created and these are easily observable by a satellite altimeter; but, as we shall see, the much more modest slopes associated with the general circulation of the oceans are more difficult to detect against the undulations of the geoid and other uncertainties.

Satellite altimetry works on very simple principles requiring very accurate measurements. By measuring the distance between the spacecraft and the sea surface it is enough to know the satellite's orbit and the altimeter will have revealed the topography of the sea surface. We saw earlier that in the absence of forces acting on the ocean the sea surface is the geoid. Again if we know the shape of the geoid then the difference between the measured sea surface and this geoid would show the geostrophic velocity at the sea surface. The scale of vertical variations due to water movement is no more than 1-2 meters over the globe whereas as we have seen the geoidal undulations are of the order of ± 100 m. Thus for the method to work we must know the geoid to very high accuracy and this is rarely the case. However, since the geoid is time-invariant it is possible to measure differences in surface current by repeated sampling over the same track. During the last month of Seasat's operation the satellite entered into a 3-day repeat pattern which indeed revealed marked variations in sea surface topography as shown.

Having to track the satellite to an accuracy of a few centimetres and requiring to know the shape of the geoid to similar accuracy are daunting enough tasks but unfortunately other corrections arise due to forces acting on the sea surface and to varying conditions in the atmosphere. A summary of these effects and the magnitude of the corrections required are shown in table 4.

Table 4(a)

Basic Requirements

- Absolute height of spacecraft
- Shape of Geoid
- Effect of atmosphere and sea surface elevations on signal

Table 4(b)

Altimeter Path Length

Contributions from Atmosphere and Ocean Surface

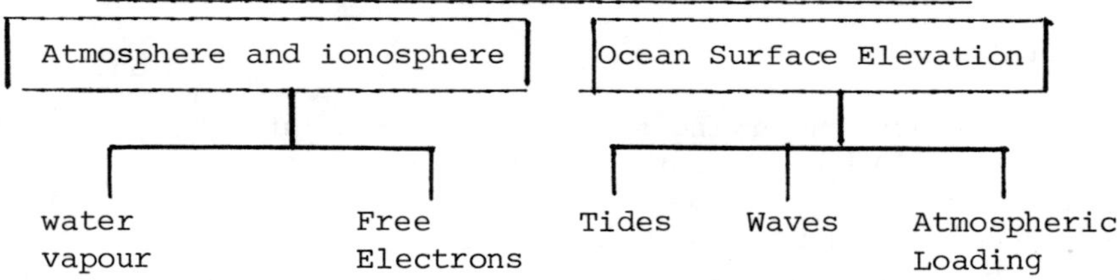

Table 4(c)

Scale of Corrections

	Amplitude of correction	Est. Error	Approx space scale
SEA SURFACE			
Tides	100 cm	10 cm	Basinwide
Waves	2-5 % H_S	10-15 cm	200-2000 km
Atmos. loading	1mb = 1cm (20-30cm)	2-3 cm	200-2000 km
ATMOSPHERIC			
WATER VAPOUR			
Dry Component	230 cm	1-2 cm	100-2000 km
Wet component	5-35 cm	3-6 cm	100-2000 km
IONOSPHERE			
Free Electrons	1-30 cm	3-5 cm	

Some of the effects can be modelled but most require direct measurement either from the same satellite or possibly from other operational satellites. The wet component of atmospheric water vapour for example, is most accurately estimated by direct vertical measurement at two microwave frequencies and for this reason an altimeter mission requires a supporting microwave radiometer besides the necessary instrumentation for accurate tracking. Another important consideration is the pattern of laying down the satellite orbits. We have seen that repeat orbits are essential since the geoid is rarely known with sufficient accuracy to identify the 'ocean' signal from a single pass. But frequent repeats can only be made at the expense of spatial coverage. Other instruments making up the payload may require quite a different orbit pattern than that required for altimetry so compromises must be made. Let us briefly examine some of the considerations that satellite agencies must take into account when planning a mission.

3. Orbits

Most earth observing satellites operate at a height in the range 700-1200 km. Seasat flew at a height of 800 km and an inclination of 108° (measured anti-clockwide from the equator). Its speed over the ground was 6.7 km/s and it completed 14 1/3 orbits per day. Examples of the coverage achieved are shown. It can be seen that if the orbit pattern is made to repeat after 3 days (i.e. if exactly 14 1/3 orbits are completed each day) then the distance between adjacent tracks at the equator is almost 1000 km. Thus, if this pattern were persisted with, then in a year each track would have been sampled over 100 times (reducing the geoid uncertainty by a factor of 10) but large areas of the sea would remain unsampled. For missions solely directed to altimetry, repeats at reasonably frequent intervals are essential (say 10-15 days) but if a synthetic aperture radar is carried - and, especially if part of the mission objective is to image land areas - then wide spatial coverage is likely to be more important than frequent sampling.

A plot of nodal separation versus sampling frequency for 1 to 4 satellites is shown (Fig. 5). It should be noted that minimal changes in the satellite's orbiting height are required to change a 3-day repeat into (say) a 17-day repeat orbit. In the one case the satellite would complete 14 1/3 orbits per day, in the other 14 6/17 orbits per day. The actual height difference at an orbiting height around 800 km is no more than a few hundred metres so that thruster motors on the satellite can be used to change from one repeat pattern to another.

Sampling strategy may be less critical for wide-swath instruments such as the scatterometer and multi-spectral scanners such

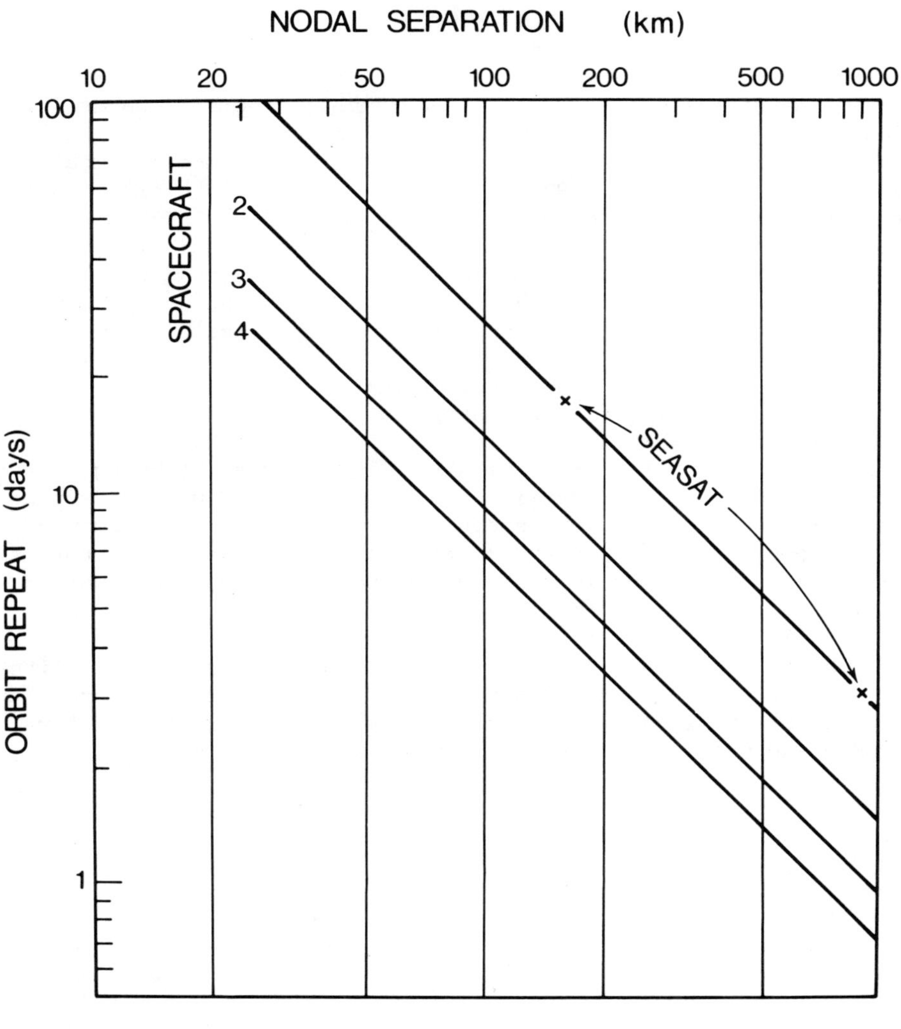

Figure 5. Plot of nodal separation at the equator versus repeat sampling period for 1 to 4 satellites. The two patterns adopted by SEASAT are shown.

as the Advanced Very High Resolution Radiometer of NOAA 6/7 and the CZCS of Nimbus-7. An example of scatterometer coverage of the Atlantic for the 400 km swath model proposed for ERS-1 is shown in Fig. 6.

4. Future Satellite Programmes

A list of proposed satellite programmes directed towards ocean surveillance is shown in Table 5. Following the success of the experimental (pre-operational) Seasat programme it might have been expected that NASA would follow up with an operational series of ocean satellites similar perhaps to the NOAA meteorological series. But budget cuts caused the cancellation of the proposed NOSS (National Oceanographic Satellite System) programme and, for the present, it has been left to Europe and Japan to build ocean programmes around the concepts proved by Seasat.

4.1 MOS 1, 2, 3. MOS-1 (Marine Observation Satellite) represents Japan's first entry into earth surveillance. It is designed to carry three sensors: a visible (MESSER), and near-infrared scanning radiometer (VTIR) and a microwave scanning radiometer (MSR). The first is primarily concerned with detecting changes in ocean colour, the second with sea-surface temperature changes and the micro-wave radiometer for all weather measurements of SST and water vapour in the atmosphere. The proposed launch date is now 1985.

The Japanese plan to follow up MOS-1 with two other satellites, MOS-2 and MOS-3. Both MOS-2 and MOS-3 will carry active microwave instruments including a radar altimeter, wind scatterometer on MOS-2 and with the addition of an imaging radar on MOS-3.

4.2 SPOT (Poseidon). The first SPOT satellite (designed by the French Centre National d'Etudes Spatiales in association with Belgian and Swedish partners) is for land use and is scheduled for launch in 1984. Plans are underway for follow-up satellites which may have dual ocean/land use. One such programme called Poseidon (SPOT-2) envisages inclusion of a radar altimeter of comparable precision to that carried on Seasat. Provisional launch date is 1986. The proposed altitude is 832 km, inclination 98° with a repeat period of approximately 28 days and a 'near-repeat' every 3 days. A vertical microwave sounder would be included.

4.3 TOPEX. At the present time the only altimetric mission designed solely for scientific research is the Ocean Topography Experiment (TOPEX) designed by the Jet Propulsion Laboratory, Pasadena (who also designed and managed the Seasat mission). The proposed launch date for TOPEX is the late 1980s'.

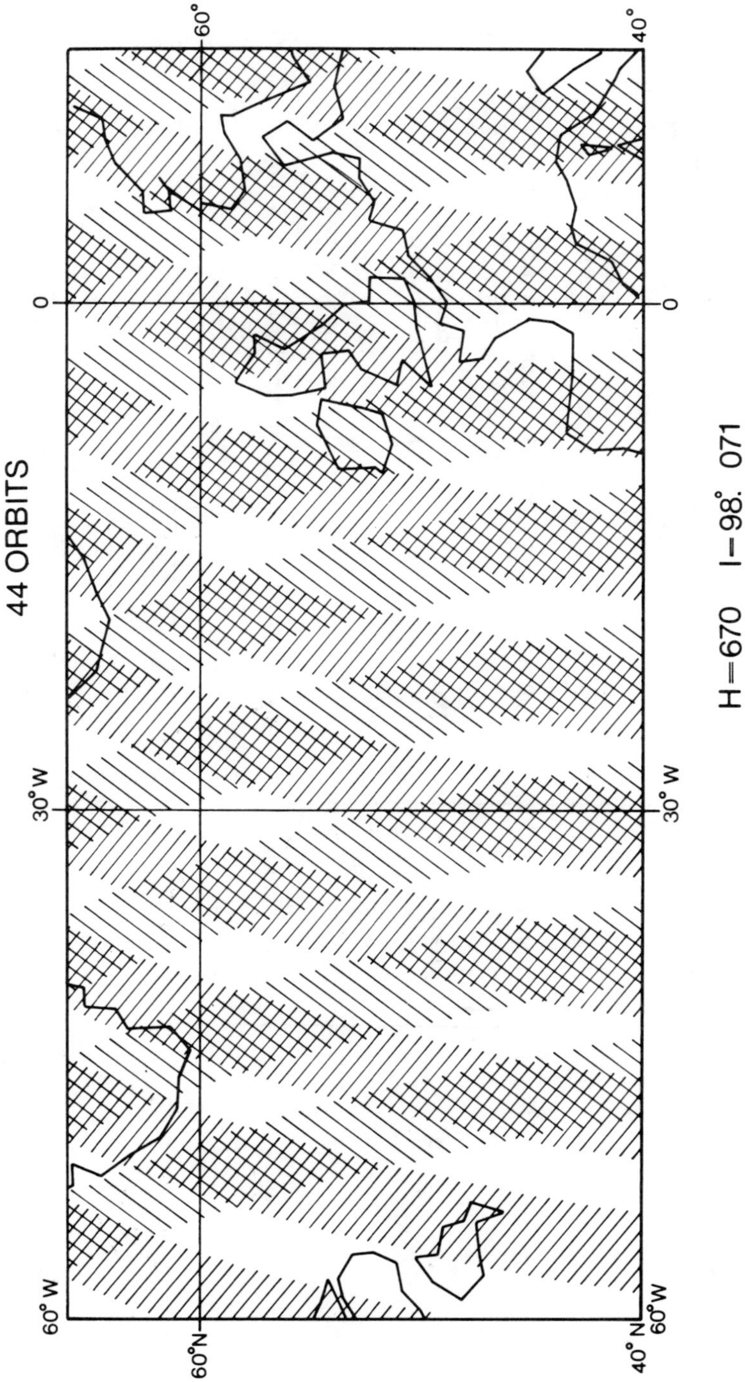

Figure 6. The coverage of wind measurements that would be achieved in 3 days over the N. Atlantic with the scatterometer proposed for ERS-1.

Table 5:

SATELLITES WITH OCEANOGRAPHIC APPLICATIONS PLANNED FOR FUTURE

Spacecraft name	Agency	Proposed launch	Sensors
ERS-1	ESA	1987-88	Altimeter, SAR, Scatterometer, VIS and IR Radiometer (ATSR)
TOPEX	NASA	1987-88?	Radar altimeter
SPOT II (Poseidon)	CNEX (France)	1987(?)	Altimeter is proposed as part of payload of SPOT II
MOS 1,2, 3	NASDA	1985-88	MOS-1 - radiometer in vis, IR and microwave bands MOS 2/3 - addition of active microwave sensors
Spacelab	NASA/ESA	Continuing	Synthetic Aperture Radar
RADARSAT	Canada	1990	SAR, Altimeter(?)

TOPEX is designed to carry dedicated microwave sounders to establish the water vapour correction. In the absence of other instruments the constraints imposed on multi-purpose missions through compromise duty cycles and orbit patterns present no difficulty for TOPEX. Unlike most satellites which require sun-synchronous orbits to power a variety of sensors TOPEX is designed to operate in a non-sun-synchronous orbit so avoiding the aliasing of tidal measurements. Its design altitude is 1300 km (to minimise air drag), its repeat period 10 days and its inclination 65°. It uses dual-frequencies to correct for the effect of variations in free electron content in the ionosphere. Final approval to build TOPEX has not yet been given but if it does fly it should attract a wide following of oceanographers around the world.

4.4 ERS-1. Earlier this year the nations contributing to the European Space Agency gave their approval to proceed with Phase B of the proposed ERS-1 (European Remote Sensing) mission. The ERS-1 payload is adapted from that of Seasat which proved so successful. Its microwave package consists of

- Radar altimeter
- Active Microwave Instrumentation (AMI)

consisting of

i) a wind scatterometer mode
ii) a wave scatterometer mode
iii) a synthetic aperture radar

In addition an infrared radiometer (the Along Track Scanning Radiometer designed by the Rutherford-Appleton Laboratory) carrying also two vertical microwave channels for altimeter water vapour corrections is included in the payload, probably together with a satellite tracking system.

The AMI will operate at C-Band (5 GHz). Its syntethic aperture radar will cover a 100 km swath as for Seasat and it will have a similar resolution. A wind scatterometer will be limited to a 400 km swath on one side of the spacecraft - as opposed to Seasat's 500 km on either side. In the wave scatterometer mode the SAR will image 5 km^2 patches of ocean every 100 km; in this way it is hoped to derive statistics on wavenumber spectra.

ERS-1 is conceived as an applications satellite and its primary task is the measurement of wind and waves over the ocean. There is nevertheless a consensus that the maximum of scientific research should be extracted from this mission and to this end it is possible that the satellite will be accurately tracked and other measures taken to utilise the altimeter not only for the measurement of significant wave height but also for the monitoring of ocean circulation.

5. Conclusions

More and more people are becoming aware of the potential benefits offered by remote sensing. Equally it behoves us to keep our feet on the ground in an almost literal sense since no remote-sensing measurements can stand alone without support from 'in situ' observations. It is becoming clear that far from diminishing the need for research vessels satellite missions may actually increase the amount of surface data required to interpret and validate much of the data from spaceborne sensors. This does not prevent oceanographers from being apprehensive; there is always the possibility that multi-million satellite programmes may require to be at least partly funded by reducing investment in ships. Another aspect of this very real problem is the contrasting styles between going to sea with comparatively well-trusted instruments and a few colleagues to study phenomena at local scales, and the strong project management and relatively impersonal approach required by many co-ordinated agencies before the relevant satellite data arrives at the oceanographer's desk.

It is perhaps too simplistic for oceanographers to write off space programmes as expensive luxuries supported in the main by military, political and commercial objectives. Equally space scientists and engineers must not be surprised that many oceanographers remain to be convinced that satellite programmes are cost-effective (in terms of the extra ships and buoys that could be bought with the same money). To some extent space programmes have suffered from being over-sold in the 60s'.

Oceanographers and agencies must adapt to each other and this may take some time.

REFERENCES

1) Allan, T.D. (1979) Monitoring the sea surface. Proc. Tech. Conf. 'Use of data from meteorological satellites. ESA SP-143, ESA, Paris, 202-215.

2) Allan, T.D. & Guymer, T.H. (1983) Seasat measurements of wind and waves on selected passes over JASIN. Int. J. Rem. Sens. (in press).

3) Chelton, D.B., K.J. Hussey and M.E. Parke (1981) Global satellite measurements of water vapour, wind speed and wave height Nature, 294, 5841, 529-532.

4) Webb, D.J. (1981) A Comparison of Seasat altimeter measurements of wave height with measurements made by a pitch/roll buoy. J. Geophys. Res., 86, C7, 6394-6398.

5) Wunsch, C. (1981) The promise of satellite altimetry. Oceans, 24, 3, 17-26.

INDEX

A

A contd.

B

B contd.

C

C contd.

C contd.

D

D contd.

E

E contd.

F

F contd.

G

G contd.

H

H contd.

I

I contd.

J

K

L

L contd.

M

M contd.

N

P contd.

Q

R

R contd.

R contd.

S

S contd.

S contd.

T

T contd.

U

V

V contd.

W

W contd.

X

Y

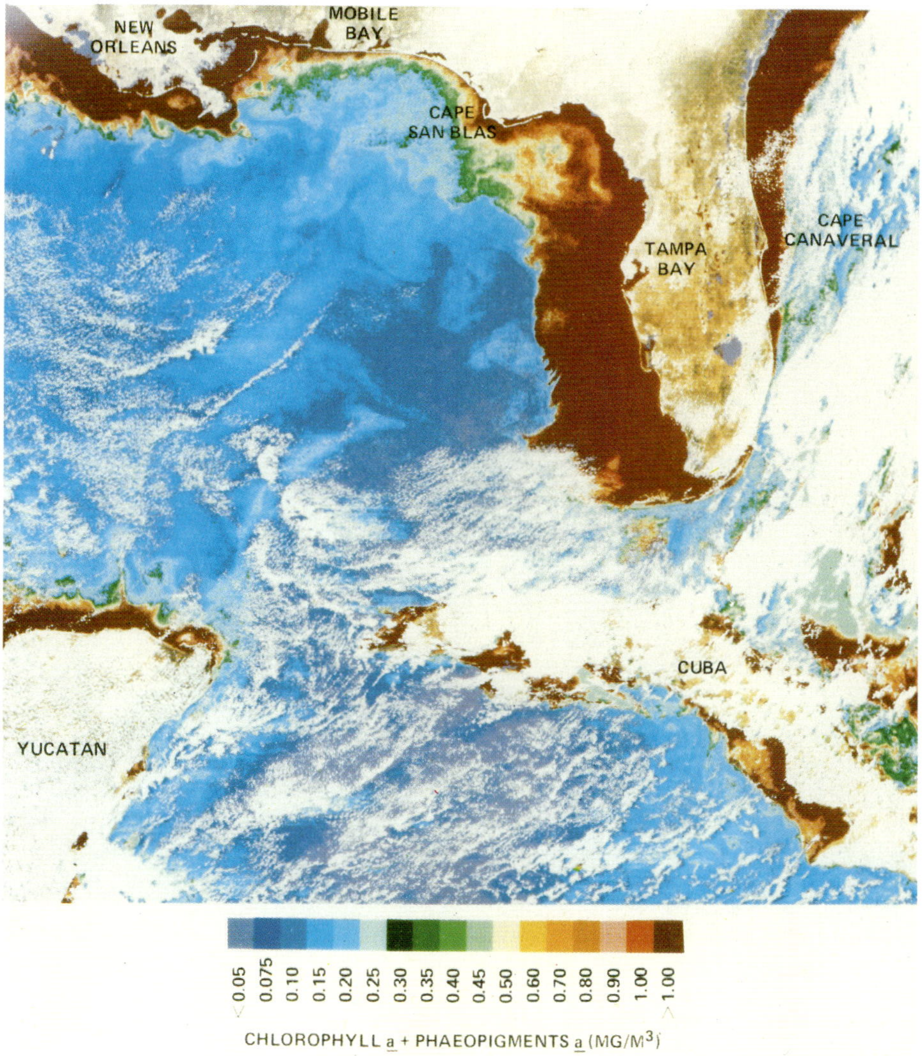

J. R. Apel: Remote Measurement of the Ocean – An Overview. Fig. 2. Colour coded image from Coastal Zone Colour Scanner (CZCS), flown on NIMBUS-7 satellite. See p. 10 for ref.

B. Sturm: Selected Topics of Coastal Zone Color Scanner (CZCS) Data Evaluation. Fig. 7.8 – Calciferous algae bloom in western British Channel, CZCS orbit 17997, Date: May 18, 1982. False color image of Ch1 → blue, Ch2 → green, Ch3 → red. Note the white signature of the algal bloom indicating high reflectance in all three channels in contrast to green algae which generally show a decrease of reflectance in Ch1. (Image is north-south inverted) See p. 165 for ref.

B. Sturm: Selected Topics of Coastal Zone Color Scanner (CZCS) Data Evaluation. Fig. 7.9 - Upwelling off shore Senegal (Dakar). Orbit No. 11875, Date: March 1, 1981. False color image of subsurface upwelling radiances in Ch1, 2 and 3 (Ch1 → blue, Ch2 → green, Ch3 → red). Note the dark water zone near the coast indicating phytoplankton blooms (Low Ch1 radiance). See p. 165 for ref.

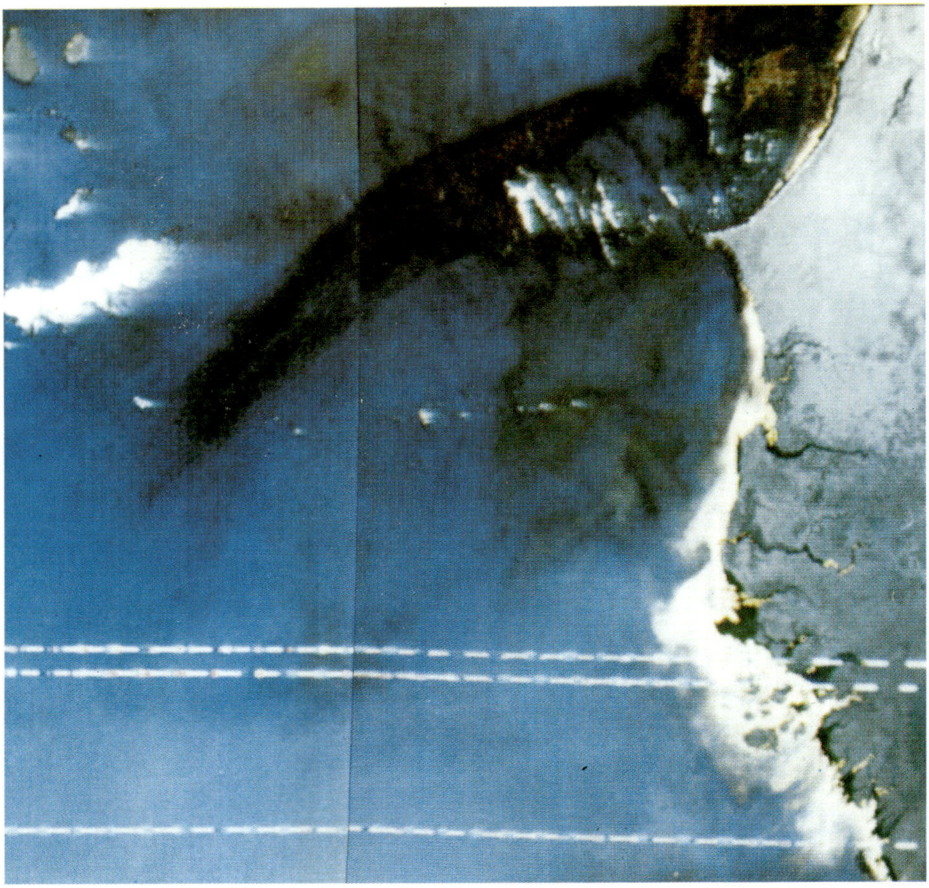

B. Sturm: Selected Topics of Coastal Zone Color Scanner (CZCS) Data Evaluation. Fig. 7.10 – Upwelling off shore Senegal (Dakar). Orbit No. 16533, Date: February 1, 1982. False color image of subsurface upwelling radiances in Ch1, 2 and 3 (Ch1 → blue, Ch2 → green, Ch3 → red). The broad dark streak north west of Dakar peninsula is due to a Sahara dust stream resulting in very high aerosol path radiance $L_A > 1.5$ mW/cm² sr/u. Note the upwelling south of Dakar and the very turbid waters along the coast of Guinea. See p. 165 for ref.

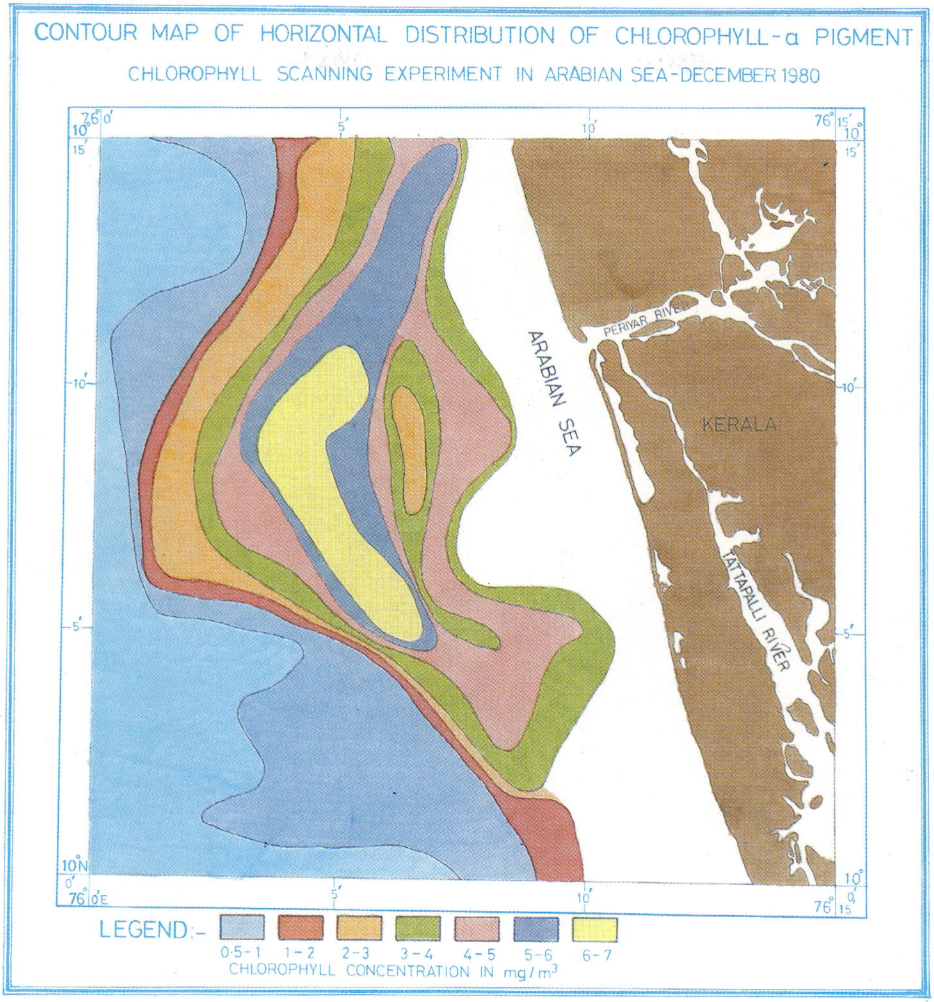

I. V. Muralikrishna: Ocean Color Studies in Arabian Sea. Contour map of horizontal distribution of chlorophyll. See p. 313 for ref.

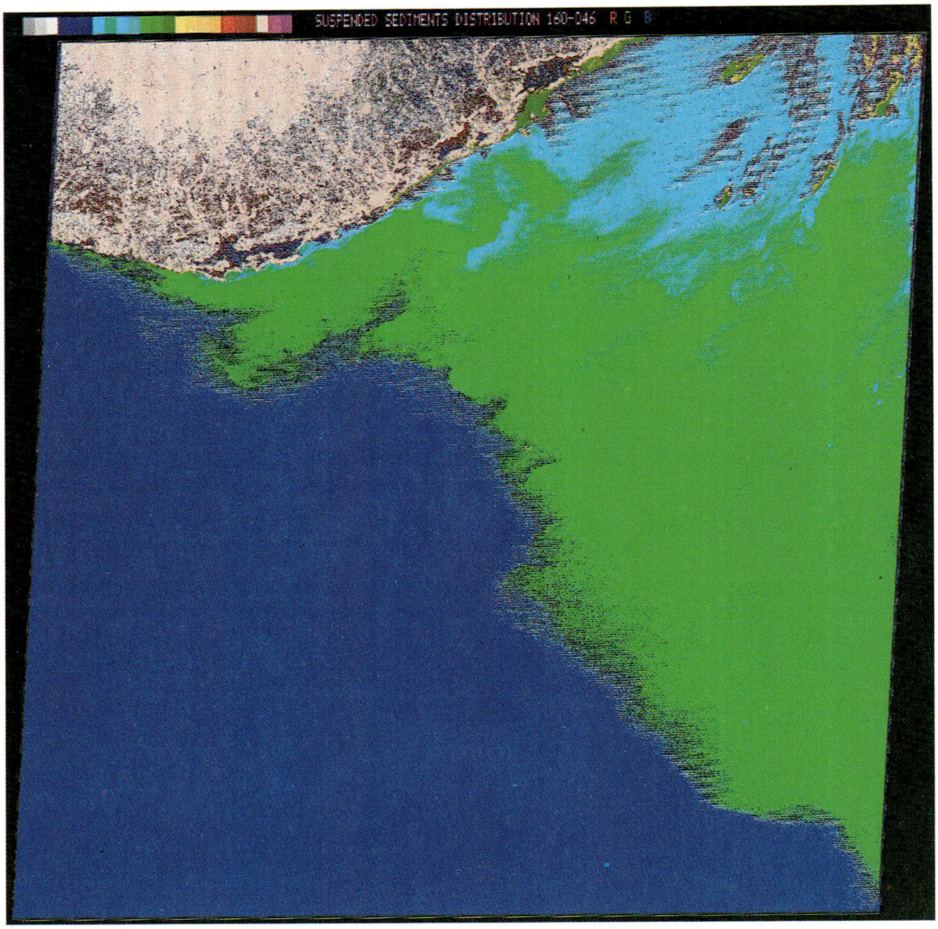

I. V. Muralikrishna: Landsat Application to Suspended Sediments Evaluation. Contour map of suspended sediments distribution. See p. 321 for ref.

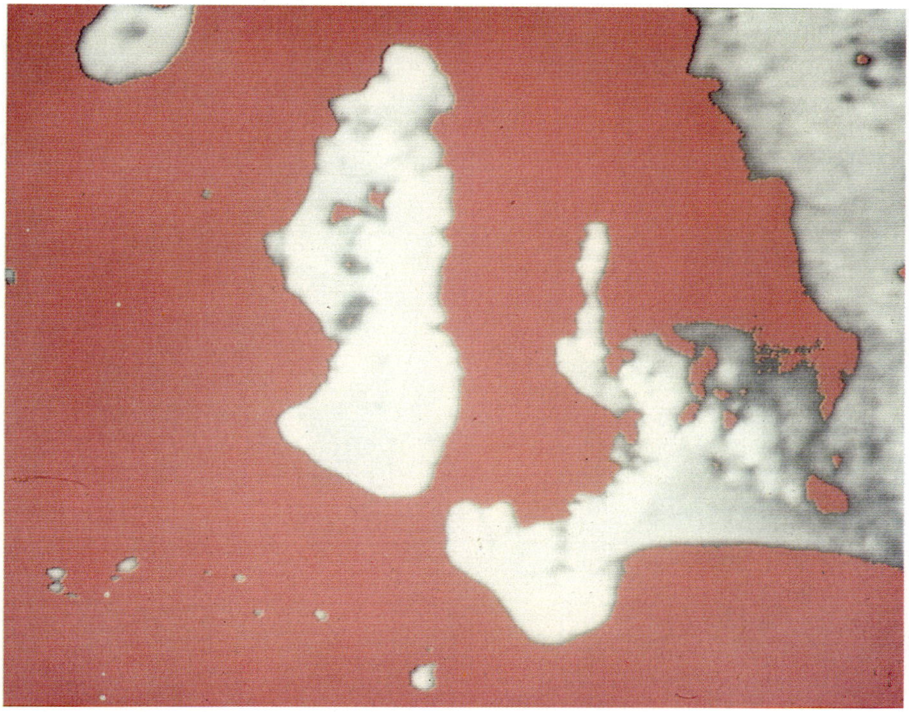

R. K. Bullard: Detection of Marine Contours from Landsat Film and Tape. Colour plate 1. Band 4 Landsat image of parts of the Red Sea and Southern Yemen Republic together with Kamaran Island. See p. 376 for ref.

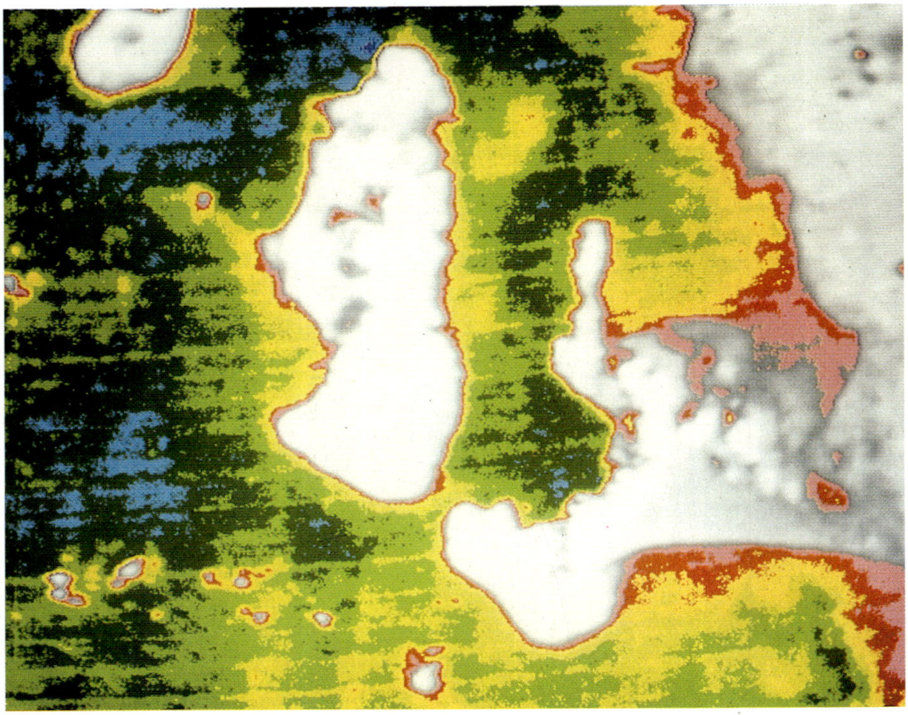

R. K. Bullard: Detection of Marine Contours from Landsat Film and Tape. Colour plate 2. Density slice of the band 4 image shown in Colour plate 1. See p. 377 for ref.

LIST OF PARTICIPANTS

- ALLAN, T.D.	Institute of Oceanographic Sciences, Brook Road, Wormley, Godalming, Surrey, GU8 5UR, United Kingdom.
- ALPERS, W.	Max Planck Institute, Bundesstrasse 55, 2000 Hamburg 13, Germany.
- ALVES, M.L.G.R.	Oceanography Group, Department of Physics/Geophysical Centre, University of Lisbon, Rua da Escola Politecnica, 58, 1200 Lisboa, Portugal.
- ANDERSON, J.M.	Carnegie Laboratory of Physics, University of Dundee, Dundee DD1 4HN, Scotland, United Kingdom.
- APEL, J.R.	Applied Physics Laboratory, The Johns Hopkins University, Johns Hopkins Road, Laurel, Maryland 20707, U.S.A.
- BAYLIS, P.E.	Department of Electrical Engineering and Electronics, University of Dundee, Dundee DD1 4HN, Scotland, United Kingdom.
- BERRIT, G.R.	Orstom, 24 Rue Bayard, 75008 Paris, France.

- BOWYER, P.A.	Department of Oceanography, University College, Galway, Ireland.
- BULLARD, R.K.	Department of Land Surveying, North East London Polytechnic, Forest Road, London E17 4JB, United Kingdom.
- CALLISON, R.D.	Carnegie Laboratory of Physics, University of Dundee, Dundee DD1 4HN, Scotland, United Kingdom.
- CALMET, D.P.	DPR/SPS/LPEP, Cenfar BP No 6, 92260 Fontenay aux Roses, France.
- CHARMASSON, S.	Laboratoire d'Ecologie Numerique, Campus de Luminy, 70 Route Léon Lachamp, 13288 Marseille Cedex 9, France.
- CLEMENT, F.	Avenue des Combattants, 7, 1340 Ottignies/Lln, Belgium.
- CRACKNELL, A.P.	Carnegie Laboratory of Physics, University of Dundee, Dundee DD1 4HN, Scotland, United Kingdom.
- DE LEFFE, A.	ESA, 18 Avenue Edouard Belin, 31055 Toulouse Cedex, France.
- DEMARCQ, H.	C.T.A.M.N. Ecole des Mines, Rue Claude Daunesse, Sophia Antipolis, 06565 Valbonne Cedex, France.

- DE SIMONE, M.D. — Department of Land Surveying,
N.E.L.P.,
Forest Road,
Walthamstow,
London E17 4JB,
United Kingdom.

- DUPOUY, C. — Laboratoire d'Optique Atmospherique,
BAT P5,
Université des Sciences et
Techniques de Lille,
59655 Villeneuve d'Ascq Cedex,
France.

- ECONOMIDES, B. — 142 Draycott Avenue,
Kenton,
Harrow,
Middlesex HA3 OBZ,
United Kingdom.

- FLIKKE, A.J. — State Pollution Control Authority,
P.O. Box 8100 Dep,
Oslo 1,
Norway.

- FONT, J. — Inst. de Investigaciones Pesqueras,
P. Nacional S/N,
Barcelona - 3,
Spain.

- GASPAR, P. — Inst. d'Astronomie et de Geophysique
Georges Lemaitre,
Chemin du Cyclotron, 2,
B-1348 Louvain-La-Neuve,
Belgium.

- GOODMAN, M. — Physics Department,
University College Dublin,
Stillorgan Road,
Belfield,
Dublin 4,
Ireland.

- HABERÄCKER, P. — MBB Apparate,
Postfach 80 11 49,
8000 München,
Germany.

- HOFFMANN, G. C/o Erno Raumfahrttechnik GMBH,
Dept PRV-5,
Hunefeldstr. 1-5,
2800 Bremen,
West Germany.

- JAYKO, K. P.O. Box 71,
W. Kingston, RI 02892,
U.S.A.

- JOLLY, J.B. 5416 California Ave S.W.,
Seattle,
WA 98136,
U.S.A.

- JORGE DA SILVA, A. Inst. de Desenvolvimento Pesqueiro,
P.O. Box 1723,
Maputo,
Mozambique.

- KERR, Y. Image Processing Division,
C.N.E.S. OT-TI-TH,
18 Ave. Edouard Belin,
31055 Toulouse Cedex,
France.

- KIHLBLOM, U.G. SWECO,
P.O. Box 5038,
10241 Stockholm,
Sweden.

- KURTER, A. Marine Sciences and Geographical Institute,
University of Istanbul,
Muskule Sokak. Vefa-Istanbul,
Turkey.

- LACEY, S. Department of Civil Engineering,
University of Southampton,
Highfield,
Southampton,
Hants.,
United Kingdom.

- LAMBERT, P. La Loubassane,
M1,
13100 Aix En Provence,
France.

- LANGOT, F.
L'Harmattan,
Impasse de la Paquerette,
13090 Aix En Provence,
France.

- LAWRENCE, D.H.
Department of Communication Engineering,
Faculty of Technology,
Plymouth Polytechnic,
Drake Circus,
Plymouth PL4 8AA,
United Kingdom.

- LIN, S.R.
The Second Institute of Oceanography,
Hangchow,
Zhejiang,
The Peoples' Republic of China.

- MACEDO, M.E.
Direccao de Servicos de Hidrologia,
Rua Bernado Lima No. 5, 40,
1100 Lisboa,
Portugal.

- MacFARLANE, N.
Oceanography Department,
The University,
Southampton SO9 5NH,
United Kingdom.

- McHARDY, C.P.
Carnegie Laboratory of Physics,
University of Dundee,
Dundee DD1 4HN,
Scotland,
United Kingdom.

- McMANUS, J.
Department of Geology,
University of Dundee,
Dundee DD1 4HN,
Scotland,
United Kingdom.

- MAGNIN, F.
21, Avenue de Lambesc,
13840 Rognes,
France.

- MAHMOOD, K.
House No. 480 Bylal Gung,
Near City Church Sahiwal,
Pakistan.

- MALMGREN-HANSEN, A.	Water Quality Institute (VKI), 11 Agern Alle, DK-2970 Horsholm, Denmark.
- MARACCI, G.	Scientific Officer, J.R.C. Euratom, 21020 Ispra (Varese), Italy.
- MAWDSLEY, J.	10 Victoria St. W., Amherst, Nova Scotia, Canada B4H 1B8.
- MISKOLCZI, F.	Institute for Atmospheric Physics, H-1675 Budapest, P.O.B. 39, Hungary.
- MURALIKRISHNA, I.V.	National Remote Sensing Agency, No. 4 Sardar Patel Road, P.B. 1519 Secunderabad 500-003, Andhra Pradesh, India.
- NACU, A.F.	Natural Resources Management Centre, 8th Floor Triumph Building, 1610 Quezon Ave, Quezon City, Philippines.
- NEEDHAM, B.	NOAA, Envir. Data and Information Services, National Climatic Center, Satellite Services Division, Washington DC 20233, U.S.A.
- NILSEN, V.	University of Tromso, Institute of Mathematical and Physical Sciences, P.O. Box 953, N-9001 Tromso, Norway.
- NYFORS, E.G.	Helsinki University of Technology, Radio Laboratory, Otakaari 5A, SF-02150 Espoo 15, Finland.

- OFFILER, D. — Meteorological Office,
Bracknell RG12 2UR,
Berkshire,
United Kingdom.

- QUADER, O. — SPARRSO,
Mohakash Biggyan Bhaban,
Agargaon, Sher-E-Bangla Nagar,
G.P.O. Box 529,
Dacca,
Bangladesh.

- REID, G.S. — Department of Geology,
University of Dundee,
Dundee DD1 4HN,
Scotland,
United Kingdom.

- RIPLEY, H. — Maritime Resource Management Service,
P.O. Box 310,
Amherst,
Nova Scotia,
Canada B4H 1B8.

- ROTHAMEL, J. — Scientific Technical Services Dept.,
Bremen University,
D-2800 Bremen,
West Germany.

- SAFAK, M. — Hacettepe University,
Department of Electrical and
Electronic Engineering,
Beytepe Ankara,
Turkey.

- SATHYENDRANATH, S. — Laboratoire de Physique et
Chimie Marines,
B.P. 8 La Darse,
06230 Villefranche-Sur-Mer,
France.

- SCHARFEN, G.R. — WDC-A for Glaciology,
Cires/Campus Box 449,
University of Colorado,
Boulder, CO 80309,
U.S.A.

- SCHROTER, J. — Max Planck Institute für Meteorologie,
Bundesstr. 55,
2000 Hamburg 13,
West Germany.

- SENGELE, N. — Inst. National d'Etude et Recherche,
Agronomique (I.N.E.R.A.),
B.P. 1513,
Kisamgami,
Zaire.

- SINGH, S.M. — Carnegie Laboratory of Physics,
University of Dundee,
Dundee DD1 4HN,
Scotland,
United Kingdom.

- STOREY, B.E. — Carnegie Laboratory of Physics,
University of Dundee,
Dundee DD1 4HN,
Scotland,
United Kingdom.

- STOTT, A.P. — Geography Department,
University of Manchester,
Oxford Road,
Manchester M13 9PL,
United Kingdom.

- STURM, B. — Joint Research Centre, EURATOM,
Ispra,
21020 VA,
Italy.

- SUN, Y-Y. — Anhui Institute of Optics and
Fine Mechanics,
Academica Sinica,
P.O. Box 25,
Hefei, Anhui,
China.

- TACCINI, S. — Via Torino, 18,
56100 Pisa,
Italy.

- TOOKE, A.O. — Carnegie Laboratory of Physics,
University of Dundee,
Dundee DD1 4HN,
Scotland,
United Kingdom.

- TSOLAKIDIS, C.	90, 28th October St., Rhodes, Greece.
- UDIN, V.I.	Swedish Meteorological and Hydrological Institute, Box 923, S-601 19 Norrkoping, Sweden.
- UYSAL, T.	Marine Science Department, Middle East Technical University, P.K. 28, Erdemli, Icel, Turkey.
- VALERIO, C.	C.E.T.E. Division Laboratoires, Z.I. Les Milles, Rue Einstein BP 39, 13290 Les Milles, France.
- VALEUR, H.H.	Danish Meteorological Institute, Lyngbyvej 100, 2100 Copenhagen O, Denmark.
- VAN HETEREN, J.	Deltadienst, Van Alkemadelaan 400, Den Haag 2597 AT, Netherlands.
- WARREN, D.	Department of Electrical Engineering and Electronics, University of Dundee, Dundee DD1 4HN, Scotland, United Kingdom.
- WHYTE, G.R.	Carnegie Laboratory of Physics, University of Dundee, Dundee DD1 4HN, Scotland, United Kingdom.
- WILSON, A.	Department of Physics, University of Lancaster, Lancaster LA1 4YB, United Kingdom.

- WYATT, L. Department of Electronic and Electrical Engineering,
University of Birmingham,
Birmingham B15 2TT,
United Kingdom.

- YENTSCH, C.S. Bigelow Laboratory for Ocean Science,
McKown Point,
West Boothbay Harbor,
Maine 04575,
U.S.A.

- YOUNG, W.M. Carnegie Laboratory of Physics,
University of Dundee,
Dundee DD1 4HN,
Scotland,
United Kingdom.